INVITATION TO
CONTEMPORARY PHYSICS

Q. Ho-Kim
Université Laval, Canada

N. Kumar
*Indian Institute of Science and Jawaharlal Nehru Centre for
Advanced Scientific Research, India*

C. S. Lam
McGill University, Canada

World Scientific
Singapore • New Jersey • London • Hong Kong

Published by

World Scientific Publishing Co. Pte. Ltd.

P O Box 128, Farrer Road, Singapore 9128

USA office: Suite 1B, 1060 Main Street, River Edge, NJ 07661

UK office: 73 Lynton Mead, Totteridge, London N20 8DH

Library of Congress Cataloging-in-Publication Data

Ho-Kim, Q. (Quang), 1938-
      Invitation to contemporary physics/Q. Ho-Kim, N. Kumar, C.S.
Lam.
           p.   cm.
      ISBN 9810207239. -- ISBN 9810207247 (pbk.)
      1. Physics. 2. Astrophysics. I. Kumar, Narendra, 1940-
II. Lam, Harry C. S. III. Title.
QC21.2.H6   1991
530--dc20   91-26392
                                                                          CIP

#24218297

Printed in Singapore by Singapore National Printers Ltd.

INVITATION TO
CONTEMPORARY PHYSICS

TABLE OF CONTENTS

PREFACE

This book deals with seven of the most exciting areas of modern physics: lasers, superconductivity, chaos, symmetry, stars, particles, and cosmology. Together, these topics cover a large spectrum of physical concepts and phenomena — from the tiniest objects to the whole universe, from the hottest to the coldest systems, from abstract principles to ideas with immediate practical implications, from the beauty of symmetries in Nature to the baffling aspects of complex systems. We hope, from this mosaic, a picture will emerge that can convey to the reader all the richness and the enchantment of modern physics. The choice of these seven areas is necessarily subjective to some extent, but multiple authorship has reduced this risk to an acceptable level.

Our goal is to make some parts of physics comprehensible to the general reader not having much prior training in physics, yet intellectually alert and armed with a keen interest in science. We are motivated to write because of a shortage of books of this kind. Quite often we ourselves would like to learn about some area of contemporary intellectual activity outside our own field, but find it difficult to locate a suitable and convenient source of reading material. Textbooks and scientific review articles tend to be too technical, while popular books and magazine articles are often long on facts and short on explanations, as such they might act like analgesics merely to dull the pain of incomprehension. We realize from our own experience that we want to know the essential facts and ideas in the area of interest, short of all the nitty-gritty technical details that a specialist in the field must worry about. But most of all, we want to *understand* the basic principles behind these facts and ideas, the logic in the organization of concepts and in the working of things. In other words, we would like to be told in plain English *why*, and not just *what* and *how*. We believe that the greatest joy in learning a subject is to *understand* what is really going on, rather than simply to register all the relevant facts. We have attempted to transmit our own joy in learning physics through this book, by exploring with the reader a variety of physical phenomena and trying to understand them in terms of a few basic physical ideas.

Since we feel that such a book is of use to all those who love physics and to all those who are more than just simply curious about science, we have written for a diverse readership. The book can be used as a reference text for a university course aiming at a general survey of modern physics; in fact this book grew partially out of a course one of us (QHK) teaches. On the other hand, it can also be read by an individual who wants to understand the essence of one of the topics treated without having to be bothered by the detailed formalisms. Each of the seven chapters can be read independently. Each starts from the very beginning, then progresses into more modern and, perhaps deeper, areas in the field.

This *Invitation to Contemporary Physics* is therefore an invitation to an excursion into a marvelously varied country. There are many roads, some easier than others; but the traveler willing to take unfamiliar paths, to climb hills and mountains will be amply rewarded by beautiful sights. A novice at a first reading may find some of the later sections of a chapter rather difficult to master, and a more mature reader may find the earlier sections far too elementary. In other words, although the book has been written in such a way that any intelligent reader with a determination can understand, it is not to say that some of these topics can be understood without some effort on the part of the reader. If such difficulties are encountered, we advise the reader to skip the section or the rest of the chapter, and come back to it at a later date when some of the new concepts have sunk in. On the other hand, the more mature reader would perhaps prefer to gloss over some parts and concentrate mostly on those sections or chapters of interest to her or him. Footnotes are sometimes used to explain finer points or to give quantitative results. The more complex sections are recapitulated in short summaries. Finally, some of the fundamental concepts and facts of physics commonly used in the main text are described in three Appendices.

We are indebted to a great number of people for their valuable suggestions and comments. In particular, we wish to thank our friends and colleagues at McGill University: Abdelhamid Bougourzi, Cliff Burgess, John Crawford, Subal Das Gupta, Al-Amin Dhirani, Charles Gale, Terry Gannon, Noureddine Hambli, David Hanna, Jonathan Lee, Zhishun Li, Rob Myers, Bob Sharp, and Douglas Stairs; at Université Laval: Sea Leang Chin, Veronique François, Michel Gagnon, Eduardo Hardy, Roger Lessard, and Serge Pineault; and at the Indian Institute of Science: N. Mukunda, T.V. Ramakrishnan, Rahul Pandit, Sriram Ramaswamy, and H.R. Krishnamurthy.

Quang Ho-Kim
Narendra Kumar
Chi-Sing Lam

May 1991

CHAPTER I

LASERS AND PHYSICS

CONTENTS

1. INVITATION TO THE NEW OPTICS

The laser may turn out to be one of the most significant inventions of our times. A product of modern quantum mechanics, it generates a light endowed with many remarkable properties and qualitatively very different from the light hitherto available to us from conventional sources. This form of light, which gives us a completely new tool for probing Nature, already transforms and broadens to an extraordinary extent the ancient science of optics. It gives us a radically new power of control of light that opens up seemingly limitless applications in arts and sciences, in medicine and technology. Physicists have used lasers to study minute details of the structure of atoms and molecules, to catch atoms in flight, and to perform delicate experiments to test the very foundations of quantum mechanics. Biologists have used lasers to study the structure and the degree of aggregation of various biomolecules, to probe their dynamic behavior, or still to detect constituents of cells. Mathematicians actively involved with nonlinear complex systems have been intrigued by the possibility that their ideas could be tested by observing the dynamical instabilities exhibited by some lasers. And not only scientists or engineers — artists and dentists, soldiers and spies have also been touched by this invention.

The term *laser*, which is an acronym for *light amplification by stimulated emission of radiation*, is an apt description of the device. The principle on which the laser is based can be traced back to a work by Albert Einstein who showed in 1917 that 'stimulated', or induced, radiation could be obtained from an atom under certain conditions. But the actual invention of the laser did not come until 1960 when Arthur Schawlow and Charles Townes demonstrated that it was possible to amplify this kind of radiation in the optical and infrared regions of the spectrum. Soon thereafter, the first laser beam was obtained.

The laser is a device for producing a very tight beam of extremely bright and highly coherent light. To appreciate these remarkable properties of laser light that make the laser the unique tool it has come to be for research and applications, we will begin by considering light as it is normally found and discuss its characteristic features, with the view of contrasting them with the distinctive properties of laser light. There exist now many types of lasers, which use different substances as active media, achieve atomic excitations through different pumping techniques, and generate light at widely different wavelengths. They all share, however, the same basic principles. We will describe many applications of lasers, not only in technology, but also in basic research in many branches of physics where the use of this tool has led us to new directions, taken us to new frontiers.

---

2. CONVENTIONAL LIGHT SOURCES

2.1 Light and Electromagnetic Radiation

Light is ordinarily produced in hot matter. In 1704, in his second great work, *Opticks*, Isaac Newton wrote:

> Do not all fix'd Bodies, when heated beyond a certain degree,
> emit Light and shine; and is not this Emission perform'd by
> the vibrating motions of their parts?

Light can be described as a perturbation of space of the kind one may observe in the vicinity of two electrically charged plates or near the path of a rapidly moving charged particle. Once we have observed how iron filings spread around the plates shift about to align themselves into a regular pattern, or watched a small probe charge attached to the end of a very thin thread respond to the particle's motion, we are left in no doubt that space nearby is pervaded with a certain distribution of force that is called *electromagnetic field.* The precise way in which this field varies in space and time is described concisely by a set of differential equations, due to James Clerk Maxwell (1873), which replaced and generalized all previous empirical laws of electricity and magnetism. According to the theory encapsulated in these equations and later confirmed by experiments, the electromagnetic field is a disturbance that propagates from point to point in all accessible directions and behaves at large distances from the source as a wave. For this reason we may refer to it simply as an *electromagnetic wave*. One of its basic properties is that it can convey energy through empty space without the transfer

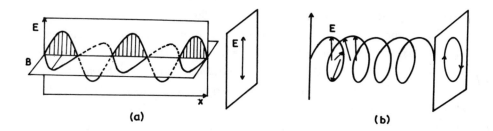

FIG. I.1 *Representation of an electromagnetic wave as space and time variations of electric field* **E** *and magnetic field* **B**. *In (a)* **E** *stays at all times in a plane passing by the propagation line, and the polarization is said to be planar. In (b)* **E** *changes direction as it evolves, and the polarization is nonplanar.*

of matter and that it is always moving at high speed, indeed at the speed of light ($c = 300000\,\mathrm{km\,s^{-1}}$ in empty space), and when it stops moving, it ceases to exist. We implicitly refer to this kind of energy transfer whenever we use the term '*electromagnetic radiation*'. An electromagnetic wave has all the characteristics, except visibility, of light: it can be reflected, refracted, or diffracted. Light, in effect, is the *visible* form of electromagnetic radiation.

A light wave, like any other electromagnetic wave, is described by the variations in space and time of two vectorial quantities, namely, an electric field **E** and a magnetic field **B**. These vectors remain at all times perpendicular to each other and perpendicular to the direction of propagation. Thus, the wave in question is a traveling transverse wave, much like the ripples on a disturbed water surface. The direction of the electric field is called the direction of the *polarization* of the wave. Over a period of time, the electric field vector defines a vibration pattern which may be projected in the **E-B** plane as a line segment, a circle, or an ellipse (Fig. I.1). The states of polarization of a light wave can be easily discovered: let light pass through a polarizer — a tourmaline crystal or a Polaroid filter — and observe the output as you slowly rotate the polarizer. A typical polarizer is a substance composed of long straight molecules aligned perfectly parallel to one another which strongly absorb the electric component parallel to the molecules but let the perpendicular component pass on through with almost no absorption. (The rope-and-picket-fence analogy is off by 90°).

Electromagnetic waves differ from one another in their characteristic *wavelengths*, λ, the distance from one peak to the next. The whole range of electromagnetic wavelengths, called the *electromagnetic spectrum* (Fig. I.2), covers values from the very small (high-energy gamma rays) to the very large (electric waves, radio waves). A small portion of it forms the *optical*, or *visible spectrum*, from 4×10^{-7} m (violet) to 7×10^{-7} m (deep red). Instead of wavelength,† one may equivalently speak of wave *frequency*, ν, which refers to the rate of vibrations, and is measured in cycles per second, or Hertz (Hz), or of angular frequency, ω, which is just the ordinary frequency multiplied by 2π, and so is given in radians per second. In practice, we can characterize a given radiation by its wavelength, λ, or by its frequency, ν or ω. For example, orange light has wavelength $\lambda = 0.6 \times 10^{-6}$ m $= 0.6$ microns (μm), or frequency $\nu = 5 \times 10^{14}$ Hz, or $\omega = 3 \times 10^{15}$ rad/s.

In many respects light behaves as a stream of quantum particles, which we call the *photons*. The production of an electric current by a sheet of copper when irradiated by intense light (a process known as the *photoelectric effect*) and the absorption or scattering of light by atoms are just two examples of phenomena that are more naturally explained in terms of the *corpuscular model* than the *wave model of light*. The photon has all the attributes of a particle: it has a mass and a charge, both of which are exactly vanishing; it has an intrinsic angular momentum (or spin) equal to 1 (meaning loosely that it can be described by a vector quantity, such as $\mathbf{E}$). Of all the elementary particles known to physicists (Chapter VI), the photon is the only particle that can be directly seen by those two marvelously sensitive detectors of ours — our eyes. The photon, of course, travels with the speed of light and has two other attributes of a particle, energy and momentum. The simple proportionality relations between the energy (or momentum) of a photon, on

---

† $\lambda = c/\nu$, where λ and ν are the wavelength and the ordinary frequency of the wave; the angular frequency ω is defined as $\omega = 2\pi\nu$. A brief discussion of the concepts of waves and fields can be found in Appendix A.

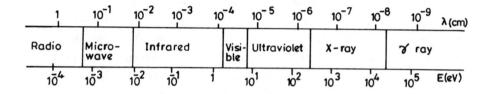

FIG. I.2 *Electromagnetic spectrum. Upper chart shows wavelengths in centimeters (cm); lower chart shows corresponding energies in electron volts (eV). Names associated with different parts of the spectrum are also indicated.*

one side, and the frequency of the electromagnetic wave that we may associate with the photon, on the other, provide the links between two apparently diverging pictures.‡

2.2 Spontaneous Radiation

An atom, like any other stable quantum system, is characterized by a set of discrete energy states; the atom may exist as a stable system in any one of these allowed states but in no others. In normal circumstances, the state of lowest energy, or the ground state, is the state in which the atom is found, and to which it ultimately returns after being excited. When an atom is placed in an electromagnetic field, the field acts on the atomic electrons, transferring to them energy. If an electron can convert the energy gained into potential energy for itself, then the electron, and therefore the atom to which it is bound, may be lifted to a higher energy, or excited, allowed state. This transition (a photon absorption) can occur only if the energy added

---

‡ $E = h\nu = \hbar\omega$, where E is the energy and momentum of the photon, ν (or ω) the frequency of the corresponding wave, h a numerical constant called the Planck constant and $\hbar = h/2\pi$. Thus, a radiation can be characterized by its wavelength λ, or by its energy E. Other aspects of the wave-particle duality are discussed in Section 6 of this Chapter and Appendix B.

will just raise the energy to one of the allowed values; otherwise there is no absorption.

Once an atom has reached an excited state, it stays there only for a short lapse of time — typically, about ten nanoseconds ($\tau = 10\,\mathrm{ns} = 10^{-8}\,\mathrm{s}$) for a transition in the visible region — then gives up its excess energy by dropping spontaneously to a lower energy level, emitting electromagnetic radiation (*i.e.*, photons) in the process. This transition is therefore called a *spontaneous emission*. The word 'spontaneous' refers to the fact that the transition is not provoked by the action of any external force, but is rather the result of an interaction of the atom with the all-pervasive electromagnetic field present in the medium itself. The frequency of the radiation emitted, ω, and the energies E_2 and E_1 of the initial and final levels are related by Planck's Law: $\omega = (E_2 - E_1)/\hbar$. Since the allowed energies of an atom are discrete, the corresponding emission frequency is discrete and is called a *spectral line*. The series of such lines forms a *spectrum*.

In other situations, a continuous, broad range of frequencies is observed: atomic transitions to an unbound state (whose allowed energy is nondiscrete), radiation of an incandescent liquid or solid (where atoms are packed closely), and radiation of a hot body (where frequent collisions cause loss of energy to the medium).

The light that emerges from any such source is nondirectional, non-monochromatic, and incoherent.

It is nondirectional because each point of the source radiates isotropically, with equal probability in any accessible direction in space. One may attempt to obtain radiation in a selected direction by placing a screen with a small hole in it some distance from the source, or by focusing the light output into a narrow beam with a mirror or a lens. But, obviously, part of the light will fall outside the collecting angle and will be lost. Even with the best available point sources, such as arc lamps, the resulting beam will nevertheless spread.

Perhaps you think that light waves behave as perfectly sinusoidal curves, oscillating rythmically and indefinitely over long distances. They do not. Ordinary light actually comes in a jumble of very short wave trains. The light vibration may change in shape and orientation over the duration of the wave train, depending on a large number of perturbations that might have affected the radiating atom during the emission process. What the eye perceives in the short lapse of time needed to make an observation is an average of the effects produced by an enormous number of unrelated, tiny wave packets it receives. Thus, even if the beam may have some dominant pattern of vibration at some given instant, the vibrations of its field components keep continually changing, favoring now one pattern, now another. Natural light emitted by ordinary light sources, by the Sun or any other star behaves in this way; it exhibits no long-term preference as to vibration pattern: it is *unpolarized*.

The lack of coherence in ordinary light results from both the finite size of the wave packets and the spatial spread of the radiating points in the light source. Suppose that two such points, separated by a small distance, emit at fixed time intervals short identical wave packets along two intersecting paths. Perhaps you agree that two such wave packets can meet and cause interference only if they overlap at the intersecting point. Moreover, only if the wave trains are sufficiently long does the interference pattern remain stable long enough to be seen. Consider now several such point sources radiating identical wave packets at random. Optical interferences could be observed on a screen some distance away. If the point sources are closely spaced, the interferences are almost identical for all sources, and a stable pattern appears. If, on the other hand, the emitting points are distributed over a relatively substantial volume — as they are in a conventional source — the wave packets follow paths of very different lengths, and produce at any point on the screen a succession of interferences which fluctuate in brightness one instant to the next. Such a pattern is very unstable and can hardly be visible to the eye. This is not necessarily an unwanted virtue: you would not want to read by a coherent light.

An idealized source that can emit infinitely long sinusoidal waves at a fixed frequency — *e.g.,* in radiative transitions between two infinitely stable and extremely sharp levels — is said to emit a *monochromatic* radiation at that frequency. Such manifestly is not the case with sources of continuous radiation; neither is it with ordinary sources of discrete radiation, as we now discuss.

Suppose that a certain number of atoms (or ions, or molecules, etc.) have been raised somehow, via some (pumping) mechanism, to some upper atomic level E_2. These atoms will spontaneously decay, or relax to lower

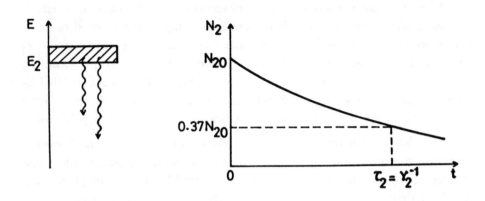

FIG. I.3 *Spontaneous relaxation of a population of N_2 atoms from atomic level E_2.*

energy levels, giving up their excess energy in the process. The energy release can be done in two ways. One is *radiative* relaxation, in which the radiation energy materializes as photons, directly measurable with a suitable photodetector. The other, called *nonradiative* relaxation, occurs mainly in solid-state materials; in this mechanism, the energy goes into setting up mechanical vibrations of the surrounding crystal lattice and not into producing electromagnetic radiation.

Any spontaneous transition can be characterized by its energy-decay rate, denoted by γ_2, which in general is the sum of a radiative part and a nonradiative part. The decay rate γ_2 tells us how fast atoms spontaneously relax downward from level E_2 according to an exponential decay law (Fig. I.3). This decay law simply says that after each lapse of time given by $\tau_2 = 1/\gamma_2$ the atomic population in level E_2 decreases by 63%. A discrete transition $E_2 \rightarrow E_1$ gives rise to an exponentially decaying signal oscillating with the frequency of the transition, $\omega_a = (E_2 - E_1)/\hbar$. Alternatively, we can describe such fluctuations with time at a given point as a superposition of an infinite number of periodic oscillations having different amplitudes and frequencies. Thus, the radiation carried by such a wave, although most intense at frequency ω_a, exists at other frequencies as well. The distribution of its intensity over all possible frequencies, called the *frequency spectrum* of the signal, is a bell-shaped curve centered at ω_a and has a linewidth $\Delta\omega_a = \gamma_2$. When γ_2 includes the contributions from the decay rates to all lower energy levels, it represents the *lifetime line-broadening* of level E_2. Its reciprocal, $\tau_2 = 1/\gamma_2$, is the *lifetime* of level E_2.

Other mechanisms may accelerate the relaxation process of atoms from a given level, and contribute to further broadening of that level. In gases, the most important of all such mechanisms is collisions between the radiating atoms and other atoms or molecules in the medium. In solid-state materials, the major contributing factor is the modulation of the atomic-transition frequency by lattice vibrations of the surrounding crystal lattice. All these line-broadening mechanisms act on all of the members of the system in the same way, so that the response of each member is equally and homogeneously broadened.

In many other situations, however, different atoms in a system of identical atoms may have their resonance frequencies unequally shifted, such that the resulting values of the resonance frequencies for individual atoms, ω_a, are randomly distributed about some central value ω_{a0}. When a signal is sent through the medium, it cannot pick out distinct responses from individual atoms, but it will receive a cacophony of overlapping responses from all the atoms present. This gives the effect of a broadening of the line transition, an effect generically referred to as the *inhomogeneous broadening* (Fig. I.4).

The prime example of such a broadening is the Doppler broadening of resonance transitions in gases, where the random change of the resonance frequencies arises from the *Doppler shift*, an effect akin in nature to the apparent change in pitch of the whistling from a passing train. Atoms in a gas have, in addition to internal motion, random kinetic motion, also called thermal motion. When an atom moving with a velocity v_x along x-direction interacts with an electromagnetic wave of frequency ω traveling with velocity c along the same direction, the frequency of the wave as seen by the atom is shifted to a new value given by $\omega' = (1 - v_x/c)\omega$. This means that the applied signal can resonate with the atomic transition only when the atomic transition frequency, ω_{a0}, coincides with the Doppler-shifted signal frequency, ω'. Alternatively, we may say that, seen in the laboratory frame, the atomic resonance frequency ω_{a0} appears to change to a new value, $\omega_a = (1+v_x/c)\omega_{a0}$.

From results of thermal physics (Appendix C), we know that the thermal

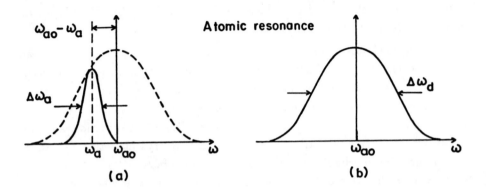

FIG. I.4 *(a) Individual atomic response with homogeneous broadening* $\Delta\omega_a$ *is Doppler-shifted from* ω_{a0} *to* ω_a. *(b) Inhomogeously-broadened atomic transition.*

velocity v_x of an atom of mass M in a gas at temperature T has a mean-square value given by $< v_x^2 > \approx kT/M$, where k is a numerical constant. So, the Doppler shift for that individual atom, and hence also the Doppler broadening for the whole system of agitated atoms, will depend (weakly) on both temperature and atomic mass (*i.e.*, as $\sqrt{kT/Mc^2}$). Typically, the Doppler broadening amounts to a few parts per million if we take $Mc^2 = 10^9$ eV (the mass of the hydrogen atom) and $kT = 25 \times 10^{-3}$ eV (corresponding to room temperature). For visible light with $\nu = \omega/2\pi \approx 6 \times 10^{14}$ Hz, it is $\Delta\omega_d/2\pi \approx 2 \times 10^9$ Hz.

In short, for reasons rooted in atomic and molecular physics, ordinary light sources are bound to generate light over a broad range of frequencies without supplying much power at any particular frequency.

Summary

Light is the visible form of electromagnetic radiation; its range of frequencies covers only a small part of the whole electromagnetic spectrum. In conventional sources (heated matter), light is produced in radiative collisions and when excited atoms in the source materials spontaneously relax downward to lower energy levels, releasing their excess energy in the process. The light that emerges from such sources emanates in all accessible directions of space, and is a mixture of very short unrelated wave trains.

---

3. LASER LIGHT

Lasers differ from ordinary light sources in the extent to which the radiation of the surplus atomic energy is controlled. We have seen that when an atom absorbs a photon, it converts the energy of the photon into internal energy for itself. The atom is then inevitably raised to an excited quantum state (Fig. I.5). After a short but unpredictable interval of time, the atom spontaneously decays, falling back to a lower level and radiating energy in the form of a photon. This spontaneous emission of photons and radiation in collisions are the sources of conventional light. The question is, can we coax an excited atom to release its excess energy at a moment of our choice? The answer is yes: while still in the excited state, the atom can be induced to decay and emit a photon if it comes in interaction with a photon that has an energy precisely the same as the energy that would be released in an allowed transition. (This photon may, for example, be part of an incoming beam that covers a range of frequencies, or the product of a previous transition of the same kind in the medium). In this process, known as the *stimulated emission*, the incoming photon is not absorbed; it merely triggers off the emission of a second, exactly identical photon. The waves carrying the two outgoing photons travel along the same direction, have the same frequency and polarization, and are *in phase*, that is, they rise and fall rythmically in step. This extraordinary behavior, observed with no other stable elementary particles but the photon, is the basis of the laser operation, and holds the key to our understanding of the laser light's many remarkable properties.

In matter at normal temperatures, atoms are never at rest. Their nervous, random motion produces at any point within and at the surface of the system a kind of pressure we call heat. It also takes them in a collision course with one another, constantly shifting about their energies and changing their states. But when the system reaches *thermal equilibrium* — when heat is uniformly distributed throughout — there are as many particles coming into each state as there are leaving it. The atoms will then be distributed among all the allowed atomic energy levels with a profile uniquely determined by the

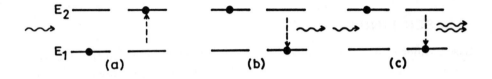

FIG. I.5 *Spontaneous and stimulated transitions in atoms.*
(a) Atomic excitation upon absorption of a photon.
(b) Emission of a photon in spontaneous downward relaxation of atom.
(c) Stimulated emission occurs when the excited atom is induced to decay
by a photon of frequency equal to the atomic transition frequency.

ambient temperature (the *Boltzmann distribution*). At a typical finite temperature, the atomic population in different allowed levels decreases smoothly and rapidly as the level energy increases (Fig. I.6). In a system at room temperature, practically all the atoms will occupy the ground state. And in this case, incoming photons are more likely to be absorbed by the multitude of atoms in the ground state than to provoke the few existing excited atoms to tumble down and radiate. We should expect there are far more photons lost through absorption than gained through stimulated emission. As a result, light gradually loses its intensity in its initial direction of propagation as it passes through the medium.

The above description of light propagation in an ordinary medium suggests a possible way to achieving light amplification: namely, by placing a great number of atoms in a selected excited state (the *active level*), thereby breaking the thermal equilibrium that prevails in the medium. Let us suppose, for example, that a suitable optically active system — perhaps a ruby crystal or a flask containing carbon dioxide — is injected with electromagnetic energy from a flash light or an electric discharge. Atoms in the system are then raised to a level or a band of levels from which they immediately decay to a lower energy level, the level one actually wishes to activate. The active level is, as a rule, chosen to be *metastable*, that is, unusually long-

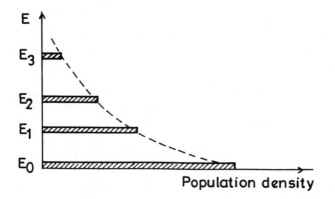

FIG. I.6 *Boltzmann distribution of the population densities of atomic levels in an ensemble of atoms at a given temperature* T.

lived; atoms can stay in this state hundreds of millions of times longer than in an ordinary excited state because their normal relaxation may be impeded by certain quantum laws. The population of a metastable state can thus be built up in excess of that of any lower levels. We describe this thermally unstable situation by saying that there is an *inversion of level population* in the system. Before this point is reached, spontaneous fluorescence is the dominant process, and emission is completely incoherent. At this point, however, the light thus spontaneously emitted has a high probability of inducing the remaining excited atoms to relax together to a lower, more sparsely populated level (*the terminal level*) and to generate massive waves of identical photons. The radiative field in the system changes radically in nature, then disorderly and chaotic, now orderly and coherent. A fundamental change of state has occured, rather similar to the phenomenon of phase transition, observed, for example, when vapor changes into liquid, water into ice, or in special circumstances, when fluid becomes superfluid (see Chapter II).

We refer to the quantum transition from the active level to the terminal level of an active medium as the *lasing transition*. Of course, simultaneously, some incident energy may be lost in reverse transitions between the same two atomic levels. If this process of absorption is much weaker than the stimulated emission, light will gain in intensity, and amplification will be achieved.

A suitable medium can certainly amplify light but, by itself, it cannot sustain a continuous production needed to make a useful laser beam. To transform an active medium into a generator of light, one encloses it in a resonant cavity, which is essentially a narrow cylinder closed at both ends by small, slightly curved mirrors facing each other. To obtain a resonant feed-back, the cylinder length L is to be equal to an integral number n of the radiation half-wavelengths, $L = n\lambda/2$ (Fig. I.7).

Atoms, when excited by an ordinary energy pump, radiate in all directions. Most of this radiation passes out of the cavity and is lost to the system. However, those photons emitted in the lasing transitions and that happen to move within the medium for some distances can interact with the atoms still on the active level, and induce them to give up more photons of the same frequency and the same phase. A coherent wave with that frequency and phase is then gradually built up with more and more identical photons as the wave sweeps through the medium. When it finally reaches a mirror at one end of the cavity, it is reflected back into the medium where it induces further radiation. Slight imperfections in the mirrors at both ends of the cavity may scatter away some of the impinging wave, but these losses are more than compensated for by the gain through repeated emission by the medium. To keep the medium radiating, it is necessary to maintain the population of the active level above that of the terminal level. Therefore, the atoms that have radiated and fallen to the ground state are continuously pumped back to the active level where they are again available to further stimulation. The steady, coherent wave that now travels in a direction precisely parallel to the axis of the cavity is reflected back and forth between the mirrors, and grows in amplitude with each new passage. The energy output by individual atoms — which are distributed over a relatively large volume and yet radiate with the same phase — adds up coherently to yield a powerful radiation. When its intensity has reached a suitable level, this radiation can be extracted through one of the mirrors made semi-transparent for this purpose. The emerging beam is very bright, very tightly collimated and, being made up of rows and rows of identical photons, behaves like very long waves of temporally and spatially coherent light.

Temporal coherence of light is, as we already know, measured in terms of the lengths of its wave trains or, equivalently, the widths of its spectral bands.

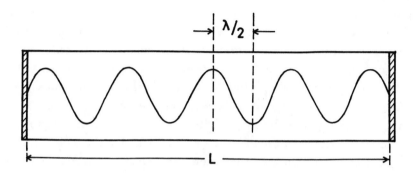

FIG. I.7 *Resonance condition in a standing-wave laser cavity of length L.*

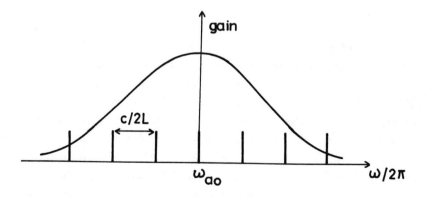

FIG. I.8 *Gain profile of a typical laser system. The available axial-mode frequencies are equally spaced, separated by c/2L, where L is the cavity length. The centrally-located frequency coincides with the natural atomic transition frequency, ω_{a0}.*

Radiation emitted by a laser, just as ordinary light, could take place through-out the Doppler-broadened distribution of frequencies (Fig. I.8). However, two factors contribute to make the linewidth of a laser gain considerably narrower. First, stimulated emission is more likely to occur at the lasing transition frequency, at the center of the spectrum, than at any other fre-quencies. Secondly, if the laser operates with a resonant cavity, only radiation whose wavelength is such that a half-integral multiple of wavelengths fits be-tween the mirrors can be supported as a cavity axial-mode, and subsequently amplified. Within the range of allowed frequencies, a large number of such modes may be supported by long cavities (since the mirror spacing, L, is usually much greater than the radiation wavelength, λ); the emission is then said to be *multimode*. On the other hand, for a short cavity, only one mode may lie within the gain bandwidth, and the laser emission is *single-mode*, producing an *extremely monochromatic* light output. For example, a helium-neon laser — a type of gas laser commonly used for reading bar-codes at supermarkets — can be so designed that the light it emits emerges in wave packets a hundred kilometers long with frequencies within a spectral band only a kilohertz wide. The monochromaticity quality of laser light, like its coherence, arises primarily from the resonant-cavity properties of the laser resonator rather than from any of the quantum properties of the radiating atoms.

We can readily demonstrate this striking property (Fig. I.9) by directing the beams from two such lasers on a photomultiplier (which generates a cur-rent by the photoelectric effect) connected to an oscilloscope (which displays visually the changes in the varying current). We assume that the two lasers are programmed to emit simultaneous pulses with two slightly different fre-quencies, ω_1 and ω_2. The two beams are then combined to give signals with frequencies ranging from $\omega_1 - \omega_2$ to $\omega_1 + \omega_2$. A filter is used to remove all but the lowest frequencies so that the signal actually received by the photomul-tiplier is a single modulated sinusoid of frequency $\omega_1 - \omega_2$. The intensity of the signal is large when the superposed waves interfere constructively. It de-creases and vanishes when they interfere destructively. Then, again, it grows larger. With the passing of time, the recurring pattern of enhanced and re-duced interferences moves along with the wave velocity. This phenomenon is known as *beats*. Beats cannot be observed with ordinary, incoherent light;

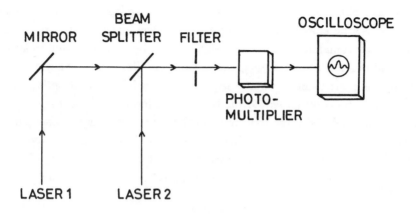

FIG. I.9 *Experimental set-up to observe optical beats. The beams from two lasers of nearly identical frequencies are directed by a mirror arrangement to a photomultiplier connected to an oscilloscope. The resulting signals have many compound frequencies. If all but those having a frequency equal to the difference between the input frequencies are filtered out, beats can be observed.*

they result from a stable interference of two long waves of nearly identical frequencies. Musicians often use acoustical beats to bring their instruments in tune at the beginning of a concert. They listen to the beats, and adjust one instrument against another to reduce the beat frequency until it disappears.

Spatial coherence, on the other hand, is sensitive to the size of the source of light. Light emerging from the aperture of a laser system diverges slightly. But when focused through a suitable lens, it always gives a point image as if emitted by a point source. This remarkable property can be demonstrated by repeating the two-slit interference experiment, first performed by Thomas Young in 1806. In this experiment, light passes through two parallel slits and falls on a screen placed some distance away. When the light waves emerging from the two openings fall in phase (crest on crest) at some point on the screen, they produce a bright spot. When they arrive out of phase (crest on valley), they cancel out, producing a darker spot. This superposition of waves gives rise to a pattern of alternating bright and dark lines on the screen, each indicating a half-wavelength difference in lengths between the two intersecting optical paths. The spacings between the fringes depend on the relative obliquity angle of the interfering incoming waves: the greater this angle is, the closer the fringes are. When the experiment is performed with a conventional light source, the source must be small or made to appear small, and must be placed some distance from the slits so that the wave fronts reaching them are as nearly plane as possible. If neither condition is met, no pattern of useful contrast will form. But when a laser is used as the source of light, it can be placed directly in front of the slits and a clear, stable pattern can be seen. More remarkable still, observations of interference of waves from different lasers that emit long wave trains of well-defined frequencies become practical, a feat evidently not possible with ordinary light sources.

Summary

A laser device has three essential components:

(1) a laser medium consisting of a collection of atoms, molecules, or ions in gaseous, liquid, or solid state;

(2) a pumping mechanism injecting energy into this medium to excite its component particles to higher energy levels; and

(3) a resonant structure that allows a beam of radiation to pass back and forth in the laser medium; in a laser amplifier, however, the beam passes only once through the medium.

The light produced by a laser device, though basically of the same nature as the light generated by any source of electromagnetic radiation, has a few significant differences with ordinary light. Laser light is both spatially coherent (spatially in phase) and temporally (or spectrally) coherent. Moreover, individual lasers can be, among other things, very powerful, very widely tunable, and very frequency-stable.

4. TYPES OF LASERS

The first working laser model was announced by Theodore Maiman in 1960; soon, lasers of different types were built. They all incorporate the three basic components described above: an *active medium* that amplifies light by stimulated emission, a *pumping mechanism* that injects energy into the system and thereby causes an inversion of population in the medium, and, finally, a *resonant structure* that transforms the emitted waves into a steady, directional and coherent beam having a few selected frequencies. Lasers use a variety of substances as active media and are capable of producing intense light at different frequencies, ranging from the infrared to the ultraviolet, and beyond.

4.1 Solid-state Lasers

In Maiman's laser, ruby was used as the medium. Here, as in many other solid-state lasers now in operation, the crystal that serves as the active medium is typically machine-tooled into a cylindrical rod about five centimeters long and half a centimeter across. Its ends are polished flat, parallel, and are partially silvered. It is placed at the center of a coil of electronic flash tube that can produce intense light. Ruby is aluminum oxide (sapphire) in which one of the aluminum ions has been replaced by a chromium ion (Cr^{3+}). The chromium ion absorbs green and yellow light, and lets the blue and red pass through, which gives ruby its characteristic color. When excited by visible light, the Cr^{3+} ion is excited to a broad band of energy levels which it quickly leaves to relax downward to a relatively long-lived level lying immediately above the ground state. This transition is non-radiative: it sends out no observable signals, the small surplus energy released is absorbed by the surrounding crystal lattice. As more and more chromium ions throughout the crystal reach the metastable state, the population of this state rapidly exceeds that of the lowest energy level, and we have an inversion of population. Soon, a few ions spontaneously fall to the ground state. The photons that are thus released go on striking other still excited chromium ions, triggering off a cascade of photons, all having the same frequency (Fig. I.10).

To keep the ruby laser in operation, it is necessary to pump at least half of the chromium atoms in the crystal to the active level. Such an effort consumes a good deal of energy. It can be significantly reduced in substances that allow a laser transition to end not in the ground state itself, which is always densely populated, but on a level that lies at some energy above it. The population of such a level is normally sparse and, provided a suitable metastable state exists, population inversion can be achieved with only a small energy expenditure by placing a relatively small number of atoms in the metastable state. Neodymium is an example of such a substance. Traces of neodymium are inserted in yttrium-aluminum oxide. When energy is injected

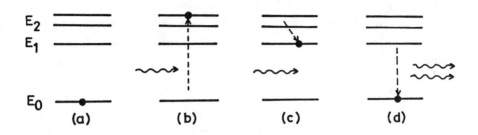

FIG. I.10 *Principle of operation of the ruby laser. Chromium ions are raised from the ground state (a) to a band of levels upon absorption of a photon (b). The atoms relax downward to a metastable state (c), from which they decay and radiate by stimulation (d).*

into the doped oxide sample, the neodymium atoms are excited to a band of levels, from which they drop to the ground state in three steps. The first and the last transitions are fast and spontaneous, and do not contribute to the laser beam. The intermediate transition, between a metastable state and a terminal level lying just above the ground state, proceeds by stimulation and produces coherent light.

4.2 Gas Lasers

Whereas most solid-state lasers operate in the pulsed mode, that is, they generate short, intense bursts of power, gas lasers in contrast are capable of continuous operation. A gas laser consists of a tube filled with atomic or molecular gas and placed in a resonant optical cavity. The pumping energy is provided by a high-voltage electric current. Energetic electrons are injected into the tube by the electric discharges and, through collisions, boost the gas atoms or molecules to excited quantum states. What makes gas lasers highly efficient is that there always exist in gaseous media many transitions capable of laser emission that terminate on levels above the ground state. It is essential that the atoms, once they have fallen back on such terminal levels, leave them as fast as they arrive and return instantly to the ground state, where they are again available for regenerating the populations of the active levels. Deexcitation of the terminal levels can be, and usually is, accelerated by adding to the laser medium some other appropriate gases, which provoke

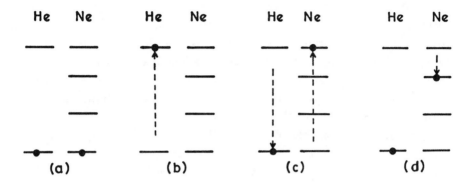

FIG. I.11 *Principle of the helium-neon laser. Both atoms are initially in the ground states (a). Helium atoms are excited by electron bombardment (b) and subsequently transfer their surplus energy, via collisions, to neon atoms which are excited to high-energy levels (c). When stimulated by an incoming photon, the neon atom fluoresces, contributing a photon to the laser beam; it then falls back to the ground state in steps (d).*

more frequent collisions and convert the surplus internal energy more quickly into kinetic energy. The high rate of depopulation of the terminal levels and a continuous repopulation of the active levels are the two factors that contribute to make gas lasers highly efficient continuous sources of light.

The first gas laser used atomic neon (Fig. I.11). It generated a continuous beam with excellent spectral purity, but its power output was low compared to the power output of solid-state lasers. The situation radically changed with the advent of molecular gas lasers. The carbon dioxide laser, the prime example of this type of lasers, is capable of producing beams several kilowatts strong. A carbon dioxide molecule consists of a carbon atom and two oxygen atoms arranged in a linear, symmetric pattern. The spectrum of its energy levels is far richer than the spectrum of each of its atomic components. Besides the familiar electronic excitations, the molecule, without altering its electronic configurations, can change its internal energy through oscillations of its component atoms about their mean positions, or through rotations of the system as a whole. Thus, each electronic level is associated with a set of vibrational levels, each in turn accompanied by a band of rotational levels. The spacings of electronic levels in molecules are comparable to those in atoms — a few electron volts — but the spacings of vibrational and rotational levels are smaller by a factor of ten and one hundred, respectively. In carbon dioxide, the transitions giving rise to laser light occur between rotational levels belonging to two different vibrational bands. These bands are produced by vibrational excitations of the molecule in the same electronic configuration. The difference in the energies of the two vibrational states determines the frequency of the light emitted, usually in the infrared. The vibrational levels responsible for laser oscillations in carbon dioxide are particularly long-lived: they can effectively store energy for about a millisecond. Their radiation can be controlled by a simple device, the *Q-switch*, which is simply a rotating mirror replacing one of the usual resonant cavity mirrors. The switch normally interrupts the path of the laser beam in the resonator. When the rotating mirror lines up with the opposite stationary mirror, the path is restored and the coherent beam passing through the column of gas touches off a massive avalanche of photons. Operating with such a scheme, a carbon dioxide laser can produce sharp, nanosecond pulses of energy which can reach peaks a thousand times greater than the average power it normally produces in a continuous operation.

Even shorter pulses can be achieved by '*mode-locking*'. As we have seen in the previous section, lasers usually allow within their oscillation bandwidth many axial-mode frequencies separated by a frequency spacing that depends on the length of the resonant cavity. The radiation of these modes is uncorrelated in phase, which results in a randomly fluctuating laser field. If, however, the modes could be correlated and made to oscillate with comparable amplitudes and a fixed phase difference from one mode to the next, they would interfere regularly in step to produce periodic pulses. The oscillations of the fluorescence thus produced could be visualized as a stable pulse that propagates back and forth between the cavity mirrors. The duration of the resulting laser pulse will vary in inverse proportion with the total oscillation width or the total number of modes locked in phase: the larger the laser gain bandwidth is, the shorter the pulse produced. Thus, gas lasers with bandwidths of the order of 10^{10} Hz are limited to pulses no shorter than 0.1 nanosecond. To obtain picosecond (10^{-12} s) pulses, the bandwidth must be greater than 10^{12} Hz. For this purpose, *dye lasers* (in which the active medium consists of solutions of organic dye compounds in liquids) are the ideal choice. Considerable progress in generating ultrashort laser pulses has been achieved: in 1981, pulses shorter than 100 femtoseconds (10^{-13} s) were observed; in 1987, scientists at Bell Laboratories in the U.S.A. produced pulses six femtoseconds in duration. Just imigagine: it takes light one second to circle the globe seven times, 50 femtoseconds to cross a distance shorter than the thickness of a hair!

4.3 Semiconductor Lasers

Solid-state lasers usually operate at a single frequency or, at best, at a few frequencies. Gas lasers and especially molecular gas lasers are more flexible; they can generate power at a very large number of discrete wavelengths lying within a narrow band. (For example, carbon dioxide can oscillate at about a hundred different wavelengths between nine and eleven microns; carbon monoxide has even more lasing lines on various vibrational-rotational transitions within a limited range of wavelengths). Still, their radiation frequencies cannot be continuously varied and controlled or, as one says, tuned. This may be a drawback for some applications. Fortunately, fully tunable lasers are available; they are based on semiconductors.

Atoms within solids are generally arranged in a stable, regular pattern. Although in perpetual oscillations about their equilibrium positions even at the lowest possible temperatures, they remain within the range of their mutual electromagnetic interactions. The electrons in the inner orbits are strongly bound to their respective atoms and are largely unaffected by outside forces, with essentially the same energies as in isolated atoms. On the other hand, the electrons in the outer orbits, farther from the centers of forces, are more susceptible to external perturbations. In fact, the clouds of such electrons in adjacent atoms may even overlap. As a result, the outer electrons interact strongly with one another and, therefore, are not limited in energy to a few discrete levels but may have a broad range of energies which tend to cluster in *bands* around the sharply defined levels that would be otherwise observed in isolated, non-interacting atoms.

In general, the bands are fully occupied from the lowest energy all the way up to a certain limit (the *Fermi level*), above which the electron population of the bands drops off abruptly to zero. Such an energy spectrum

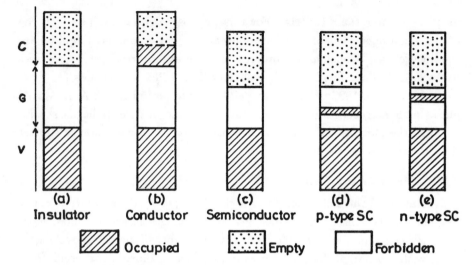

FIG. I.12 *Valence bands (V), conduction bands (C), and forbidden band gaps (G) for (a) an insulator, (b) a metal, (c) a pure semiconductor, (d) a p-type semiconductor, and (e) an n-type semiconductor.*

indicates that the electrons remain firmly in their atomic orbits. In most metals, however, electrons in the external orbits are loosely bound. Some of them may even escape their orbits and move about freely in interatomic space. Thus, normally occupied states become empty —'*holes*' are created — and normally empty states become occupied. The presence in metals of free electrons, unattached to any atomic sites, makes such substances good *conductors* of electricity and heat. In an electrical *insulator*, such as sodium chloride or calcium fluoride, all electrons stay in their orbits. In this case, the lowest-energy empty band (the *conduction band*) is separated from the highest-energy occupied band (the *valence band*) by an energy region that has no allowed quantum states (the *band gap*); electrons cannot have an energy lying in this interval (refer to Fig. I.12).

Semiconductors are intermediate between conductors and insulators. A pure semiconductor (silicon, selenium, gallium arsenide, etc.) has an energy structure similar to that of an insulator with a small gap between the valence band and the conduction band. In practice, semiconductors are often doped with impurity atoms to modify their energy spectra and hence their electrical properties. Interactions with the foreign substances may remove some

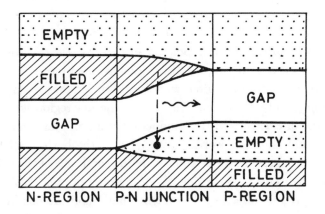

FIG. I.13 *Principle of operation of a semiconductor laser. Layers of p-and n- semiconducting materials are separated by a thin layer of the same, but undoped material (p-n junction). Electrons are pumped into the n-region. At the junction, the electrons drop into empty states of the p-region, emitting photons. This recombination radiation is amplified by the geometry of the junction plane, the mirrors at both ends of the junction, and a proper choice of the refractive index such that the junction can act as a wave-guide.*

electrons from the valence band and introduce some in the conduction band, creating a condition more favorable to conduction. So the situation normally observed in metals can be re-created in some solids in a way that may better suit our purpose. In a *type-p* (positive) semiconductor, the impurity atoms have empty electron levels just above the top of the valence band of the pure semiconductor. These empty electron levels can be readily reached via thermal excitations by valence electrons, leaving holes in the valence band, which then act as carriers of positive electric charges. In a *type-n* (negative) semiconductor, the impurity atoms have electron levels, each filled with an electron, just below the conduction band of the pure semiconductor. The electrons that occupy these levels can be excited to the conduction band, where they act as negative carriers of electricity.

The *semiconductor laser* in its simplest form is a junction diode formed by the juxtaposition of a type-n semiconducting crystal and a type-p semiconducting crystal (Fig. I.13). The opposing faces of the crystals are polished flat and parallel. They are separated by a thin layer of undoped semiconducting crystal. An electric current can be fed into the diode by connecting its p-component to the positive pole of an external electric source and its n-component to the negative pole. Electrons injected into the system move from the n-region to the p-region, whereas the holes present in the crystals move in the opposite direction. Thus, the junction region is quickly filled with an excess of electrons in the conduction band and an excess of holes in the valence band. In other words, levels lying at the bottom of the conduction band receive many new electrons while the top levels of the valence band lose many of their customary residents and become empty. There exists then a population inversion between the valence band and the conduction band, exactly the situation required to produce a sustained light emission when the conduction electrons, which exist in abundance all along the plane of the junction, are stimulated to drop into the empty states of the valence band. This (recombination) radiation steadily grows as it propagates between the two reflecting inner faces of the crystals which act as wave guides. The amplified wave finally emerges as a laser beam at one end of the junction region.

Semiconductor lasers are highly efficient sources of energy because every electron fed into the system contributes a useful photon (nonradiative decay being negligible) and no radiation is wasted in non-coherent transitions. Their output is, moreover, tunable. For instance, lasers based on gallium arsenide emit light at a wavelength around 9000 Angstroms ($1\text{Å} = 10^{-10}$ m) at room temperature. As the ambient temperature is gradually lowered, the emission wavelength continuously decreases to 8400 Å. We can further widen the range of available wavelengths by inserting impurities in the semiconducting medium. For instance, lasers based on gallium-indium arsenide can operate between 8400 and 31000 Å. The emitted light can be tuned over this considerable range of wavelengths simply by varying the temperature of operation. Another practical advantage of semiconductor lasers is that they can be miniaturized. Thus, you may have a diode laser, composed of several layers of semiconducting materials, as small as a grain of salt.

4.4 And all the Others

The above descriptions give but a small sample of the huge variety of media capable of sustaining laser action. The range of wavelengths of laser radiation now available extends well beyond the visible region, and covers all radiation between the far infrared and the soft X-ray regions of the spectrum. If we include in our considerations the *maser* — a device preceding the laser and generating not visible but microwave radiation — then the wavelengths of coherent radiation available range from 5 centimeters to 5 nanometers, two extremes separated by a factor of 10^7. Power outputs range from a few milliwatts in low-power continuous lasers to hundreds of kilowatts in high-power continuous lasers, and up to 100 terawatts in pulsed lasers. There is also an enormous choice of pulse widths, anywhere from the millisecond $(10^{-3}\,\text{s})$ level to the femtosecond $(10^{-15}\,\text{s})$ level. The physical dimensions of different types of lasers also vary widely; the low-power semiconductor laser diodes are miniatures with dimensions less than a millimeter, while the high-power neodymium-glass laser called NOVA is a mammoth many stories high.

Summary

Since the first laser was presented to the public in 1960, many types of laser systems have become available. Coherent light is now obtainable in wavelengths extending from the far infrared (at wavelengths around 0.5 mm) to the near ultraviolet (at wavelengths around 1000 Å). Outside this interval, microwave radiation is readily available from masers, and soft X-rays have been generated in laboratories by a new class of lasers. Laser action can take place in many media. Some of the main types are:

(1) solid-state lasers, where the active elements are often impurity ions (e.g., transition metal ions, notably Cr^{3+}, or rare earth ions, notably Nd^{3+} or Er^{3+}) introduced into an ionic crystal; optical pumping is the commonly used excitation technique;

(2) gas lasers, where the active elements are atoms, molecules, and ions in gas or vapor phase; pumping is achieved by passing a strong electric current through the gas;

(3) semiconductor lasers, using a variety of semiconductors as active media (notably GaAs), which operate on the principle of stimulated emission of recombination radiation.

Our list gives only a sample of the most common lasers. Many more lasers exist, such as liquid lasers (organic dye compounds in liquids, gels, vapors, or plastics), free (i.e., unbound) electron lasers, and X-ray lasers.

5. APPLICATIONS OF LASERS

5.1 Microelectronics and Microsurgery

Many of the most familiar applications of lasers derive from the extreme brightness of the laser light, which can be many orders of magnitude greater than the output obtained from the best conventional light sources. The laser beam can be captured with a suitable lens and focused to an extraordinary density into a very small image whose size, perhaps a few microns across, depends only on the resolution of the lens and not on the laser aperture.

This tremendous concentration of energy can be used to heat, weld, melt, or cut small areas of any substances in operations that require a high precision and a good control of the amount of power to be applied. In microelectronics, collimated beams of laser light are now routinely used to make microcapacitors by cutting meander paths through a conducting film vapor-deposited on some substrate, or to make highly specialized microcircuits by performing discretionary wirings on general purpose circuits. The unique properties of lasers have also attracted, early on, the interest of practitioners in the medical field. A well-focused laser image, flashing intense pulses less than a millisecond in duration, makes it an ideal surgical tool: it can make precise small cuts and cauterizes as it cuts; it can stop blood circulation in a small volume of tissues; it can melt away the constricting plaque inside blood vessels that could lead to a heart stroke. A laser beam can be carried inside the human body on optical fiber light guides, and used to attack ulcers and tumors in internal organs. Because of the absence of any contact, it ensures a perfect asepsis and an excellent cicatrization. Specially designed, compact lasers have now widely replaced the more traditional tools in ophthalmology to weld torn retinas to their support by coagulation and to remove the degenerative blood vessels that cause diabetic retinopathy, in dermatology to treat angiomas or tumors of blood vessels, and in odontology to selectively destroy caries.

5.2 Surveying and Fitting

Surveying and machine-tooling are just two examples of fields that specifically exploit the high intensity and directionality of laser beams. The continuous power output of a gas laser, such as the helium-neon laser, can be focused through a telescope to give a beam that remains constant in cross-section, one millimeter across, over a large distance. The directionality and the considerable length of the beam make measurements at intermediate points unnecessary, even in works done over wide areas such as required in building roads, tunnels, or in laying pipelines.

For short distances, up to 50 m, a frequency-stabilized helium-neon laser is used together with interferometric techniques. The laser beam is split into a reference beam and a measurement beam. The first is reflected by a fixed mirror while the second by a mirror fixed to the object being measured. The two reflected beams are then recombined so as to interfere on some suitable electronic device that can count the interference fringes. When the position of the object changes in the direction of the beam by half a wavelength of the laser light, the interference signal will shift from a maximum to a minimum, or vice-versa, and the distance of the object relative to a given initial position is measured, often with an accuracy of one part per million. For greater distances, perhaps up to 1 km, the beam of a helium-neon laser or a gallium arsenide laser is amplitude-modulated (the amplitude of its output is modified in a well-defined way), and is reflected on the object being measured. The distance is then determined by the phase difference between the direct and the return beams. This technique is common in geodesy and cartography. For even greater distances, the distance is determined by the time of flight of a nanosecond pulse emitted by a Q-switched carbon dioxide or neodymium laser and reflected on the object. Thus, a routine ranging of satellites can be done, and the distance to the Moon can be measured with a precision of some ten centimeters.

The directionality of laser light also makes it ideal for alignment purposes. The laser beam serves as a straight reference line, for example, in assembling large sections of aircraft or in civil engineering works. The beam is sent toward four solid-state detectors disposed in quadrants; it is perfectly centered when the detectors, upon receiving parts of the beam, send out photocurrent signals of exactly equal strengths; otherwise the beam is readjusted until perfect alignment is obtained.

5.3 Interferometry

Interference is used to measure optical paths and to detect their small variations. In particular, it is a standard technique for controlling the shapes of polished surfaces and for detecting small surface defects. For example, to test the planarity of a polished glass or metallic plate, a parallel beam of light is directed on a reference, perfectly plane glass plate and a sample surface placed just behind. The rays reflected from the two surfaces are then combined to produce an interference pattern on a screen. The interference fringes are expected to be rectilinear and parallel if the test surface is perfectly plane. If the test surface is spherical, the fringes are circular and concentric, and their radii can be related to the radius of curvature of the sample surface. Using a laser, rather than a conventional source, has several advantages: its output is adequate both for visibility and for producing interference, and the long coherence length of the laser light frees the experimenter from the impractical need of exactly matching each time the reference path length with the reflected path length. Thus, in testing planarity, the two plates need not be parallel nor be at some definite distance apart. Likewise, a polished lens can be tested for sphericity and its radius of curvature can be measured with a reference lens of any curvature.

5.4 Communications

The advent of the laser marked a new era in long-distance *communications*. Light, it is true, had been used for signaling since time immemorial. But it had been handicapped in its development as a practical means of communication by the noisiness and feebleness of existing sources. Consequently, it was replaced by the turn of this century by the more efficient and more versatile electric techniques. With the invention of the laser, light has become, once again, the focus of interest as a vehicle for long-distance, high-volume communications. This renewed interest is justified both by the inherent superior capacity of light for transmitting information and by the special properties of the laser light itself.

A long-distance communication system based on electromagnetic waves consists of five basic components: an oscillator to generate a pure, smooth information-carrying wave, a modulator to encode information on this carrier wave by altering some of its properties (frequency, amplitude, or phase) in a controlled way (Fig. I.14), a medium to convey the modulated wave, a detector to receive the signals, and, finally, a demodulator to extract information from the signals.

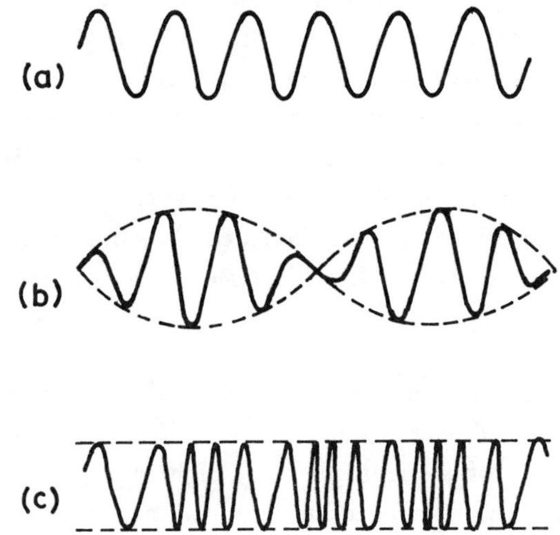

FIG. I.14 *Two common methods of wave modulation. In amplitude modulation (b), the amplitude of a perfectly sinusoidal carrier wave (a) is modulated according to some lower-frequency wave. In frequency modulation (c), the frequency of the carrier is modified according to some definite pattern.*

The capacity of a communication system is measured by the maximal information it can transmit in a unit interval of time. This capacity depends crucially on the *frequency of the carrier wave*: the higher the frequency is, the more rapidly the oscillations follow one another, and so the more numerous bits of information can be imprinted on the carrier wave. In the lower frequencies, around one megahertz, used by ordinary radio broadcasts, only voice and music can be transmitted; but at fifty megahertz, electromagnetic waves can carry the complex details of television pictures. Imagine the amount of information that visible light, at frequencies some ten thousand times higher, could transmit! Over the past decades, radio engineers have extended the usable radio region of the spectrum to all frequencies between 10^4 Hz and 10^{11} Hz — from the navigation bands to the bands reserved for use by microwave relays and radar stations. But with the insatiable needs of modern society, the available capacity for transmission of the electromagnetic spectrum has almost reached its point of saturation. The visible region of the spectrum, which ranges in frequencies from 4×10^{14} Hz to 7×10^{14} Hz, could in principle support ten thousand times more transmission channels than all the present radiowave and microwave portions combined, and could satisfy our communication needs for many years to come. In addition, the infrared and far infrared light, which is available in many types of gas or semiconductor lasers, holds another attraction: it suffers little loss in transmission on Earth and out into space because the Earth's atmosphere is partially transparent to that light.

Carrier waves used in long-distance transmission of information — much like sheets of paper used in written communications — play a key role in determining the quality of the transmission. Light waves generated by lasers are ideal for use as carrier waves in intercity communications for at least two reasons. First, as we have already discussed, the *spatial coherence* of the laser light makes it possible to have a highly-directional, well-focused beam over very large distances. The narrowness of the beam means that a large fraction of the radiation output can be coupled to the transmitting medium. The power output, already susbstantial to begin with, suffers little loss, and so can provide perfect conditions for transporting broadband information over great distances. Secondly, the *monochromaticity* of laser light is a decisive advantage because it helps to preserve the integrity of the information transmitted. When a light signal containing a mixture of colors travels through a dispersive medium — an optical glass fiber, for example — its spectral shape is inevitably distorted. Waves of different colors travel at different velocities: the higher the frequency, the lower the velocity. Blue light falls behind, red light gets ahead, and the pulse spreads out unevenly. The distortions, which are appreciable in pulses with broad bandwidths, could pose severe limits on transmission. Consider for example the high-intensity light-emitting diodes commonly used in light wave communications. Their output has a spectral bandwidth of about 350 Å, centered in the infrared part of the spectrum. A pulse from these sources would spread over 65 cm/km, which would limit the signal rate to about 1.5×10^8 pulses per second. In contrast, a pulse from a typical diode laser, which has a bandwidth of about 20 Å, would suffer a wavelength dispersion of only 4 cm/km, allowing a transmission rate twenty times better.

5.5 Holography

The laser has also been instrumental in the rapid development of another field, *holography*. Invented in 1947 by Dennis Gabor, holography is a photographic process which does not capture an image of the object being photographed, as is the case with the conventional technique, but rather records the *phases* and *amplitudes* of light waves reflected from the object. The wave amplitudes are readily encoded on an ordinary photographic film, which converts variations in intensity of the incident light into corresponding variations in opacity of the photographic emulsion. Recording the phases is another matter since the emulsion is completely insensitive to phase variations. But here comes Dennis Gabor with a truly ingenious idea: why not let the light reflected from the subject interfere with a reference coherent light on the photographic plate so as to produce interference patterns, which are then visible to the film emulsion?

To understand Gabor's idea, let us first consider a simple arrangement in which the object being photographed is just a point (Fig. I.15). A beam of perfectly coherent light is divided into two parts: one goes on to illuminate the object before being reflected away, the other — the reference beam — goes directly to a high resolution emulsion plate. The two halves of the beam recombine on the plate and produce a clear pattern of alternating dark and bright circular fringes. The spacings of the fringes depend on the angle between the reflected and direct waves as they together strike the plate. The greater this angle is, the more closely spaced the fringes are. The pattern tends to be coarser at the center of the plate, directly facing the illuminated object, than near the edges, where the reflected beam makes a greater angle with the direct beam. Therefore, the variations in the spacings of the fringes give an exact measure of the phase variations of the reflected waves. Similarly, local variations in the amplitude, or intensity, of the reflected waves translate into local variations in the contrast of the fringes. In other words, the perfectly coherent waves of the reference beam act as carrier waves on which is impressed the information transmitted by the light reflected from the object. The waves thus modulated are then recorded by the photographic emulsion. Once the plate is developed in the traditional way, it has the 'whole picture' of the object photographed: it is a *hologram*.

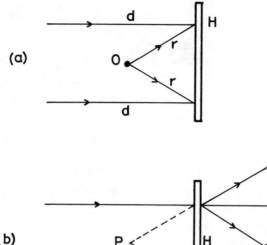

(a)

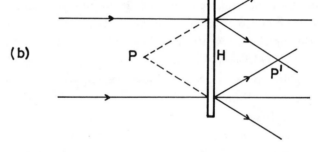

(b)

FIG. I.15 *(a) Making of a hologram. Interference patterns are produced and recorded on a photographic plate H when the direct (d) and the reflected (r) beams interact on the plate.*
(b) Viewing of the hologram. When the developed hologram H is illuminated by the same coherent light as used in the recording, two images are formed: a virtual image P' and a real image P'' at points symmetric with respect to H.

When such a hologram is illuminated by a collimated beam of coherent light, it shows the same properties as a grating surface. The transparent slits on the negative let the light pass on through and effectively act as sources of radiating cylindrical waves. These waves reinforce each other in certain directions and produce diffractions of varying degrees of intensity. For example, the two directions of strongest reinforcement can be constructed — as suggested by Christiaan Huygens, one of the first proponents of the wave theory of light in the seventeenth century — by drawing lines tangent both to a wave front emerging from each slit and to the wave fronts emerging a period earlier from the two adjacent slits. We can draw in this way two se-

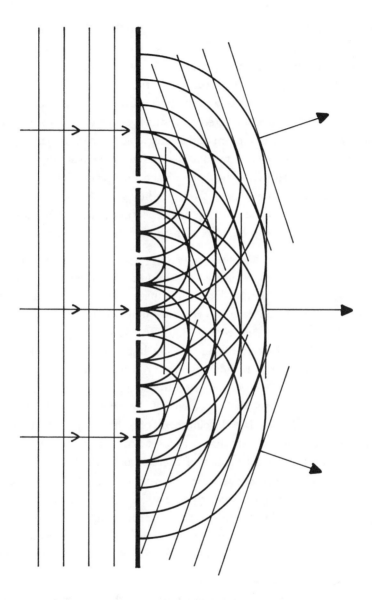

FIG. I.16 *Diffraction pattern as could be formed in a hologram reconstruction. The figure shows an unscattered wave parallel to the direction of the incident light, and two diverging diffracted waves which could be prolonged backward to meet at the virtual image point.*

ries of parallel lines, representing moving wave fronts, going away from the
hologram in two diverging directions (Fig. I.16). Each direction defines a
diffracted wave. Its obliquity depends on the separation of the slits: the finer
the grating spacings, the greater the diffraction angles. Since the fringes become more closely spaced as one moves away from the center of the pattern,
it is evident that the two diffracted waves have reverse curvatures. One wave
diverges away from the hologram and seems to emanate from the point where
the real object was placed; it produces a *virtual image* visible to an observer
placed on the opposite side, in front of the hologram. The other, in contrast, converges as it moves away from the hologram to form a *real image* at
the point symmetric to the position of the virtual image with respect to the
hologram. This image can be seen with the eye or recorded with a camera.

Now, if we take as the subject of our experiment a realistic, complex
three-dimensional object, we still can obtain an interference pattern by making the split beam of coherent light recombine on a photographic plate, just
as before. Of course, the data recorded are much more complex. Each point
on the surface of the object reflects light to the entire photographic plate;
conversely, each speck of emulsion receives light from all reflecting parts of
the object. So the local variations in opacity and spacings of the interference
fringes on the plate are directly related to the irregularities in the impinging
waves and, ultimately, to the complexity of the reflecting surface. When the
hologram obtained from the development of a film exposed in this way is
placed in a beam of coherent light, two sets of strong diffracted waves are
produced — each an exact replica of the original signal-bearing waves that
impinged on the plate when the hologram was made. One set of diffracted
waves produces a virtual image which can be seen by looking through the
hologram. It appears in a complete three-dimensional form with highly realistic perspective effects. In fact, the reconstructed picture has all the visual
properties of the original object and, for optical purposes, can be as useful
as the original object.

As we already mentioned, each point on the hologram received light
reflected from every part of the illluminated object and, therefore, contains
the complete visual record of the object as a whole. Any fragment of the
hologram, no matter how small, can be used to reconstruct the whole image.
The only limitation here is that resolution deteriorates and the perspective

effect is reduced as the fragment becomes smaller.

Another interesting property of the hologram with considerable practical potential is the possiblity of enfolding many wave patterns in a single plate by exposing it successively to light with a different frequency for each recording. (Compare this process with a simultaneous transmission of many radio messages by carrier waves of different frequencies). Each image is then unfolded with a laser beam having the same frequency as the light used in the construction process.

Several recent developments have contributed to make this photography-by-reconstruction-of-light-waves an exciting field of research. One of them is the introduction of three-dimensional holography. An ordinary hologram is an essentially two-dimensional recording of wave fronts as variations in opacity of a thin photographic film. A three-dimensional hologram is a thick plate of high resolution emulsion that can record interference fringes throughout its thickness. This is possible when the reference beam and the reflected beam make a large angle between them as together they strike the plate; the interference fringes will then be much finer than the thickness of the emulsion layer. The plate, when developed, acts much like a three-dimensional grating or a crystalline lattice. When exposed to a beam of light, it diffracts light in the same way as a crystal would diffract X-rays. In contrast to an ordinary hologram, which diffracts light of all colors whatever its incidence angle, a three-dimensional hologram is more discriminating. For a given angle of illumination, only waves of a specific color, reflected from successive emulsion layers, can fall in phase and lead to a strong diffraction. It follows that a volume hologram can be illuminated by ordinary white light, a mixture of all colors, and still can give a good reconstruction of the recorded image: the diffraction-in-depth 'knows' which proper band of wavelengths to pick out from the incident light. Another practical possibility is to make multicolor holograms by superposing volume holograms of different colors.

Summary

Many of the more familiar practical applications of lasers take advantage of the unique qualities of the laser light. We have briefly mentioned some such applications in the medical field, microelectronics, surveying, fitting, and interferometry. We have also discussed in greater detail the use of coherent light, of which lasers are excellent sources, in communications and holography.

6. QUANTUM OPTICS

In the previous sections we have described how a purely theoretical idea of Albert Einstein's has contributed to the invention of the laser and the subsequent development of one of the most important technologies of our time. The laser technology in turn has dramatically stimulated not only the field that has spawned and nurtured it but also many other scientific endeavors, and even has initiated many new, unexpected lines of research. Using the laser as a tool, physicists, chemists, biologists, and medical researchers have pushed their respective fields to new frontiers. In what follows, we will focus on physics and discuss how physicists make use of the light newly available to them to scrutinize matter from various angles — its structure, its bonding, its interaction with light — and to explore many fundamental aspects of physics — the particle-wave duality, the reality of quantum-mechanical entities.

6.1 Atomic and Molecular Spectroscopy

Atoms and molecules are quantum-mechanical systems which can exist, in contrast to classical objects, only in a certain number of discrete states and no others. These states are directly determined by the composition and the dynamics particular to each system and the general laws of quantum mechanics. Each quantum state (or energy level) is defined uniquely by a set of physical characteristics, called quantum numbers, two examples of which are the energy and the angular momentum (or spin). The full set of such states is unique to the system (atom, molecule, etc.), and so can serve as its signature. It is essential to know exactly the physical properties of the atom in its various states because by comparing them with theoretical calculations, the physicist can identify the system, learn about its structure and the forces shaping it.

One of their favorite approaches is to measure, whenever possible, the radiative transitions between levels and, from information on levels already identified and knowledge of the electromagnetic force governing such processes, extract bits of facts from unknown levels. Unfortunately, the signals they detect for a given quantum transition are, as a rule, spread around the expected transition energy (or frequency) rather than located precisely at this energy itself, as one would anticipate from general conservation of energy. In Section 2, we mentioned that this kind of radiation distribution, or line broadening, arises mainly from two effects. First, quantum levels in real physical systems are fuzzy rather than sharply defined (because of their finite lifetimes, various perturbations they are subjected to, or whatever), so that any transitions may take place within these uncertainties. Second, the frequencies of all these transitions are further altered by the Doppler shift arising from the thermal motions of atoms or molecules; this effect is most pronounced in high frequency transitions in gaseous samples involving light mass particles at high temperatures. What one observes then is not a single emission mode but many transitions, closely related yet differing in frequencies. The upshot is that the object of the experimenter's quest, the energy level, is hidden somewhere beneath this background of noise.

One of the most significant contributions of the laser to atomic and molecular spectroscopy is to reduce this noise by eliminating the (first order) Doppler broadening, which is (except in transitions involving very short-lived excited states) the major factor limiting the resolution ultimately attainable in conventional spectroscopy. Another notable success of the laser is that, with its high intensity energy output, it makes accessible to the experimenter's scrutiny transitions that can only be reached by processes involving more than one photon at a time.

• *Single-photon transitions*

In a typical spectroscopic experiment making use of a laser light source, the light output from a tunable dye laser with very stable lines and very small widths is beamed into an atomic (or molecular) gas sample, traverses it once, then is reflected back into the gas from the other end. The intensity of the reflected beam is measured for different laser frequencies and, at the peak intensity, the laser is tuned to the frequency of an atomic transition unaltered by any Doppler effect.

In this experiment, atoms moving in the gas generally see the direct and reflected beams at Doppler-shifted frequencies, one up, the other down. If the laser light is sufficiently intense to excite a large number of atoms, the velocity distribution of gas atoms at ground states will be depleted at some definite velocity and we will see two dips in the distribution, one produced by each beam, at *different* atomic velocities (Fig. I.17). The separation of

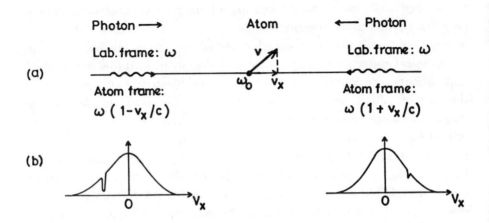

FIG. I.17 *(a) Doppler effect for an atom moving obliquely to two counter-propagating laser beams: photon frequency ω is shifted to $\omega(1\pm v_x/c)$. (b) Velocity-distribution of atoms. Dips in the distribution are produced by the two beams tuned to resonate with obliquely moving atoms, and occur at $v_x = \pm c(1 - \omega_0/\omega)$, where ω_0 is the atomic transition frequency.*

the two dips gives a measure of the atoms longitudinal velocities. Whenever the two dips merge together, it means that the *same* sets of atoms interact with the two beams. This can happen only when their longitudinal velocities vanish, *i.e.*, when the Doppler effect is absent. Then, provided the laser light is strong enough to excite a sufficiently large number of atoms, the direct beam will bleach a path through the sample, completely depleting it of a set of atoms moving at right angles to the beams over a range of velocities comparable to the radiative transition linewidth. It means that less of the reflected beam will be absorbed, resulting in a greater intensity than at any other frequencies, and its strong signal will display the spectrum of the transversely-moving atoms *free of Doppler broadening.*

With this (saturation one-photon) technique, experimenters were able to study the fine structure of the famous red line of atomic hydrogen at wavelength 6565 Å. As you know, this line is one of the lines first observed in 1885 by Jakob Balmer in the absorption spectrum of the Sun that were later shown to be part of the spectrum of the hydrogen, and eventually became known as the 'hydrogen Balmer series'. Balmer showed empirically that the frequencies of these lines obeyed a very simple formula, something like R/n^2, where $n = 2, 3, 4, \ldots$ and R is a constant, called the Rydberg constant. This remarkable formula remained a mystery until 1913 when Niels Bohr derived it from his quantum model. The red line which concerns us here corresponds to fluorescence emission between states $n = 2$ and $n = 3$. But what appeared as a single line to Balmer and other physicists up to recent days is actually a set of seven closely-spaced lines corresponding to seven possible transitions between component states with $n = 2$ and $n = 3$ but different values of energies and spins. This is what we meant by 'fine structure'. Most of this fine structure is obscured in conventional spectroscopy by the very large Doppler broadening found in the hydrogen atoms, but the Doppler-free laser method succeeds in displaying details unseen before. With the hydrogen lines resolved into individual, well-defined components, the absolute wavelengths can also be measured accurately, which in turn leads to significantly improved precision in the value of the Rydberg constant, a cornerstone in the determination of other fundamental constants in physics. The most recent measure by laser spectroscopy has yielded a value for R with the accuracy of one part in 10^9.

• *Multiphoton transitions*

So far we have discussed electromagnetic transitions involving only one photon at a time. They are practically the only ones possible in weak fields. Generally, the probability for a radiative transition to occur is sensitive to two factors which vary in opposite directions with the number of photons involved. First, it is proportional to the response of the system to the electromagnetic field, a factor that decreases by about a hundredfold for each additional participating photon; thus multiphoton excitations are normally weaker by many orders of magnitude than allowed one-photon transitions. Secondly, it depends strongly on the intensity of the incident radiation, so that multiphoton transition probabilities may become significant when a sufficiently strong source of light is used. In other words, in the strong field created by an intense laser light, multiphoton processes will become observable.

To illustrate, let us consider the excitation of state i to state f by absorption of two photons. The nature of the electromagnetic interaction is such that only one photon is absorbed or emitted at a time. In our case, the system — an atom or a molecule — absorbs a photon and passes to some allowed intermediate state m, then jumps to the final state f on absorbing the

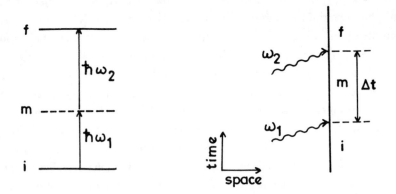

FIG. I.18 *Two-photon absorption process via an intermediate state. (a) Energy level diagram;* $\hbar\omega_1 = E_m - E_i$, $\hbar\omega_2 = E_f - E_m$. *(b) Space-time representation of the sequential two-photon absorption. For a transition via virtual state m,* $\Delta t \approx 10^{-15}$ *s.*

second photon (Fig. I.18). In general, there are many such states m accessible through one-photon absorption, all contributing to the process. Their contributions to the total transition probability interfere destructively with deep minima when waves following different available routes come up badly out of step, and constructively with high maxima when the radiation frequency matches the frequency of a resonant intermediate state. If there exists an accesssible *resonant* intermediate state in the system, that is, if there is an observable state with a well-defined lifetime in the range $10^{-6} - 10^{-9}$ s, then the second photon needs to arrive only within this lapse of time after the first absorption for the process to complete successfully. The transition is being carried out in two steps, an absorption completed before the next starts. But if there are no accessible intermediate resonant states, the transition still can go through, provided the second photon arrives within the flyby time, say, 10^{-15} s, of the first. The state m is not observable, it is a *virtual* state, a very fleeting situation created by a temporary distortion of the system under the applied force. It is in this particular circumstance that the laser plays a unique role.

Of course, if one-photon transitions between two levels can occur at all, they will dominate over any other transition mode. But such transitions are not always physically allowed; whether they are or not depends on the nature of the interaction and the conservation of energy, spin, or other symmetries relevant to the system. Those considerations are conveniently summarized in a set of rules called the *selection rules*. Consider, for instance, the possibility of exciting the hydrogen atom from its ground state, called 1s, to an excited state, called 2s. No observable states exist between the two. When in either of these, the atom has a spherically symmetric configuration. But the interaction primarily responsible for one-photon transitions, the electric dipole mode, is represented by a mathematical object that has the symmetry of the vector joining the positions of two electric charges; for instance, it changes sign under inversion (an operation that flips the signs of all position vectors). When acting on a symmetric state it changes the state to one having the same symmetry as itself, and so cannot lead to a symmetric state: the transition 1s→2s cannot occur via one-photon absorption, it is said to be *forbidden*. But if a second photon with the right attributes (*e.g.,* angular momentum) happens to be there, while the system is still distorting after a dipole absorption, it can act in the reverse direction to bring the system back to a symmetric configuration: the transition 1s→2s is *allowed* in the two-photon mode.

In fact, this transition, which is of great importance in physics and astrophysics, has been observed in a recent experiment using a tunable laser as the driving force for atomic excitations. Photons at half the transition frequency are sent toward the gas sample in two parallel beams from opposite directions with opposite circular polarizations. When a moving atom absorbs in rapid succession two photons, one from each beam, it gains an energy exactly double of the photon energy and devoid of any Doppler effect.† Thus, the transition frequency can be measured with enough accuracy to determine the Lamb shift of the hydrogen ground state, a purely quantum effect, which plays a crucial role in verifying the validity of relativistic quantum theory (see Chapter VI).

Even when one-photon transitions are allowed, the multiphoton methods may still offer practical advantages. For one thing, the Doppler-free technique

---

† The two photons have energies $(1 - v_x/c)\hbar\omega$ and $(1 + v_x/c)\hbar\omega$ in the atom frame, so together they contribute an energy of $2\hbar\omega$ to the atom. Here, ω is the light frequency in the laboratory frame, v_x the atom velocity in the direction parallel to the beams.

introduced in the above paragraph allows all particles in the sample to take part in the experiment, and so is more efficient than the saturation one-photon absorption technique (described in the previous subsection) which singles out a set of particles having just the right velocities. For another, with multiphoton transitions involving high-lying states, the energy required for excitations can be equally shared among several photons, so that the applied field need not be of particularly high frequency. The density of excited states in an atom or a molecule increases rapidly at energies near the ionization limit. Excitations to such high energy states require intense vacuum-ultraviolet (vuv) radiation for one-photon absorption. But absorption of two near-ultraviolet or three visible photons is equally possible, and provides an attractive alternative.

Multiphoton excitation allied with Doppler-free techniques has been used extensively in atomic and molecular spectroscopy to increase within a very short time, both in extent and detail, our understanding of the structure of gaseous atoms and molecules. From these advances, physicists have built a vast knowledge of the nature of bonding in molecules as well as of the way they dissociate. This new body of facts is essential for progress in other fields of research — in studies of combustion processes, of gas-phase reactions in the upper atmosphere, of gas-surface interface, and so on.

6.2 Nonlinear Optics

So under intense light, atoms and molecules undergo complicated mutations in their private little worlds. How can this affect our world? What does it do to the medium as a whole? And how does light behave through all that? A short answer to these questions is, in effect, intense radiation elicits from the medium a collective response quite unlike that observed with weaker light, which in turn alters its own properties, and even produces an output with novel features.

• *Harmonic generation*

As we have seen, when an intense electromagnetic field is applied on an atomic or a molecular medium, photons are absorbed one by one, either singly or multiply, and individual atoms or molecules are carried to excited states. This picture of the microscopic world translates into the appearance of *nonlinearities* in the optical bulk properties (*e.g.*, refractive index or susceptibility) of the medium.

Normally the presence of a weak field does not affect the medium; the substance acts as an inert background through which the field propagates. Its response will merely vary linearly with the field intensity. But in a strong field, the medium itself is modified in a field-sensitive way. Its response depends on both the applied field and the modified medium, so that the overall dependence, direct and indirect, on the applied field is rather complicated. To illustrate, consider the force acting on a playground swing to bring it back to the vertical position. For small deviations, the restoring force is proportional to the deviation angle; for larger deviations, it can be expressed as a finite combination of successive powers of the angle. But for still larger angles, it will vary in a fairly complicate manner. In the same manner, we may say that atoms always respond nonlinearly to external forces, but nonlinearity, normally overshadowed by the dominant linear effects, becomes significant and its effects become detectable only when the applied field is not much smaller than the interatomic field. The Coulomb electric field in an atom is of the order 10^{11} volts/m; fields of 10^6 volts/m from a typical laser at $1\,\mathrm{MW\,cm^{-2}}$ intensity, though appreciably smaller than the typical electric field in an atom, are already sufficient to induce nonlinear effects discernible to highly sensitive detectors.

To be specific, let us consider the effects of the electric component of an intense electromagnetic field on a dielectric (nonconducting) material. We assume the field to be monochromatic at frequency ω. As it propagates through the medium, the applied field displaces the positive and negative charge distributions in each atom in opposite directions, inducing a time-dependent electric field, called the induced *dipole moment*. This effect on the whole collection of atoms can be expressed in terms of a macroscopic quantity, the electric dipole moment per unit volume or, simply, the *electric polarization*. It is a space-time dependent vector field, and can be expanded in powers of the applied field. In contrast to the applied field, the induced polarization is anharmonic; it oscillates with more rounded peaks and deeper valleys than the field itself, and can be decomposed into components of various frequencies, called *harmonics*. These include, besides the frequency of the applied field, all of its whole multiples. The simplest term, linear in the field, is responsible for classical weak-field phenomena, such as ordinary refraction or absorption;

its oscillations faithfully retrace the field vibrations with the same frequency but only scaled down in amplitude. The higher-order terms, in higher powers of the field, arise from processes involving several photons, and can be visualized as anharmonic oscillations of less intense higher harmonics.

The polarization vector has the same symmetry as the electric field vector which, let us recall, changes sign under inversion. If the medium also has that symmetry, even-order terms must vanish because they have the wrong symmetry, they do not change signs under inversion. Examples are isotropic systems (crystals, liquids, or gases); they support only odd-order processes. However, for molecules and lattice sites in crystals without inversion symmetry, both odd and even powers of the field may appear. Since for a field of reasonable strength, contributions decrease in importance with increasing orders, the dominant nonlinear term in a nonisotropic medium is generally the second power term. Quartz, calcite, and anisotropic crystals are materials of this kind.

The induced polarization in a *linear medium* has a very simple frequency dependence. If the applied field has a single frequency, this is precisely the frequency of the induced oscillations; if the field has components at several different frequencies, their contributions to the polarization are additive, and the response of the medium is simply proportional to the sum of all different components. Nothing new so far. Now, if you have an anisotropic *nonlinear medium* (*e.g.*, quartz), then a laser operating at frequency ω induces a polarization which radiates at both ω and 2ω. In other words, you send red light in through a piece of quartz, you get out both red and violet light (of course, at lower intensities). Similarly, if the crystal is shined on by two beams of frequencies ω_1 and ω_2, it will generate waves of frequencies $2\omega_1$, $2\omega_2$, $\omega_1 + \omega_2$, and $\omega_1 - \omega_2$. Finally, for an example of polarization in an isotropic medium, when the applied field contains components with three distinct frequencies ω_1, ω_2, and ω_3, the dominant nonlinear components will have twenty-two different harmonics with frequencies of the types: $3\omega_1$, $2\omega_1 + \omega_2$, $\omega_1 + \omega_2 + \omega_3$, $2\omega_1 - \omega_2$, $\omega_1 + \omega_1 - \omega_3$, and ω_1. If the input frequencies are equal, with common value ω, the polarization radiates only at ω and 3ω.

The practical implication of the above discussion is clear: you can use a suitable nonlinear medium to boost to higher values the frequency of an available strong field. In fact this idea has been applied to produce tunable vuv radiation by generating third harmonics from a dye laser beam. The nonlinear medium is chosen so that the third harmonic of the field (or any other combination of input frequencies you wish, for that matter) falls close to a resonance of the medium. This harmonic will then dominate over all others. Let us take an example. The output of a pulsed dye laser, tunable over the wavelengths $3610 - 3710\,\text{Å}$, is focused into a sample of krypton gas. Now the krypton atom has a known allowed three-photon transition to a state, call it 5s, at wavelength $1236\,\text{Å}$. Thus, three light quanta will be absorbed almost simultaneously by the atom, which on decay emits a single photon at a frequency triple the input frequency. Other harmonics are suppressed. Third harmonic generation in krypton and other noble gases is a proven method of producing vuv radiation of narrow bandwidths, high intensity, and good tunability. The technique has even been applied to molecular gases (CO, N_2, C_2H_2, and HCl, just to name a few) to produce vuv and extreme ultraviolet radiation. Many enthusiastic practitioners of this art are convinced that, given the considerable number of molecular gases exhibiting sharp, high-energy levels that favor, even require, three-photon absorption, it will be possible to generate any wavelength in the range $1000 - 2000\,\text{Å}$ by the one-color three-harmonic generation process and a judicious choice of molecules and input light.

• *Phase conjugation*

Let us return to the general example of harmonic generation in a nonlinear isotropic medium under the action of a three-component strong field, which we briefly mentioned above. We will refer to this process as the *four-wave mixing*. We want to study the nature of a particular radiation produced in this process by three incident beams (of frequencies ω_1, ω_2, ω_3 and momenta k_1, k_2, k_3) through nonlinear frequency mixing. Of all the possible harmonics generated, we focus on the radiation component at frequency $\omega = \omega_1 + \omega_2 - \omega_3$ and momentum $k = k_1 + k_2 - k_3$. If now we let the three beams have a common frequency, the fourth (radiated) wave will also have this frequency, ω. Its momentum is determined by the geometry of the incoming beams: we let the first two (called the *pump beams*) counterpropagate, $k_1 + k_2 = 0$, so the radiated beam has momentum opposite to that of the third (the *probe beam*), $k = -k_3$. This means that the fourth beam is exactly identical to the third, except it travels in the reverse direction, retracing the steps of

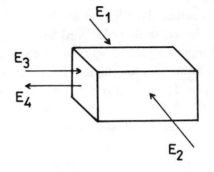

FIG. I.19 *Phase conjugation by nonlinear four-wave mixing. Two strong pump beams (E_1 and E_2) counterpropagate into a nonlinear medium; at the same time, a probe beam (E_3) travels into the medium and generates a fourth wave (E_4) having the same frequency but moving in the opposite direction.*

the incident wave all the way to its source (Fig. I.19). For this reason it is also called a *time-reversed* wave, being the time-reversed replica of the probe wave, or, more correctly, a *phase-conjugate* wave because of its dependence on the complex conjugate of the probe wave.

The nonlinear medium, excited by the pump beams to states at twice the pumping energy, acts as a perfect *reflector* for the probe beam, which is turned back precisely along the retrodirection. When you gaze into such a mirror you will not see your face but just the dots of light scattered by the corneas of your eyes. If the probe light is distorted or scrambled along its way, say, by a frosted glass plate, the distortions will be gradually but completely undone as the light wave retraces its path through the glass to its source. The medium may also act as an *amplifier* of the incident light if the pumping fields are sufficiently strong; the reflective index of the medium depends directly on their intensities. Thus, the light reflected on a phase-conjugate mirror may even be brighter than the light beamed in.

The reader may have noticed a certain resemblance between the production of phase-conjugate waves and conventional holography. In holography the reference beam and the light reflected from an object interfere on an emulsion plate to produce a hologram. After the film is developed, the static hologram can be decoded with the same reference beam and a realistic image of the object obtained. In four-wave mixing the nonlinear medium acts as the photographic emulsion, the probe beam as the object beam, and the pumping beams as reference beams. The probe beam and each of the pumping beams interact to produce a wave pattern in the medium — a sort of dynamic, real-time hologram. The phase-conjugate beam is radiated when the other probe beam is reflected from the hologram.

The remarkable optical properties of phase-conjugate wave systems — perfect retro-reflectivity, perfect homing ability, cancellation of aberrations in wavefronts, and amplification — all point to applications in both practical and fundamental domains. For example, a high quality beam can be transmitted through a turbulent atmosphere, collected and possibly amplified by a phase-conjugate laser system, and sent back to its point of origin free of degradation (provided the intervening atmosphere does not change appreciably in the interval). Such beams can be used in tracking satellites, self-targeting of radiation in fusion, processing images (comparing fingerprints, identifying cells and their mutations, and so on), or in realizing novel classes of ultrasensitive detectors, sensitive enough to respond to gravitational waves. Not to be neglected are the potential benefits for spectroscopic studies: the properties of the observed conjugate phase beam could be used to probe the nonlinear medium that produces it, giving us further insight into matter and its interaction with light.

6.3 Is Quantum Physics Real?

It is very significant that the laser, a product of quantum theory, has played an essential role in the ongoing process of clarifying some deep and disturbing questions still remaining in that theory. Let us hasten to say that the problem does not concern any technical aspects of the theory nor its overall validity. Its spectacular success in the last sixty years leaves no doubts about that. It is a superb theory which describes the physical world to a level of precision unprecedented in sciences. Who could seriously think of dismissing its predictions on the atomic and molecular spectra, the thermal properties of radiation, the creation and annihilation of elementary particles, the existence of superconductivity and superfluidity, and much else? And if more concrete proofs are needed, just have a look at an electron microscope, a transistor, and, certainly, a laser. No, the problem has rather to do with the interpretation of quantum mechanics,† how to reconcile its strange character with our intuitive, common sense perception of nature.

We know that light has characteristic wavelike properties; it can be made to produce diffraction and interference patterns. We also know it can knock electrons out of metals to produce a photoelectric current, an effect representative and indicative of a particle-like behavior. The co-existence of the wave and particle properties in the photon — its *wave-particle duality* — is one of the first basic realizations of the theory, with implications that defy an easy interpretation.

Consider again the now familiar two-slit interference experiment. The interference pattern that one observes in this experiment is a clear proof of the wavelike nature of light. But one can also argue just as convincingly in terms of quanta: the corpuscular photons, each arriving at a definite point on the screen and each leaving its own speck, all cooperate to build up a mosaic of spots which gradually takes the form of a regular interference pattern by the law of averages of large numbers. This is the first surprising conclusion: before the advent of quantum theory, the world was simpler, it was completely predictable; now it looks as if events in the quantum world are only known in a probabilistic sense. In Young's experiment, when one of the two slits is plugged, only a bright spot on the screen marks the image of the open slit: no traces of interference fringes. We would certainly fail if we attempted to reconstruct the interference pattern by superimposing the patterns obtained separately with each individual aperture acting alone. The photons behave quite differently from the way they did before; they 'know' that this time only one hole is open and pass on through the aperture. The argument is in no way based on a perceived collective wavelike behavior of the whole group of quanta but only on the inherent character of each individual. How do they know, if they are independent, indivisible particles? In classical physics, a particle moves along well-defined paths. Not so in quantum physics. Suppose a photon is in a certain state at a certain time, and you want to calculate the probability to find it in some other state at a later time. You simply allow the photon to go wherever it wants to go in space and time, provided only that it starts and ends in the two given fixed states. You obtain the required probability by adding together the contributions from all possible paths. In

---

† See Appendix B for a brief review of quantum theory.

this way, the photon keeps itself informed about what is going on all around itself. Perhaps. But a direct, concrete proof would greatly ease our sense of discomfort.

The wave-particle duality is not confined to the photon alone. Electrons, atoms, particles of matter, quanta of energy, and even cats and dogs, all have both wave and particle behavior. Wave-particle duality, probability of events, and undetermined paths, all this implies an inescapable degree of *indeterminacy* in the entire quantum world that is not due to experimental limitations but rather inherent in nature. Such imprecision is a consequence of the formalism of quantum mechanics, and is perfectly compatible with the best accuracy obtainable in experimental measurements. Suppose we want to observe an electron under a 'microscope' by illuminating it with a strong radiation. At the instant when the position of the electron is measured, *i.e.,* just when the quantum light is diffracted by the electron, the latter makes a jump, changing its momentum discontinuously. So just when the position is determined, the momentum of the electron is known only up to a certain latitude which corresponds to the discontinuous change. The greater this change is, the smaller the wavelength, and hence the more precise the position measurement. Conversely, if we want to measure the momentum of the particle accurately, its location becomes unavoidably uncertain.

Let Δx denote the latitude within which a coordinate x is determined and Δp the latitude for the x-component of the conjugate momentum p in the same experimental arrangement. The indeterminacy in the values of position and momentum is then given by the inequality: $\Delta x \Delta p \geq \hbar/2$. This equation means that no matter how hard we try, we *cannot* know the values of x and p to a better precision than indicated: $\hbar/2$ is the best we can do for the combined deviations. There exist similar inequalities involving other pairs of conjugate variables, such as energy and time of a particle, or amplitude and phase of a wave. All those formulas, referred to as the *Heisenberg uncertainty relations*, describe the irreducible level of uncertainty in our knowledge of those pairs of variables when they are observed together.

In quantum mechanics, a particle (or more generally, a system) is described, not by its trajectory as in classical mechanics, but by its wave function, or *state function*, which contains complete information about the particle. Another basic tenet of quantum mechanics is the *superposition of states*, which asserts that from any two independent quantum states of a system other states can be formed. For example, states in which the polarization vector of a photon points in any direction can be given as a definite combination of quantum states with the photon's polarizations perpendicular to each other. We cannot say for sure which of these polarization states the photon actually is in; all we know is such a photon state exists with such a probability. More generally, we would say that a physical system resides at each moment in some arbitrary state, which can be described as a superposition of the infinitely many states of some given set. Left to itself, the system evolves normally in this indeterminate, aimless state, much like the chatter in a theater at intermission, waves in open seas, or any isolated wavelike system. But, if any sort of measurement is made that leads to a result, then suddenly, one of the ghostlike component state stands out, very real, with its definite wave pattern and other characteristic features, while all others disappear, leaving no trace.

The principle of superposition leads to many disconcerting conclusions, not only at the invisible, quantum level but also at the macroscopic, real life level as well. To dramatize the kind of philosophical problem one might encounter with the superposition principle, Erwin Schrödinger, one of the founders of quantum mechanics, devised the following thought experiment. Suppose a cat is penned up in a steel chamber along with a poisoning device that has equal probability of releasing or not releasing a deadly poison within one hour. As long as the box remains sealed and as far as we know, the poor animal is neither live nor dead — it is *both* equally live and dead! And it will remain in this uncertain state until we open the box and have a look, at which time it either jumps out fully alive or drops really dead, either way with one-hundred percent probability. If you smile with skepticism, you are not alone. You would perhaps ask, along with many other physicists, is the moon there when nobody is looking?

This sort of spookiness of the quantum world has led Niels Bohr to advise us to make a clear distinction between 'the measurement instruments, the description of which must always be based on space-time pictures, and the objects under investigation, about which observable predictions can in general only be derived by the non-visualizable formalism'. Ernest Mach, the leader of the positivist school, went even further and categorically de-

nied any reality to any concepts not directly perceivable by the senses.† The
question is, are atoms and molecules, photons and electrons mere products
of our imagination, or can they be made 'objects of sensuous contempla-
tion' endowed with an objective reality, independent of the subjective act of
observing?

• Delayed-choice experiment

An experiment that could sharpen the concept of wave-particle duality was
proposed some time ago by John Archibald Wheeler. It is basically a modern
version of the classic Young interference experiment in which the two slits
are replaced by the two arms of an interferometer (Fig. I.20). A pulse of laser
light, so severely attenuated that at any time it carries only one photon into
the apparatus, is split by a beam splitter (BS1) into two beams (A and B).
These beams are later deflected toward the lower right of the set-up by two
mirrors (M). A detector is placed at the end of each of the two light paths.
Two situations could be considered. In one, a second beam splitter (BS2) is
placed at the point of crossing of paths A and B. With a proper adjustment
of the lengths of the two arms of the interferometer, interference signals can
be recorded by the two detectors. This result would be evidence that the
photon came by both routes, thereby showing its wavelike property. In the
other, the second beam splitter is removed, and the detectors will indicate
whether the photon came along one of the two possible paths, A or B, thereby
revealing its particle-like property.

Wheeler then asks whether the result of the experiment would change if
the experimenter's decision for the mode of observation — with or without
the second beam splitter, *i.e.*, detecting wavelike or particle-like properties —
is made *after* the photon has passed the first beam splitter. Theoretically, one
would decide to put in or take out the second beam splitter at the very last
moment. A photon will take 15 nanoseconds to travel a distance of 4.5 meters,

---

† Ernest Mach, experimental physicist and philosopher of the early twentieth century, assumed that the laws of science were only a convenient way
of describing relations between observed facts; in particular, he rejected the
existence of atoms, being a mere creation of the mind. Nevertheless, he made
very significant contributions to mechanics and was considered a precursor
of the general theory of relativity.

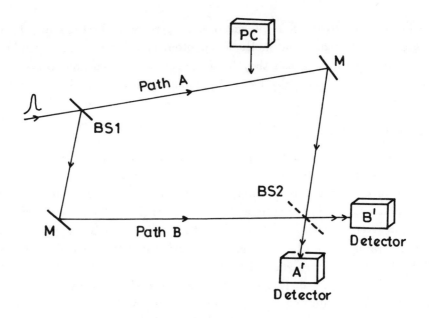

FIG. I.20 *Modern interference experiment. A single-photon pulse enters an interferometer via a beam splitter (BS1) and the two beams may follow paths A and B. In the absence of the second beam splitter (BS2), detectors A' and B' will reveal the route taken by the photon, either path A or path B. With BS2 in place, this particle-like information is lost, and the two detectors will record a wave interference signature. In the delayed-choice version of the experiment, the second beam splitter is put in place, and one of the light paths may be interrupted by actuating a Pockels cell switch (PC) installed for this purpose.*

which is the length of each route in a typical experimental arrangement. This would not give enough time for an ordinary mechanical device to switch between the two modes of measurement. In practice, the experimenter's choice is made possible with a switch called a Pockels cell, which can respond in six nanoseconds or less. The switch is installed on one of the light path, which can be interrupted by applying a voltage to the Pockels cell. In an experiment recently performed, measurements are made in two ways. In the *normal* mode of operation, the Pockels cell is open when the light pulse reaches the first beam splitter and remains open during the whole transit time of light through the apparatus. In the *delayed-choice* mode, the cell is normally closed, and is flipped open a few nanoseconds *after* the pulse has passed through the beam splitter, and thus, has been well on its way to the detectors. For data taking, the operation is switched back and forth between the two modes, normal or delayed-choice, with each successive light pulse, and the photon counts are stored in different multichannel analyzers. Note that in either mode, the data come from many single-particle events, and the information obtained results from a time average, not an ensemble average as in the usual many-particle experiments. The results of the experiment show that there is no observable difference whatsoever between the interference patterns obtained in the normal and delayed-choice modes.

Thus, the observations are completely consistent with the mainstream understanding of quantum physics: the photon behaves like a wave when undulatory properties are observed, and like a particle when corpuscular properties are measured. In other words, the photon in the interferometer resides in an ambiguous state that leaves many of its properties indefinite until a measurement is made. An indefinite property becomes definite only when it is observed, the transition from indefiniteness to definiteness is performed by some 'irreversible act of amplification'. Or, as Wheeler puts it, 'In the real world of quantum physics, no elementary phenomenon is a phenomenon until it is recorded as a phenomenon.'

• *To catch an atom*

Perhaps the most convincing proof of the reality of the quantum world would be to capture some of its creatures and hold them in place for all to see. This has become quite feasible with the technique of laser cooling.

When light interacts with matter, it transfers to the medium some of its own momentum. This momentum transfer manifests itself as a mechanical force acting on the atoms. This force is not constant; it fluctuates over time because the photons scatter at random times and because the recoil of the atoms following excitation and spontaneous emission is also random. However, an average force can be defined; it is parallel to the direction of propagation of light and equal in size to the product of the photon momentum and the photon scattering rate. It is largest when light is resonant with an atomic transition. Let us consider a jet composed of atoms having a strongly allowed transition close to the ground state. A laser beam with a frequency slightly below the atomic resonance frequency is directed against the atomic jet. As an atom is moving against the beam, it sees light Doppler-shifted toward resonance, and so becomes subject to a maximum scattering force which can effectively brake the motion of the atomic jet. Once the atoms are sufficiently slowed down and cooled, perhaps down to near one degree Kelvin, they can be captured and confined in a small region of space by an appropriate configuration of applied electromagnetic field. For ions, trapping poses no problem. But for neutral atoms, trapping is more delicate but still feasible. The technique used would be based on the fact that a neutral atom, with its weak magnetic dipole, interacts with a magnetic field, and so could be made immobile.

Laser cooling has a potential role in many interesting experiments. It can be used to study collisions between very cold atoms and ions. Such interactions could be observed with excellent energy resolutions. The cooling of atomic motions may allow the collective aspects of the particle dynamics to emerge. In particular, statistical properties at the particle level in various atoms and ions may manifest themselves at the macroscopic, observable level. Finally, now that the small thermal motions are largely eliminated, quantum effects could appear free of interference. Observations could be made in single particle systems rather than in collections of many particles. An interesting example of the latter class of experiments is the observation of *individual quantum jumps* in a single ion. In this experiment, atoms are cooled, trapped and confined in a region of space, and excited by a laser field. Two transitions are possible, driven by two different lasers, one to a dominant, strongly fluorescing level with a normal, short lifetime, the other to a metastable state lasting several seconds. Normally, the strong emission is easily detected; but whenever the atom makes a transition to the metastable state, this strong fluorescence ceases, and a period of darkness follows. This period of darkness ends when the atom decays from the metastable state to the ground state, at which point the strong transitions and the subsequent spontaneous radiation resume. The atom will flash on and off like a tiny lighthouse, signaling each time the absence or the occurence of the weak transition.

Summary

The laser has played a significant role in recent advances in atomic and molecular spectroscopy and in nonlinear optics. It has also been instrumental in clarifying certain fundamental aspects of quantum mechanics.

The major contribution of the laser to spectroscopic studies is to eliminate the Doppler broadening of transition linewidths, particularly in atomic and molecular gases where this effect has severely limited the resolution attainable in experiments using conventional electromagnetic sources for excitations. The high intensity monochromatic light of lasers makes accessible states that are normally closed to one-photon excitations — the ones usually available by conventional means — but are allowed to transitions via multiple photon absorptions. Multiphoton excitation has been used together with Doppler-free techniques in spectroscopy to deepen our understanding of the structure and the dynamics of gaseous atoms and molecules.

Under the intense radiation of a laser output, atoms and molecules are massively excited to high energy states. These strong and abundant interactions correspond, at the macroscopic level, to the appearance of nonlinearities in the collective response of the medium. In other words, the optical bulk properties of the medium will depend on the applied field strength not only in the first power but in higher powers as well. The irradiated medium will in turn generate various characteristic harmonics, which can give us further insight into the medium itself or can be put to other uses. One such use is the generation of a phase-conjugate beam by a nonlinear medium under the combined action of three appropriate laser beams.

Recent advances in detection techniques and the availability of high quality light from lasers have made many experiments involving single-particle events possible. Information gathered in such experiments results from a time average of successive single-particle events rather than from an ensemble average over a collection of many particles. Noteworthy for their implications in physics are the delayed-choice interference experiment and various works on the cooling and trapping of atoms.

---

7. LOOKING BEYOND

Following the steps of illustrious optical instruments of the past, the laser has radically transformed optical science and technology. More than that, it has made notable contributions to many branches of physics, and is having an important impact on other disciplines as well. We have discussed above several of its applications in technology and fundamental research. Yet the list is far from complete, and could not possibly be: quantum optics, from which the laser is inseparable, is a relatively young science and still evolving. Discoveries of new properties leading to new phenomena leading to new applications occur almost every month. Before we end this excursion through quantum optics, let us pause and look briefly into two new and promising directions.

Recently, it has been realized that the view we have thus far of the laser light, as an extremely ordered state of radiation, does not give us a complete picture. Lasers, upon careful examination, can in fact exhibit a rich variety of nonlinear dynamic behavior which includes unstable and even chaotic features.

A full description of a lasing system is based on a dynamic interplay between the applied electromagnetic field, on the one hand, and the various atomic properties of the lasing medium, on the other. In the simplest case, examples of the latter are the polarization of the material and the population inversion of excited atoms. For simplicity, let us consider all these three quantities as real variables. Now, if coherent emission is to occur at all, one or both of the material variables must respond quickly enough to the slowly varying field to maintain a proper phase correlation; their relaxation times must be much shorter than the cavity lifetime. In that case, they can simply be replaced by their equilibrium values and lose their status as dynamical variables. With only one variable (the field variable) driving the evolution of the system, the laser always operates in a stable mode. On the other hand, with the field and one material parameter acting as dynamic variables, the output may have short-term oscillations under certain conditions, but these oscillations quickly damp out to insignificance. In either case, the light emission has predictable long-term stability. But now suppose that all *three* dynamical variables participate actively in the process, and consider the situation where the medium polarization is very large, or where the atomic population inversion is allowed to reach a certain high limit. The system will then display a completely unpredictable behavior. When represented in the abstract space of the dynamical variables, its lasing emission will show complex aperiodic variations with time, spiralling around several centers of attraction and jumping at random from one center to the next — features reminiscent of other nonlinear systems (see Chapter IV). Lasers are ideal systems for studying deterministic chaos because they are simple to construct, stable in operation, and rich in diversity. The possibility of a quantum description of the observations and the potential practical applications of the sources of instabilities add further intriguing and attractive features to an interesting problem.

Laser light, in spite of its remarkable purity, is still subject to fluctuations, not the classically describable jiggling of ordinary light, but rather the inherent, unavoidable random fluctuations of quantum origin. They are a source of noise that sets apparently ultimate limits on the accuracy of ultra-sensitive measurements.

To understand quantum fluctuations, recall that according to the Heisenberg uncertainty principle two conjugate variables cannot be simultaneously measured with perfect precision; the product of their uncertainties satisfies a fundamental lower bound (namely, $\hbar/2$). Applied to an electromagnetic field operating in a single mode, this condition implies that two conjugate field variables — e.g., the amplitude and phase of the field or, alternatively, the amplitudes of the sine and cosine field components (quadratures) — are never simultaneously known with complete certainty. Ordinary light shows huge field fluctuations well above the lower bound imposed by the Heisenberg relation. Even in vacuum, the field also fluctuates, but with equal variances whose product saturates the uncertainty limit; these quantum fluctuations manifest themselves as the observable zero-point energy of the field. Coherent light emitted by lasers is another example of such a minimum uncertainty state with the minimum uncertainty equally divided between the conjugate variables; in this case both of the average field values are nonvanishing.

The Heisenberg uncertainty relations set limits on products of variances of conjugate variables, not on individual variances. Therefore, one could in principle reduce the fluctuations of one of the quadrature field components to near zero, making that component noiseless, at the cost of having a correspondingly greater indeterminacy in the other component. The intensity of light in a given mode can be determined with great accuracy, but the phase of that mode is completely indeterminate. Light in a minimum uncertainty state but with fluctuations unequally divided between the two quadratures is said to be 'squeezed'.

The production of quadrature-squeezed light is based on the nonlinear interaction of light and matter. In one scheme, the experimenter uses the three-wave mixing technique in which an incoming photon of frequency ω is split into subharmonics of frequency $\omega/2$ (the inverse of the frequency doubling we discussed before). The amplification process that ensues is phase-sensitive. Only that part of the initial wave at $\omega/2$ for which the phase fluctuation vanishes is amplified, while the other part, at 90° out of phase, is de-amplified. The result is an output with reduced phase fluctuations and enhanced amplitude fluctuations.

Squeezed light is a form of nonclassical light; it is qualitatively very different from both ordinary and laser light. It is a fundamentally quantum phenomenon and so can provide a new, unique insight into the nebulous quantum world. But it is more than just an interesting phenomenon. It is very likely to find many practical applications. For example, in spectroscopy and gravitational wave research, detection schemes insensitive to the noisy quadrature components could be designed to sense tiny effects which would otherwise be masked by quantum fluctuations. Or, in light wave communications, ever hampered by the presence of quantum noise in the transmitted signals, information could be encoded in the quieter component for transmission. What can be accomplished with such quiet dynamical components when used as probes of Nature, no one knows. *Chi viverà vedrà.*

CHAPTER II

SUPERCONDUCTIVITY

CONTENTS

Disordered states are all alike. But every ordered state is ordered in its own strange way. And the most strangely ordered of these is the state of superconductivity that we may not make even a mental image of it. For superconductivity, as we will see, is a truly quantum effect coherent over a large, macroscopic scale. It has manifest in it the wave nature of matter that is normally spoken of atoms and molecules, that is, in the domain of the very small. For these reasons superconductivity is somewhat hard to understand except for a preoccupied mind. But we can easily get acquainted with the superconductor by watching its behavior which is quite robust, and readily amenable to ordinary experiments. And when we do this we will be left in no doubt that we are in the presence of something very extraordinary, almost bizarre.

1. ZERO ELECTRICAL RESISTANCE

Superconductivity is the total disappearance of electrical resistance of a material at and below a sharply defined temperature (T_c) which is characteristic of that material (Fig. II.1).

This *critical temperature* (T_c) can, however, be very low, typically close to the absolute zero of temperature. A superconductor is thus a perfect conductor. And so an electric current once set up in a superconducting ring will go on circulating undiminished forever, or very nearly so. A simple minded calculation would give the time of decay of this current much longer than the age of the universe which is some 15 billion years! Very precise laboratory measurements of the decay of the supercurrent in a superconducting coil estimate the decay time to be about a hundred thousand years which is long enough. This persistent flow of electricity, the supercurrent, is just about as close as we can get to man's recurrent dream of a *perpetual motion* — of the first kind. Today a superconducting magnet with the supercurrents circulating in its winding coils is a common sight in the low temperature physics laboratories around the world as a quiet source of constant magnetic field as long, of course, as it is kept cold enough.

This amazing phenomenon of superconductivity was discovered by the

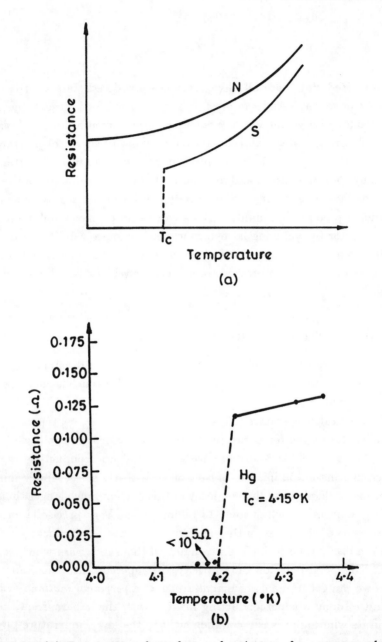

FIG. II.1 *(a) Temperature dependence of resistance for a superconductor (S) and a normal metal (N); (b) superconducting transition in mercury after Kamerlingh Onnes.*

great Dutch physicist Heike Kamerlingh Onnes way back in 1911. Kamerlingh was studying the low temperature behavior of electrical resistance of metals in his world famous low-temperature laboratory at Leiden of which he was the director. Just three years earlier, Kamerlingh had liquefied the last and the noblest of the permanent gases, helium. Helium boils at the incredibly low temperature of 4.2 K (Kelvin), that is just 4.2 degrees centigrade above the absolute zero of temperature — the lowest temperature possible as ordained by the laws of physics. This circumstance made it possible for Kamerlingh to observe things very close to the absolute zero for the first time. Kamerlingh found to his great surprise that the resistance of his sample of frozen mercury (chemical symbol Hg) dropped almost abruptly to zero, within experimental limits, when it was cooled below about the boiling point of liquid helium, 4.2 K (Fig. II.1). Kamerlingh was quick to realize that the resistance was not just low — it was essentially zero! He could observe the persistent current that flowed without an external source such as a battery. He called it superconductivity and the name stuck. For this discovery Kamerlingh won the Nobel Prize for Physics in the year 1913. His life-long preoccupation with low temperatures earned him the informal title of the 'gentleman of absolute zero'.

Since its discovery almost eighty years ago now, superconductivity has remained one of the greatest surprises of physics. Why should the zero resistance be so surprising? Let us understand this first.

1.1 Metallic Resistance

Metals are by definition good conductor of electricity. Now, even the best and the noblest of them all such as copper, silver and gold do offer some resistance to the flow of electricity. This is what causes the wasteful heating of the wire, the *copper loss*, in transmitting electrical power from one point to another. And for this, of course, we have to pay as an invisible part of our electricity bill. Remember Ohm's Law: Amperes (I)=Volts (V)/Ohms (R) and Watts (W) =I^2R. This ohmic dissipation can be reduced by going to lower operating temperatures. This is readily understandable. Much of the electrical resistance is due to the incessant thermal jiggling of the atoms or the ions (atoms that have lost one or more of their loosely attached outermost electrons) that perturbs the otherwise free flow of electricity. More precisely, the ultimate carriers of elctricity are the freely moving electrons that abound in a metal. There are roughly 10^{23} of them in each cubic centimeter of the metal. In equilibrium, that is in the absence 'of an applied electric field or potential difference, there are, on the average, as many electrons moving in any given direction as in the opposite one, and so there is no net current. In the presence of an electric field, however, there are relatively more electrons moving in the direction opposite to the applied electric field and this excess makes up the directed electric current. Now, the thermal vibrations of the background ions 'scatter' these electrons randomly in all directions and tend to neutralize this excess current causing electrical resistance. Indeed, the electron would accelerate indefinitely but for this continual scattering that offers the necessary friction forcing the electron to settle down to a steady thermal drift — much the same way as the mechanical friction or, better still, the fluid viscosity (treacliness) limits the flow of a liquid through a metal pipe or a glass capillary in spite of a head of pressure. Now, lower the temperature, lesser is the intensity of heat motion, and hence smaller is the resistance. Thus, for example, the resistance of a specimen of copper will be down by a factor of about a thousand or more as it is cooled from the room temperature of 300 K (26.85° C) to the liquid helium temperature, 4.2 K (The Kelvin (K) and the Celsius (C) scales of temperature are simply related: Degrees Kelvin = Degrees Celsius + 273.15. Thus, the absolute zero, written 0 K, corresponds to −273.15° C). But the fall is a smooth one. Ideally, then the resistance should vanish at the absolute zero of temperature where all thermal agitation halts to a dead beat. We said ideally because this would be true only if the specimen was a perfect crystal — a perfectly periodic array of atoms or ions. (It is a deep result of quantum mechanics that the electrons are not scattered by such a perfectly periodic arrangement of 'scatterers' no matter how strong the individual scatterer may be. This has to do ultimately with the wave nature of the electron). But, of course, the real materials are far from this perfect crystalline symmetry. There are the ubiquitous defects — impurities, misplaced atoms, or missing atoms (vacancies). These deviations from the perfect crystalline symmetry can scatter electrons and, therefore, offer resistance. Thus, we have to live with this *residual* resistance even at the absolute zero of temperature, which is inaccessible anyway. (Incidentally, the record lowest temperature achieved so far in the laboratory is about a nano-Kelvin, that is a billionth of a degree Kelvin. But even this is *not* quite zero.

And, of course, never mind the cost of refrigeration). Against this normal behavior consider our sample of frozen mercury that had lost all its resistance at and below about 4.2 K. Now the point is that the thermal agitation of the atoms at this low temperature, while admittedly small, is still far from being zero. Also, the randomly placed defects that were present above 4.2 K are still very much around and look just as obstructive. In fact nothing much has changed for the material by way of its chemistry or crystal structure — but the resistance has vanished completely. The scatterer has somehow lost the 'will' to scatter — it lets the electrons pass by uninterrogated. It is as if the cloud of electrons flows past these obstacles, ever adjusting, ever adapting but never quite getting perturbed — ghostlier than the ghost. The scatterer did not scatter. The dog did not bark. And that is the strange thing!

1.2 Superconductivity is Common

One may get the impression that such a bizarre phenomenon as supercon-ductivity must be a rare occurrence. But this is simply not true. A quick look at a modern *periodic table of elements* will convince you of this. Of the 92 elements prominently displayed 68 are metals, and of these at least 26 are superconductors. Then there are others that become superconducting when *pressed* hard enough. Thus, silicon which is not only not a metal but a semiconductor (that transistors and computer chips are made of), begins to superconduct under a pressure of tens of kilobars (kilo = thousand, bar = 1 atmospheric pressure). But there are exceptions. The magnetic metals, iron (Fe), nickel (Ni) and cobalt (Co) refuse to ever superconduct — ferro-magnetism seems inimical to superconductivity. Same is true of the light alkali metals such as sodium (Na), potassium (K) etc. But the most notable exceptions are the noble metals like copper (Cu), gold (Au) and silver (Ag) which are normally the best conductors of electricity. In fact, ironic as it may seem it turns out that good superconductors (the ones with high T_c such as niobium (Nb) for example) are bad normal metals. We will soon begin to see why this has to be so. Besides these elemental superconductors we have thousands of superconducting alloys (metallic mixtures), organic compounds, and now even earthy ceramics, that are found to superconduct — may their tribe increase! In Fig. II.2 we have listed some of the common superconduc-tors along with their critical temperatures marked on the absolute (Kelvin) scale. Note the crowding at the lower end of the scale. Until about four years

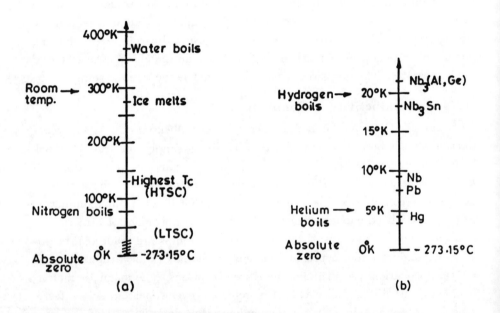

FIG. II.2 *(a) Absolute temperature scale marking well known transitions; (b) expanded scale showing T_c's of some superconductors.*

ago the highest critical temperature known was 23.2 K, for a compound of niobium and germanium (Nb_3Ge).

Thus, superconductivity is indeed very common. It is just that the critical temperatures are abysmally low. If, however, we are willing to leave our terrestrial laboratories and look elsewhere, there is high temperature superconductivity out there. There are strong theoretical reasons to believe that the interior of the neutron star is a neutronic superfluid and a protonic superconductor, with T_c of about a hundred million degrees (see Chapter V). Nearer home we have the case of the major planet Jupiter. It is again suspected that hydrogen, the major constituent of the giant planet, is crushed to a metallic density under its gravitational pressure of about a million atmospheres (megabar), and the metal so formed is a superconductor with T_c of several thousand degrees Kelvin. But here on earth, until very recently, superconductors lived only in the liquid helium cryostat (i.e., *dewar*, a sophisticated thermos or vacuum bottle that keeps cold things cold and, of course, hot things hot). It is precisely this coldness that has kept superconductors confined to the low-temperature laboratories of the world, away from the gaze of the public eye. We may call these the *liquid helium (LHe) superconductors*. More than half a century of uninterrupted worldwide research could barely push the critical temperature to a little over 20 K. So much so that one began to doubt seriously if higher T_c's were possible at all. And then came the breakthrough in 1986. J.G. Bednorz and K.A Müller of the International Business Machines (IBM) at Zurich announced in the September issue of the German journal 'Zeitschrift fur Physik' their discovery of an earthy, ceramic superconductor with T_c of more than 30 K. For this they won the Nobel Prize for Physics in 1987. This led in quick succession to superconductors with still higher critical temperatures ranging from 90 K to 125 K, thus bringing the age of the LHe-superconductors to a sudden end. It also initiated the era of the *liquid nitrogen (LN$_2$) superconductors*. Liquid nitrogen boils at a comfortable 77 K. And now there is already some responsible talk of room temperature superconductors — the *holy grail* of the solid state physicists.

These events of the last four years have altered our view of superconductivity. It is now a serious thought that in the coming decades, superconductors may revolutionize human condition more decisively than the laser, or the nuclear power or even the transistor ever could. We ought to get more than just acquainted with superconductivity.

---

2. INFINITE MAGNETIC RELUCTANCE

Zero electrical resistance is the defining property of a superconductor. It is also by far the most striking property of a superconductor. But the deciding property of a superconductor is really its infinite reluctance to admit magnetic fields in its interior. A superconductor is a perfect *diamagnet*, which as we shall see is more than being just a perfect conductor. Let us understand what all this means. Take a piece of a superconducting metal like tin (Sn) and hold it at a temperature above its T_c so that it is in the normal resistive state. Now, place it in a static magnetic field which may be conveniently produced by a permanent magnet, or by a solenoid carrying electric current. Now a normal non-magnetic metal like tin is indifferent to magnetic fields. It is almost as good as vacuum, and the magnetic lines of force run right through it undisturbed. (Of course, the act of placing the sample in the magnetic field involves initially some motion through the magnetic field which, by the *Faraday law of induction*, induces an electric field in the metallic sample. This in turn generates the *eddy currents* in the metal and the associated stray magnetic fields. But these transient effects die down rather fast, and we assume that we have waited long enough for this to happen). Now, let us cool the sample sufficiently and, lo and behold, at a certain temperature about T_c the sample turns superconducting and the magnetic lines of force (the flux) are *expelled* totally from the bulk of the superconductor. This happens unless, of course, the external field is much too strong in which case superconductivity is suppressed and the sample remains normal down to 0 K (Fig. II.3).

This dramatic phenomenon of flux expulsion or exclusion, is the famous *Meissner-Ochsenfeld* effect named after the discoverers W. Meissner and R. Ochsenfeld (1933). The process is reversible, that is the flux lines re-enter the sample if it is re-heated through the same temperature. What really happens is that in the presence of the external magnetic field, persistent supercurrents are generated in the superconducting sample of a magnitude, sense and detail which is just right so as to produce a field that cancels the external field

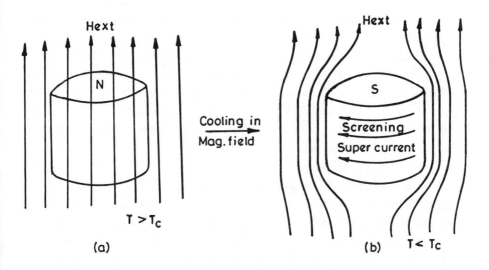

FIG. II.3 *Meissner effect: (a) normal (N) state; (b) superconducting (S) state.*

throughout the interior of the sample (See Fig. II.3). We call these the *screening currents*. They flow mostly on the surface of the sample, almost skimming it. Note that we are talking here about static magnetic fields and these *screening currents* are not to be confused with the *eddy currents* that are induced even in normal metals but by a time-dependent field (Faraday's law of induction). In a perfect conductor (infinite conductivity) these inductively induced eddy currents will be infinitely large so as to totally screen out any time-dependent magnetic field. Even in a metal such as copper which is merely a good conductor, because of this screening the alternating currents flow only on the surface up to skin depth which is about a centimeter at 60 Hz and one-twentieth of a millimeter at 1 MHz. Thus, the central core of a thick copper wire hardly carries any 'ac' current. A superconductor is not merely a perfect conductor (zero resistance) — a perfect conductor will not exclude a *static* magnetic field. On the other hand we can readily reason out why perfect diamagnetism (*flux expulsion*) must imply zero resistance. Assume to the contrary that our perfect diamagnet had a finite resistance. But then the persistent screening currents must continually dissipate energy — the I^2R loss, remember! The question now is where could this energy possibly come from. Perhaps from the energy stored in the magnetic field. But the magnetic field is given to be static (constant in time) and, therefore, it cannot supply the necessary energy. We seem to have a problem here. There is clearly no source of energy available to our system. Having thus eliminated the obvious, what remains, no matter how improbable, must be the true explanation — in this case, namely that our perfect diamagnet had no resistance to start with and hence there was no energy loss to be accounted right. We have just proved that a perfect diamagnet is also a perfect conductor. The converse is not true. It is for this reason that perfect diamagnetism is regarded a more fundamental, in fact the deciding property of a superconductor.

Perfect diamagnetism (*flux expulsion*) implies that the superconductor is repelled away from a strong magnetic field. Thus, for example, we can have a bar magnet floating above a superconducting surface (Fig. II.4). This *magnetic levitation* has led to the possibility of having ultrafast trains gliding on the frictionless magnetic cushion.

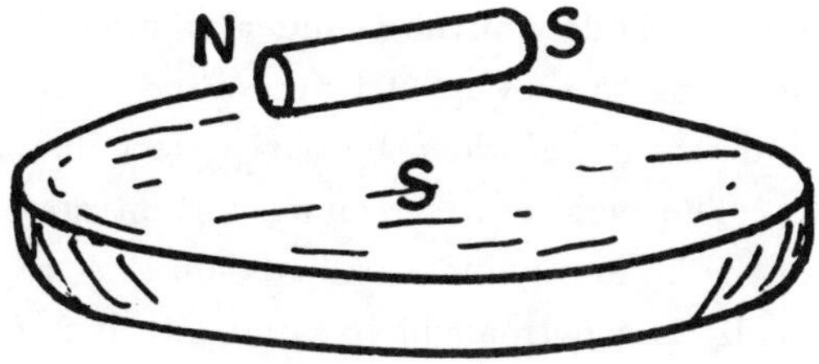

FIG. II.4 *Magnetic levitation of a bar magnet above a superconductor (S).*

3. FLUX TRAPPING

There is an interesting corollary to the Meissner effect, which is the trapping of magnetic flux by a superconductor. Let us repeat our experiment demonstrating flux expulsion, but this time with a sample in the shape of a *hollow* cylinder. As the temperature is lowered through its critical value in the presence of the magnetic field, the flux lines are again expelled from the *bulk* of the material as expected. Nothing, however, comes in the way of the flux lines threading the *hollow* of the cylinder. Persistent screening currents will flow near the inner and the outer surfaces of the hollow cylinder as shown in Fig. II.5. And now let us gradually remove the externally applied magnetic field leaving our sample all by itself. But what about the flux lines passing through the hollow of the cylinder? Surely they cannot escape sideways because in doing so they must traverse the surrounding superconducting material and this is forbidden by our perfect diamagnet. You see, the flux is trapped. The persistent screening current now circulating near the inner surface of the hollow cylinder will sustain this trapped flux. What we have got here really is a permanent bar magnet. It is robust. You may carry it around in your pocket except for the inconvenience of having to keep it cold enough. Viewed differently, you have invented a non-polluting device for storing energy — the trapped magnetic flux and the circulating screening currents form a kind of flywheel if you like.

4. WHOLENESS OF TRAPPED FLUX

The curious case of the *trapped flux* becomes curiouser if we enquire further. What is the amount of magnetic flux trapped in the hollow of the cylinder? It turns out, and we will shortly know why, that the flux thus trapped cannot have an arbitrary value. It has to be an integral multiple of a certain basic unit of flux, denoted by ϕ_0. That is to say that it must be a whole number when measured in lots of ϕ_0. Fractions, or half-measures, are not allowed! This wholeness is the celebrated *flux quantization*, and ϕ_0 the quantum of flux. This unit of flux is extremely small but still macroscopic enough. To have an idea of how small it is, imagine a circular wire loop of diameter 0.1 millimeter facing the earth's magnetic field. Then the loop will intercept

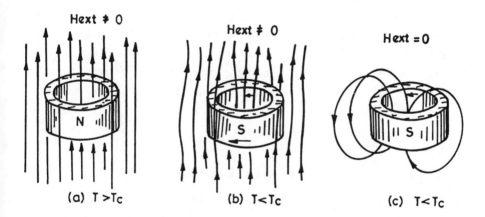

FIG. II.5 *Flux trapping by a hollow superconducting cylinder on field cooling.*

about 100 of these flux quanta! Small as it is, the trapped flux quanta can be counted by jiggling our cylinder in and out of a coil and then measuring the voltage (electromotive force) induced in the coil due to the changing flux linkages (The Faraday law of induction). This was indeed done by B.S. Deaver and W.M. Fairbank in their classic experiment in 1961 that confirmed this quantization of trapped flux. As we shall see later, $\phi_0 = hc/2e = 2 \times 10^{-7}$ gauss centimeter-squared ($= 2 \times 10^{-15}$ tesla meter-squared) where h is Planck's constant, c the speed of light and e the magnitude of the electric charge on the electron. Planck's constant gives away the hidden quantum nature of superconductivity. It is a remarkable fact that the quantum nature of superconductivity was anticipated by Fritz London in 1935, long before the fully microscopic theory of superconductivity was given by John Bardeen, Leon N. Cooper and J. Robert Schrieffer in 1957, the celebrated *BCS theory* for which the trio was awarded the Nobel Prize in 1972. In fact London had predicted flux quantization, but being far ahead of his time, he missed the all important factor 2 in the denominator of $\phi_0 = hc/2e$. We now know after BCS that it is '2e' and not 'e', and thereby hangs the tale of 'two electricities' — the *electron pairing* theory of superconductivity.

---

5. TEMPERATURE AND PHASE TRANSITION

The irresistible zero electrical resistance and the irrepressible infinite magnetic reluctance that set in at the critical temperature should leave us in no doubt that a qualitative change of state has taken place. There is a branch of physics that describes these changes of states of matter in general terms — *thermodynamics*, or *statistical mechanics* if we are interested in a microscopic treatment (see Appendix C). Temperature plays the central role here. This Section is a brief digression intended to acquaint ourselves with some simple but powerful ideas that make the change of state understandable (see Appendix C).

But first some quick remarks on the *absolute (Kelvin) scale* of temperature that we have already spoken of several times. Temperature is the *intensity of heat*. It measures the energy of random jiggling of atoms, molecules, electrons, spins, or more generally, of the *dynamical degrees of freedom* that our system may have. Absolute temperature measures it absolutely. Thus, at the absolute zero of temperature the thermal energy is zero — all motion comes to a standstill. (There is, of course an irreducible *zero-point* motion even at the absolute zero of temperature which is of a purely quantum nature, and is appreciable for the so-called quantum liquids of which we will speak later. In fact the superconductor is one such *quantum liquid*). It is clear that any property that at all depends on temperature can be used to detect changes in temperature. Thus, the common household thermometer uses the thermal expansion of mercury for this purpose. One can also use, for example, the change of electrical resistance of metals, alloys or semiconductors to measure temperature charges. Thus the Platinum (Pt) resistance thermometer is a prime standard for measuring temperatures down to $-260°$ C. But how can we meaningfully specify *equal* intervals of temperature? To say that equal changes in the length of the column of mercury in our thermometer give equal intervals of temperature is nothing more than an assertion that mercury expands equally for equal changes in temperature — clearly a circular statement empty of any objective content. For example, the equal intervals so defined may not be equal on a thermometer using alcohol instead of mercury. The question is if we can define equal intervals of temperature independently of the property of the material! The answer is Yes. As an act of almost pure reason, Lord Kelvin of Britain, one of the greatest of the classical physicist, proved in 1860 that such an absolute scale does exist and is defined in terms of the efficiency of an ideal heat engine. The absolute (or Kelvin) temperature scale (K) so defined is then conveniently graduated so that the boiling and the freezing points of water differ by 100 degrees on this scale just as on the commonly used centigrade scale (C) of Celsius. Then the absolute zero (0 K) measures $-273.15°$ C and is the lowest temperature possible. Water freezes at 273.15 K ($0°$ C), and the 'room' temperature is 300 K ($26.85°$ C). It is a deep result of classical statistical mechanics that every *degree of freedom* of a system such as a classical gas in equilibrium carries an equal amount of kinetic energy, $k_B T/2$, where k_B is the *Boltzmann constant* (*the law of equipartition of energy*).

Temperature is the single most important control parameter that determines the states of matter. A solid (ice) melts to a liquid (water) and the liquid (water) boils to a gas (steam) as the temperature is raised through the well defined melting and the boiling points. These are the commonest and perhaps the most important changes of states that have shaped our biological lives and indeed the universe itself (See Chapter VII). Yet another interesting example of change of state is the loss of magnetization when a bar magnet is heated above its *Curie temperature* — the change from the ferromagnetic to the paramagnetic state. The change from the non-magnetic resistive normal state at high temperatures to the diamagnetic superconducting state at low temperatures is also a change of state, in fact closely related to the paramagnetic-to-ferromagnetic change of state. We call these different states different *phases* of the substance. The change of state is called *phase transition*, and the corresponding temperature the transition temperature.

5.1 Order Parameter

There is a feature which is common to all phase transitions. The higher temperature phase is disordered (or less ordered at any rate) while the lower temperature phase is ordered (or more ordered). There is indeed a competition between order and disorder and temperature decides the winner. Thus, for example, the liquid state is disordered — a snap shot of the liquid state will show atoms positioned more or less at random, while the solid state formed upon freezing displays a periodic arrangement of atoms that we call a crystal. Similarly, for the magnetic case, the spins (the tiny atomic magnets) point in different directions at random in the paramagnetic phase above the Curie temperature T_c, while they align parallel on the average in the ferromagnetic phase below T_c. Indeed, one can define an *order parameter* that vanishes in the disordered phase but assumes a non-zero value in the ordered phase. For a magnet, the choice of the order parameter is obviously the magnetization. In the case of the superconducting transition, however, the nature of the order is too subtle as we will see later. Order parameter is one of the most powerful concepts in the physics of phase transition. It is an *emergent* quality. It was introduced by the great Russian physicist Lev Davidovich Landau in 1960, who gave a general theory of phase transition based on this crucial concept.

5.2 Free Energy and Entropy

What is the basic principle that determines which one of the possible phases our system in equilibrium will be found to be in? For mechanical systems with friction the answer is well known from our high school physics — the system will settle down to a state of *minimum* potential energy. Thus, a marble thrown in a bowl will eventually come to rest at the bottom-most point. This is a one-body problem. A somewhat similar *minimum* principle exists even for our many-body systems with large, almost infinitely many number of particles (or degrees of freedom) interacting with one another. Left to itself, our system too will settle down to a final state which will change no more in time — a state of equilibrium. This state will, however, correspond to a minimum of what is called the *free energy* (see Appendix C). Let us see what this means. Consider a microscopic state of our system of energy E. A microscopic state means specifying in detail the momenta (roughly velocities) and the positions of all the particles (degrees of freedom). Then the energy E is the sum total of their kinetic and potential energies. Now the fundamental principle of statistical mechanics is that all such microscopic states, which are possible at all, will occur, but with a probability proportional to $\exp(-E/k_BT)$. Next, strange as it may seem, almost all of these microscopic states of the same energy (E) *look alike* from the macroscopic (average) point of view. And it is the macroscopic point of view that matters for all practical purposes. (Indeed, even if we knew the finer microscopic details, we wouldn't know what to do with them. The fact of the matter is that the microscopic description is too fine grained while our usual probes are too coarse). Thus, for a given macroscopic state of energy E, there will be a large number of the microscopic states corresponding to the number of ways in which the energy E can be partitioned among the many degrees of freedom. Let this number be $g(E)$. Hence the probability of occurrence of the physically identifiable macroscopic state must be proportional to $g(E)$ times $\exp(-E/k_BT)$. We may re-write this as $\exp(-F/k_BT)$ with the exponent $F = E - k_BT \ln[g(E)]$. Here $\ln[g]$ denotes the 'natural' logarithm of g with respect to base $e = 2.71828....$ So the most probable state is the one that corresponds to the minimum of F, the Free Energy and not of E. The quantity $k_B \ln[g(E)]$ is the mysterious *entropy* and is usually denoted by S. Thus, we must minimize $F = E - TS$, and not just E. At $T = 0$ K, this, of course, reduces to minimizing the energy itself. This lowest energy state, called the ground state, is essentially unique for the system. For this state 'g' is unity and hence $S = 0$ (remember, logarithm of unity $= 0$). The *ground state*, the most ordered state, has zero entropy! It is clear that at a sufficiently high temperature the entropy term in F may dominate the free-energy and a different state may be preferred. In general $g(E)$ and, therefore, entropy is expected to increase with energy — there are then obviously more ways of partitioning it among the various degrees of freedom. The corresponding macroscopic state will also be the more disordered. Ordered microscopic states are fewer due to the constraints of order, and hence the corresponding ordered macroscopic state has lower entropy. (Just compare the disorderly crowd and the disciplined military and you will have the general drift of the idea). And so it happens that high temperature favors disorder. The relationship between the many microscopic states and the single macroscopic state corresponding to them is illustrated best by an analogy with the game of dice. Consider casting two dice simultaneously. Each can come face-up with a number from 1 to 6. Thus there are $6 \times 6 = 36$ possibilities. These are all the possible 36 microscopic states of our system of the two dice. Let the dice be true for simplicity. Then all the 36 microscopic states are equally probable. But suppose now that we are interested only in the *sum* of the numbers that the two dice come up with. The sum can vary from $1+1 = 2$ to $6+6 = 12$. These are then the 11 macroscopic states. Let us label them by the sums 2 to 12. Now you see that the macroscopic state $2 = 1+1$ is realized in only one way, and the macroscopic state $3 = 1+2 = 2+1$ in two ways, and so on. You can easily verify that the macroscopic state 7 is realized in 6 ways which is the maximum and hence the most probable state. If you want to push this analogy further then all you have to do is to imagine a large, almost infinite number of dice, and let the dice be *not* true. Then the most probable macrostate is all that will occur overwhelmingly. The others may be regarded as mere fluctuations about this. But a discussion of these fluctuations will take us far afield.

In the case of our superconducting material the fact that for $T < T_c$, the material is in the superconducting (S) state implies that it has a free-energy less than the normal (N) state. The difference $F_N - F_S$ is called the condensation energy and is denoted by ΔF. It is clear that ΔF is positive and a maximum at 0 K, falls off to zero at $T = T_c$ and then turns negative for $T > T_c$ when the normal state takes over.

Calculating the free energy is a horrendous task of statistical mechanics. But the principle of phase transition is now clear. The behavior of free energy determines the nature of the phase transition. It may be a discontinuous one, where the energy E changes by a finite amount even though F is (as it must be) continuous. This discontinuous change of E is the latent heat that is given out (absorbed) during freezing (melting) or condensation (boiling). We have all experienced it some time or the other, rather regretfully though — the scalding of hand exposed to condensing steam from a boiling pot. We call these *first order phase transition*. The superconducting transition, on the other hand, is a continuous transition with no latent heat associated with it. The same is true of the magnetic transition. These are called the *second-order phase transition*. Unlike the first-order transitions, the changes that take place at and near the second-order phase transition are very subtle. Several physical quantities such as the specific heat show singular behavior which is remarkably universal.

---

6. TYPE I SUPERCONDUCTORS

While superconductors are all alike electrically, namely that they all transport electricity without loss once they are below the critical temperature and at very low currents, their magnetic behavior can be really very different. The perfect diamagnetism that we have spoken of typifies a superconductor of *Type I*. Their behavior is understood quite simply. Expulsion of the magnetic field from the bulk of the superconductor requires doing some work against the magnetic pressure of the field thus expelled. It is like blowing a balloon. You may picture the magnetic lines of force as elastic strings under tension. They get stretched as they are pushed out sideways (Fig. II.3). The amount of work done on the system is proportional to the square of the field H. This raises the energy (and, therefore, the free energy) of the superconductor by the same amount. It is clear now that when this exceeds the condensation energy, the superconducting state will no longer be favorable energetically and the sample will turn normal. This defines a *critical field H_c* such that superconductivity prevails only for $H < H_c$. Inasmuch as the condensation energy decreases from a maximum at $T = 0$ K to zero at $T = T_c$, H_c too will behave likewise. Typical examples of Type I superconductor are mercury, aluminum and tin. The critical field H_c is typically 0.1 tesla (1 kilogauss).

Even below H_c, the flux expulsion is really only partial. Indeed, it is energetically favorable to allow the field to penetrate some distance into the interior of the sample. It is a kind of energy minimization by an optimal compromise as we shall see later. In fact the field diminishes exponentially as $\exp(-x/\lambda_L)$ with the depth x below the surface of the sample (Fig. II.6). The characteristic length λ_L is called the *London penetration depth*. The screening currents flow mainly within this depth. For the Type I superconductors, λ_L is typically 10^2–10^3 Å(1 Å$=10^{-8}$ cm). It is smallest at 0 K and grows to infinity (i.e., the size of the sample) as we approach T_c, where the sample turns normal and is filled with the flux lines uniformly.

---

7. TYPE II SUPERCONDUCTORS

Type II superconductors are different, and much more interesting. Discovered in 1937 by the Russian physicist L.V. Shubnikov, they are also the more important of the two types. They show the Meissner effect just like the Type I superconductors up to a *lower critical field* H_{c1}. As the magnetic field exceeds H_{c1}, something catastrophic happens. The flux rushes into the bulk of the superconductor and permeates the whole sample. But it does so in the form of filaments, or flux tubes, rather than uniformly (Fig. II.7). Each flux-tube carries exactly one quantum of flux. As the external field is increased, more and more flux tubes are formed in the sample to accommodate the increased total flux. This goes on until the flux tubes begin to almost touch each other at $H = H_{c2}$, the *upper critical field*, beyond which superconductivity is destroyed. For Type II superconductors, the upper critical field H_{c2} is typically 1 to 10 tesla (10–100 kilogauss). The flux tubes are very real. You can make them *visible* by sprinkling finely divided iron filings on the surface of the superconductor. The iron filings naturally cling to the foot points where the flux-tubes emerge from the superconductor and thus give them away. Such *flux decoration* experiments demonstrate not just the existence of the flux tubes, but also that these flux tubes are ordered in space as a triangular lattice at low temperatures. This is the so-called the *Abrikosov flux lattice*, named after the Russian physicist A.A. Abrikosov who predicted it theoretically in 1957. The flux-lattice is very real. It can vibrate elastically

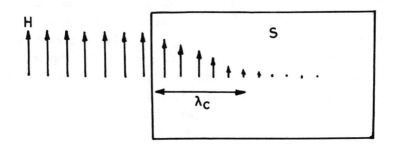

FIG. II.6 *Flux penetration in a superconductor (S). The London penetration length λ_L.*

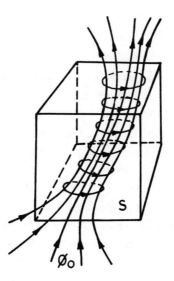

FIG. II.7 *Flux tube in a superconductor (S) carrying single flux quantum ϕ_0.*

and even melt at higher temperatures and form a flux liquid where the flux-tubes can even get entangled like the strands of melted polymers and hinder their mobility.

The individual flux tube has an interesting structure. It has a core which is in the normal state in that the superconducting order-parameter (or the condensation energy) is locally depressed to zero. The effect of this depression extends out to a distance ξ from the axis of the tube. This is the so-called *coherence length* and measures the distance over which the superconducting order is correlated — it is the minimum distance over which the superconducting order can change appreciably. Such a coherence is characteristic of all systems ordered one way or the other. Thus, the superconducting order-parameter (roughly the condensation energy) which is zero in the normal core will rise to its full value only beyond ξ, in the superconducting regions between the flux tubes. The magnetic field, on the other hand, will have its maximum value along the axis in the core region, and will fall off to almost zero beyond a distance λ_L away from the axis. Surrounding the core we will have the circulating supercurrents that do this screening of the field. Thus, the flux tube looks like a vortex and this state for $H_{c1} < H < H_{c2}$ is called the vortex state. Here the normal core region co-exists with superconducting regions intervening between the cores. For this reason the vortex state is also called the mixed state.

Let us see now what determines the type of a superconductor. The basic principle is the same — the free energy must be minimized. For a superconductor this requires compromising between two competing tendencies that operate at the interface (or the boundary) between the superconducting region of the sample and the region driven normal by the magnetic field. On the one hand it is favorable energetically to let the magnetic field penetrate the superconducting region reducing thereby the energy cost of flux expulsion. This gain in energy is proportional to the London penetration depth λ_L for a given area of the interface. On the other hand, the superconducting order (or the condensation energy) which is depressed to zero in the core, remains more or less depressed out up to a distance ξ from the axis of the flux tube. This costs an amount of energy proportional to ξ for the given area of the interface. Thus, for ξ much greater than λ_L, it is energetically favorable to reduce the total area of the interface. It is as if there is a positive interfacial surface energy per unit area (a surface tension). This will correspond to a complete Meissner effect — total flux expulsion of flux as in a Type I superconductor. For the opposite case of ξ much less than λ_L, it is energetically favorable to increase the interface area as if the surface tension is negative. This is realized by flux tubes filling the sample as in a Type II superconductor. Detailed calculation shows that $\xi \simeq \lambda_L$ is the dividing line. The basic physics here is the same as that of wetting — water wets the glass while mercury does not wet it. A more closely related situation is that of mixing of oil and water in the presence of some surfactant that controls surface tension. The mixture may phase-separate into water and oil (Type I) or, globules of oil may be interspersed in water (Type II). The Type II superconductor is indeed a *laboratory* for doing interesting physics.

---

8. THE CRITICAL CURRENT

The critical field (H_c for Type I or H_{c2} for Type II) is an important parameter that limits the current carrying capacity of a superconductor. For, a supercoducting wire carrying current generates its own magnetic field (Ampere's Law) and if this self-field exceeds H_c or H_{c2}, superconductivity will be quenched. This defines a *critical current density* J_c. Clearly a Type II superconductor with large H_{c2} up to 10 teslas, is far superior to the Type I superconductors with H_c of about 0.1 tesla. There is, however, a snag here that involves some pretty physics. Consider a flux tube threading a Type II superconductor, and let an electric current I flow perpendicular to the flux tube. Now, by the Faraday principle of electric motor, there will be a force acting on the flux tube, forcing it to move sideways perpendicular to the current and proportional to it. Physicists call it the Lorentz force. Once the flux tube starts moving with some velocity, the Faraday principle of electric generator (the dynamo principle of *flux cutting*) begins to operate — an electromotive force (potential drop V) is generated perpendicular to both the flux tube as well as its velocity so as to oppose the impressed *current I*. This leads to dissipation of energy. Remember Watts $(W) = V$ (voltage drop) $\times I$ (Amperes). We have a paradoxical situation of having a lossy superconductor! Where is the energy dissipated, you may ask. Well, you see the core of the flux tube is in the normal state and, therefore resistive. The motional electromotive force really acts on this normal region dissipating the energy. Now this would make the Type II superconductors practically useless. The flux-tubes must be *clamped* somehow. There is an ingenious way of doing it. Let us introduce some defects in the material — by adding impurities, alloying, or mechanically by cold working. This can locally depress the order parameter (or even make it zero altogether). Now it is clear that it will be energetically favorable for the flux tube to position itself such that its normal core overlaps maximally with these defects where the order parameter (condensation energy) is small anyway. Thus the flux tube is *pinned* at these pinning centers. Of course beyond a critical current, the Lorentz force will exceed the pinning force and the flux tubes will be released leading to a snap-jiggle kind of motion and hence to dissipation. Proper pinning is the secret of superconductors with high critical currents.

---

9. UNDERSTANDING SUPERCONDUCTIVITY

It shoud be clear at the very outset that superconductivity has to do with the state of the free electrons that make up our metal. The electrons repel one other and are attracted towards the oppositely charged ions that form the background lattice. The ions can collectively oscillate about their mean positions — the lattice vibrations or the sound waves called *phonons*. This is our many-body system. Thus an electron moves under the influence of all other electrons and the ions. Individually, it can easily be scattered and this is what happens in the normal resistive state. But at temperatures below T_c, the interacting electrons enter into an ordered state that somehow has the collective rigidity against such a scattering. What is the nature of this order? It is certainly not that the electrons have crystallized. With such a long-range rigid order in space they could hardly conduct, much less superconduct. No, the electrons remain a liquid, but this liquid has an order of which we may not form a simple mental picture. For, here finally we are confronted with their all pervasive quantum waviness, amplified infinitely by their indistinguishability, and reified in the stillness of absolute zero where the ions have all but gone to sleep — the single, whole *macroscopic wave function* of the many electrons. That such may be the case was anticipated by Fritz London way back 1933, almost 25 years before the fully microscopic theory of BCS was completed. Let us try to see how this may come about without getting technical.

9.1 Fermions

The electron carries an internal angular momentum (spin) which is one-half in units of $\hbar$. The spin can point either parallel or antiparallel to any direction that can be chosen arbitrarily. The quantum mechanical state of a single free electron in a metal is then specified, or labelled, by its energy, its momentum and the direction of its spin. The latter can be conveniently taken to be either up or down. As we will see in Chapter III on Symmetry, the spin-half particles (electrons) are fermions, and not more than one electron can occupy the same state. Thus at a given point in space, you can have at the most two electrons with opposite spins — the Pauli exclusion principle. This social (or rather asocial) behavior, called Fermi statistics, is the direct consequence of indistinguishability of identical particles in quantum mechanics and their half spin. Thus the ground state of a system of free electrons, that is the state at absolute zero of temperature, is obtained by placing two electrons with opposite spin directions in the lowest one-electron level, two electrons with opposite spin directions in the next higher one-electron level, and so on until we have accommodated all the electrons in the system. Thus, there will be a highest occupied state, or energy level called the *Fermi level* of energy usually denoted by E_F. (See Fig. II.8).

This distribution of occupation numbers among the allowed energy levels is called the *Fermi statistics*. This is precisely how we build up atoms. A solid is like a large extended atom. The only difference is that in an atom there is a single attractive center, the nucleus, and then the electrons are confined or localized around it like the planets around the sun. In a solid, on the contrary, there are many equivalent nuclei, and a given electron moving under their influence (potential) is as likely to be on one of them as on any other. Thus the electronic wavefunction is extended over the whole sample like a plane wave. This corresponds to freely moving electrons. Now recall that the kinetic energy of a free electron is proportional to the square of its momentum. Thus the Fermi energy E_F will define a *Fermi sphere* in the space of momenta such that all states within the sphere are occupied while the states outside are empty. Such a Fermi sphere is referred to as the *Fermi sea* and the surface as the *Fermi surface*. In a crystal the Fermi surface can

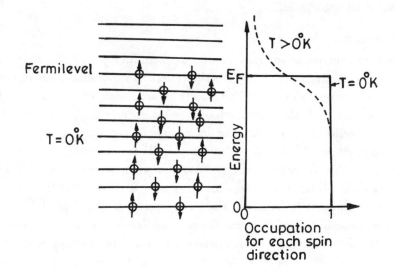

FIG. II.8 *Fermi-distribution at zero and finite temperatures.*

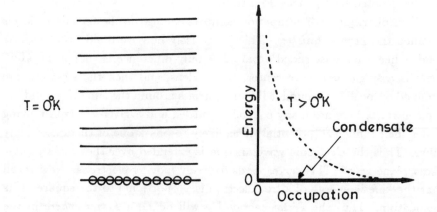

FIG. II.9 *Bose distribution at zero and finite temperatures. Thick horizontal peak at zero energy signifies Bose-Einstein condensate below the lambda point.*

have a complicated shape reflecting the symmetry of the crystal lattice.

For a macroscopic system size, the allowed energy levels are very closely spaced — they form nearly a continuum or a band. In a metal there are empty states available just above the Fermi level into which an electron can be accelerated by an external electric field, no matter how small. This is what makes a metal a good conductor — there is room at the top. (Incidentally, in an insulator, by contrast, there are no allowed higher energy levels arbitrarily close to the Fermi level into which an electron may be promoted by a small electric field, and hence no conduction. We say that the Fermi level lies in a forbidden gap between a lower filled band and an upper empty band of allowed states. A semiconductor is merely an insulator with a small band gap). In a metal, the electrons at the Fermi level (E_F) move with large speeds, the Fermi speed v_F typically 10^6 metres per second (i.e., a third of one hundredth the speed of light). Because of the exclusion principle, it is clear that only electrons lying close to the Fermi level can participate in any low energy phenomenon.

The above picture at 0 K is only slightly modified at finite temperatures. All that happens is that some of the electrons lying below the Fermi level get thermally promoted to empty states just above it. Thus the sharp step in the Fermi-distribution at 0 K gets smeared out over an energy interval of about $k_B T$ which is much smaller than E_F. Typically $k_B T_c / E_F$ is 10^{-4}.

One final remark about the electrons in a metal. The electrons, of course, repel each other by strong long-range Coulomb forces. One may wonder how we can possibly talk ourselves out of this and treat them as a Fermi gas of particles moving independently of each other, in the given potential of the background ions. Surely a moving electron creates a disturbance around it by pushing other electrons out of its way, for example. This is a complicated many-body problem — it is a Fermi liquid. It turns out, however, that these effects can be by and large absorbed in a re-definition of our particle. The *bare* electron is dressed with a cloud of disturbance around it. The *quasi-particles* so defined now move more or less independently of each other. The quasiparticles carry the same charge and spin as the bare electron. But they have a different effective mass and interact with a relatively weak, short-ranged (screened) repulsion. All we have to do then is to read quasi-particle (quasi-electron) whenever we say particle (electron). This tremendous reduc-

tion of strongly interacting *bare* particles to weakly interacting quasiparticles
is due to the great Russian physicist Landau. It is called the Landau quasi
particle picture.

9.2 Bosons

Now we turn to the other species of particles, the *bosons* that have a completely opposite social behavior — the *Bose statistics*. Bosons are particles
with spin equal to zero, or an integer. An example of direct interest to us
is that of ^{4}He, an isotope of helium having two protons and two neutrons in
the nucleus and two electrons outside of it, and with the total spin adding up
to zero, i.e., it is a *boson*. (Remember that for all phenomena involving low
energies, as indeed is the case in *condensed matter physics*, we do not probe
the internal fine structure of the composite particle like ^{4}He. It acts just like
any other elementary particle with spin zero). Bosons, unlike fermions, tend
to flock together. That is to say that any given one-particle state can be occupied by any number of the Bose particles. And this leads to a remarkable
phenomena called Bose-Einstein (B–E) condensation. To see this consider
an ideal gas of N Bose particles (i.e., non-interacting Bose particles) with N
very large, almost infinite. The ground state of the system, that is the state
at 0 K, can be readily constructed by simply putting all the N particles in
the lowest one-particle state which is the state of zero momentum and zero
energy. You see, we have a macroscopic occupation of a single one-particle
state. The occupation number is proportional to the size of the system. We
call this phenomenon *Bose condensation*. As we raise the temperature we
expect a finite fraction of the particles to be excited, or promoted to higher
energy levels (Fig. II.9) thus depleting the *condensate* partially. The gas of
excited particles in equilibrium with the condensate form a kind of interpen-
etrating two-fluid system. Finally, at and above a characteristic temperature
T_{B-E}, the condensate is depleted totally. The temperature T_{B-E} is called
the *Bose-Einstein temperature* and depends only on the number density and
the mass of the Bose particles. It increases with the increasing density and
the decreasing mass. Something drastic must happen at T_{B-E}. At and below
this temperature a finite fraction of atoms condenses into a single one-particle
state of zero momentum, and the fraction grows to a maximum (unity) as the
temperature is lowered to absolute zero. For the ^{4}He, regarded as an ideal
Bose system, the calculated T_{B-E} is about 3 K. Now ^{4}He indeed undergoes

a phase transition at 2.17 K, below which its viscosity drops abruptly by a factor of at least a hundred million. It can flow through the finest capillaries without any viscous drag. It becomes a *superfluid*! This phase is called He-II to distinguish it from the normal phase of liquid helium called Helium-I. The transition point is called the *lambda point* (T_λ) because the temperature dependence of the specific heat near T_λ has the shape of the Greek letter lambda (λ). The proximity of the λ point (2.17 K) to the Bose-Einstein temperature (3 K) is no coincidence. It is now believed that superfluidity is due to the Bose-condensation. The difference between T_λ and the ideal T_{B-E} is attributed to the fact that helium is after all not an ideal Bose gas. The atoms of ⁴He repel each other strongly at short distances due to their hard-core, and attract weakly at long distances due to the *van der Waals forces* (that act even between neutral atoms and molecules and make small particles stick together).

9.3 Bose Condensation and Superfluidity

But how come Bose-Einstein condensation should give superfluidity? The argument runs something like this. It is not just that we are allowed to put any number of the Bose particles in a given single particle state, it is rather that they tend to flock together. Thus, if a Bose particle is scattered from an initial state to a final state that is already preoccupied by N particles of its kind, then the probability of this process is enhanced by a factor $(N + 1)$ — it is like the rich getting richer. Consider now the situation where the condensate is moving with a certain velocity relative to the walls of a capillary. Let a particle be scattered out of the condensate due to its interaction with the wall. This is the kind of process that would give rise to viscous drag. Now the probability of this particle being scattered *back into* the condensate is proportional to $(N^* + 1)$, where N^* is the number of bosons in the condensate, which is a macroscopic number, almost infinite. It is clear, therefore, that the particle will relapse almost immediately into the condensate with probability unity. The condensate has a collective rigidity against scattering. Hence the superfluidity.

9.4 Phonon Mediated Attraction

But what has this got to do with the superconductivity of metals, where we have instead electrons obeying Fermi statistics? If only the electrons could somehow form bound pairs that would then behave like bosons and undergo Bose condensation! But in order to form these pairs the electrons must attract each other and not repel as they normally do. This then is the big question. It was, however, shown by H. Fröhlich that the electrons can indeed effectively attract one another in the presence of a deformable lattice of ions that is, of course, always present as the background. The Fröhlich mechanism is roughly this. An electron attracts the ions in its immediate neighborhood. The ions respond by moving, ever so slightly, towards it creating thus an excess of positive charge around it. We say that the electron has *polarized* the lattice. Another electron is now attracted towards this polarization localized around the first electron, and in doing so it is effectively attracted towards the first electron. This is very much like the water-bed effect. Imagine two persons lying on a water-bed. Then each one of the two tends to fall into the depression in the bed created by the other. You can easily demonstrate this effect by putting two marbles on a stretched linen held taut and watch them roll towards each other.

One is still left with the uneasy feeling that this, rather indirect, attraction may not be strong enough to overcome the direct repulsion between the two electrons. The argument gets somewhat subtle here and involves some pretty physics. As we have already remarked, the electrons, that is our quasiparticles repel via a short-ranged screened Coulomb potential of range on the order of the mean spacing between electrons. This is about an Angstrom or two. But what is most important is the fact that this potential acts instantaneously, that is to say that it depends only on the present positions of the two electrons under consideration. The indirect attractive interaction, on the other hand, involves the tardy movement of the ions. The ions are sluggish because of their relatively large mass which is at least a few thousand times the mass of the electron. The response time of the ions may be taken to be the period τ_D of their harmonic oscillation, and is typically 10^{-12} s. This corresponds to their typical oscillation frequency, 10^{12} Hz, the so-called *Debye frequency*. This means that the local polarization induced by the electron at a point will persist for a time τ_D even after the electron has moved away from that point. Now the electron moves with the Fermi speed v_F, typically 10^8 cm s^{-1}. It would thus have moved a distance $\tau_D v_F$ which is about 10^{-4} cm (10^4 Å) during this time τ_D. This is much larger than the mean spacing between electrons, or the range of the screened repulsive interaction. One may therefore expect a second electron to come around and feel the attraction of the persistent polarization left behind by the first electron and still be too far away from it to feel its direct repulsion. This is the essence of *dynamical screening* — a rather subtle effect. Thus the direct repulsion is strong but instantaneous, while the indirect attraction is weak but retarded, and it is this difference of the time scales that makes the weak attraction prevail over the stronger repulsion.

In the parlance of many-body physics this attractive interaction is viewed as mediated by *phonons*, that is due to the exchange of the virtual quanta of lattice vibrations. One electron emits (creates) a phonon which is absorbed (destroyed) by the other electron. Interactions between material particles by exchange of some virtual quanta in a commonplace in physics. Thus the exchange of virtual photons leads to an interaction between charged particles. We say *virtual* because the exchanged quanta exist only between the times of emission and absorption. It should be clear that these quanta must be bosonic. (Fermions have to be created and destroyed only in pairs). Thus, phonons are bosons just as photons are. There is a very transparent way of saying these processes with the Feynman diagrams (Fig. II.10). (See Chapter VI).

It turns out that the attractive interaction mediated by the exchange of phonons is maximum when the two electrons have equal and opposite momenta (velocities). This enables the two electrons to take maximum advantage of the polarization created by each other. Also, as the energy exchanged between the electrons is of the order of the energy of the phonon exchanged, which is typically the Debye energy, only electrons lying to within the Debye energy of the Fermi surface take part effectively in the process. The Debye energy of lattice vibrations is typically 10^{-4} times the Fermi energy — only a very small fraction of electrons are affected by this attractive interaction.

---

10. COOPER PAIRS AND THE BCS THEORY

Given this attraction between the electrons lying close to the Fermi surface, one is tempted to ask immediately if the electrons are going to bind to form pairs. For, the pairs so formed will be *bosons* (because of integral total spin,

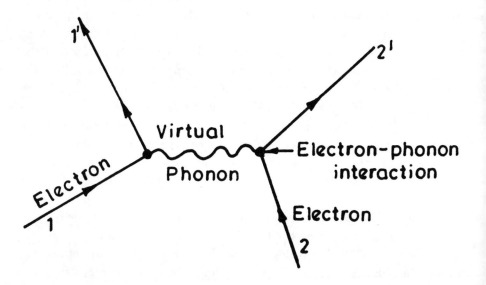

FIG. II.10 *Feynman diagram showing emission of a virtual phonon by one electron and its absorption by another electron, giving an effective electron-electron attraction.*

zero or one), and it is not hard to contemplate a Bose condensation that would lead to superfluidity (or rather superconductivity as the pairs will be charged $2e$).

It was Leon N. Cooper of the University of Illinois who first considered such a possibility. Cooper showed that two 'test' electrons having a short-ranged attraction between them and moving in the background of an impenetrable Fermi sea (*Pauli exclusion principle*) formed by the other 'spectator' electrons, will bind with opposite spins, no matter how weak the attraction is. Pairing with opposite spins means total spin zero — a singlet pairing. There was a hope. Opposite spins allow the electrons to come close enough to take full advantage of attraction. The formidable problem of pairing with all the electrons treated *at par*, gamesters and the spectators alike, was finally solved by John Bardeen, Leon N. Cooper, and J.R. Schrieffer, all at the University of Illinois at that time, in their now famous paper published in the Physical Review in 1957. This became the celebrated *BCS theory of superconductivity*. The central idea remained that of *pairing* — of the '*Cooper pairs*'. What causes the attraction is a secondary issue. Instead of phonons, other bosonic quanta may be exchanged. One should note, however, that seen from a distance, the Cooper pairs will appear and act as bosons. But, it turns out that the size of the Cooper pair is typically 10^{-4} cm (essentially the coherence length), which is much too large compared to the mean electron spacing. Indeed, millions of Cooper pairs overlap. Thus treating them as compact bosons, like ^{4}He, is an oversimplification. Still the essential physics remains the same. A direct consequence of this pairing is that it costs energy to break the pair. Thus, it turns out that unlike the normal metallic state, we need a minimum energy 2Δ to excite the superconductor. Here Δ is called the *superconducting energy gap*. It is typically $10^{-4} E_F$. That such relatively small energy scales should emerge giving a robust superconductor is an amazing consequence of quantum coherence.

The BCS pairing theory with the phonon mediated attraction explains many of the puzzles immediately. First, the electron-phonon coupling involved in the pairing mechanism is no different from that causing scattering (resistance) in the normal state. No wonder then that good superconductors are bad normal metals as we noted in the beginning. Then there is the isotope effect. Since the pairing mechanism involves motion of the ions, the ionic mass M must be a relevant parameter. Other things remaining the same, heavier the ion smaller is the displacement (polarization) and hence weaker is the pairing. Thus, replacing an ion with its heavier isotope (this being the meaning of the proviso 'other things remaining the same'), the T_c must go down. BCS predict T_c to be proportional to $1/M^{1/2}$. This is indeed observed experimentally. These and many other predictions of the BCS theory have since been confirmed. It is the correct theory of superconductivity.

The connection between superconductivity (or superfluidity) and Bose condensation seems intimate. After all, ^{4}He shows superfluidity at 2.15 K as we have remarked earlier while its isotope ^{3}He, which is fermionic (with two protons and one neutron in the nucleus and two outer electrons and thus the total spin half), shows no sign of it around that temperature. The superfluidity of ^{3}He observed at much lower temperatures in the millidegree Kelvin range is again due to pairing. But this time the pairing is with total spin one (triplet) and not zero (singlet) as for the BCS superconductors. We must caution, however, that the BCS theory goes far beyond the naive idea of Bose condensation of an ideal Bose gas.

---

11. SOME MACROSCOPIC QUANTUM EFFECTS

As we have remarked several times, particles and in particular electrons have a wave nature. Electrons may be reflected and diffracted by a diffraction grating. Indeed, electron diffraction is used to study crystal surfaces, and in the electron microscope. An electron wave can also be made to interfere (with itself) just as a light wave in the Young double-slit experiment. All that we have to remember is that these matter waves are the waves of probability amplitude — they are the quantum mechanical waves. A Bose condensate represents a macroscopic *wave function* of all the bosonic Cooper pairs in the condensate. It is, so to say, a highly coherent superposition of the individual Bosonic amplitudes — something akin to a laser beam (see Chapter I on Lasers). The wave nature of this condensate is described by the single complex wave function $\psi(r) = |\psi(r)| \exp(i\theta)$ such that $|\psi(r)|^2$ gives the density of the condensate at the point r. It has a phase which is constant all over the sample in the absence of any supercurrent. We may treat this complex wave function as the superconducting order parameter, much the same way as the magnetization $\mathbf{M}(r)$ is for a ferromagnet. The magnitude of magnetization corresponds to $|\psi(r)|$ while the phase of $\psi(r)$ corresponds to the direction of $\mathbf{M}$ (r). Similarly, all values of the phase from 0 to 2π are energetically equivalent. A fixed global value of θ for a given superconductor is the spontaneously broken symmetry akin to a fixed global direction of spontaneous magnetization (see Chapter III on Symmetry). Gradient of θ in space causes a supercurrent to flow.

11.1 Flux Quantization Revisited

With this picture of the condensate in mind, flux quantization follows straightforwardly. Considering a superconducting ring enclosing a certain amount of flux (Fig. II.11). Let us reckon the total change of phase as we go round a closed curve C deep in the material of the ring. The total change must be, of course, an integral multiple of 2π. This is because on completing the circuit we must return to the same value of ψ — it must be single-valued, we say. Now recall that $\exp(i2\pi n) = 1$ for any integer n. Thus the phase change must be $2\pi n$. Now it is known from electrodynamics that the change of phase in going round a circuit must be proportional to the magnetic flux enclosed by the circuit. In fact it is $q\phi/\hbar c$, where q is the charge on the basic entity, in our case the Cooper pair. Thus $q = 2e$. Equating the phase changes computed in these two ways, we get $\theta = hc/2e$. The flux is quantized!

11.2 Josephson Tunneling and the Superconducting Interference

The long-range nature of the superconducting coherence shows up dramatically if we interpose a thin dielectric (insulating) barrier between two superconductors. This is called a Josephson junction. We can have, for example, a tin oxide (SnO_2) film separating the two superconducting tin (Sn) electrodes. No current will flow on closing the circuit if the tin is in the normal state. If, however, the barrier is thin enough (~ 25 Å) and tin is in the superconducting state, a supercurrent can flow persistently on closing the circuit without any potential drop across the junction (Fig. II.12). There will be a phase difference θ between the two superconducting contacts on the two sides of the barrier. The supercurrent will vary sinusoidally as function of θ, i.e., $J_s \propto \sin\theta$. This dramatic effect was predicted by the 20 year old Brian Josephson at Cambridge in U.K., for which he shared the Nobel Prize for Physics in 1973. Since the maximum value of $\sin\theta$ is 1, it is clear that the

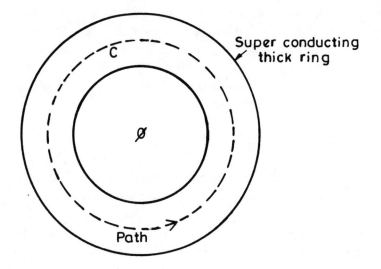

FIG. II.11 *Flux (φ) quantization through a superconducting loop. J_s is the screening supercurrent.*

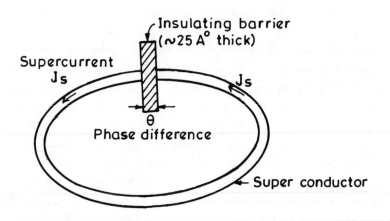

FIG. II.12 *Josephson junction with persistent supercurrent tunneling through it.*

Josephson current must be less than a critical value J_c. Beyond J_c, a voltage appears across the junction and the dc super current drops to zero!

One can use a pair of Josephson junctions in parallel and demonstrate the quantum interference effect (Fig. II.13). The relative phase of the two superconducting amplitudes along the two arms can be varied by varying the magnetic flux passing through the enclosed area. One obtains the oscillatory pattern reminiscent of the Young double-slit experiment with light. It is indeed possible to count the 'fringes' (the maxima and minima of the current) and thus measure the flux change with unprecedented accuracy. A strange thing to note here is that the effect depends only on the total magnetic flux through the area enclosed by the two parallel superconducting paths. The magnetic field need not touch the superconductors at all. This non-locality again emphasizes the non-classical nature of superconductivity. This is the basic principle of the *superconducting quantum interference device* (acronym SQUID). Magnetic fields as small as 10^{-9} gauss can be measured. This is the kind of fields produced by the tiny currents flowing in the human brain or the rusting fender of your car.

An equally fascinating phenomenon is the ac Josephson effect. If you apply a dc voltage V across the Josephson junction, the phase difference across the junction will begin to increase linearly in time making the current through the junction oscillate at a frequency $2eV/h$. Thus a dc voltage of $1\mu V$ (one micro-Volt) will produce a frequency of 483.6 MHz. You have made an oscillator! This fact can be used to measure the fundamental ratio e/h to unprecedented accuracy. Conversely, we can measure low voltages of the order of 10^{-16} Volt! The phase of the superconducting order parameter is perhaps the most important aspect of it.

As an aside let us mention that superconductivity affords us a means of addressing some very deep fundamental questions of quantum mechanics (see Appendix B). In dealing with the microscopic particles like an electron we freely superpose the wave amplitudes. Thus, if there are two possible states labelled 1 and 2 for an electron with wave amplitudes ψ_1 and ψ_2, then the electron can also be found in the superposed state with the wave amplitude $a_1\psi_1 + a_2\psi_2$. Then $|a_1|^2$ and $|a_2|^2$ give the probabilities of finding the electron in the two states 1 and 2 respectively. The question is if this is true of macroscopic objects with states which are *identifiably distinct*. This

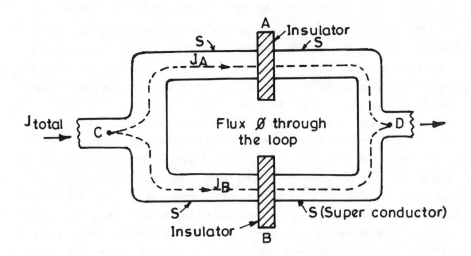

FIG. II.13 *Interference of partial quantum amplitudes through two Josephson junctions connected in parallel. The resultant supercurrent oscillates sinusoidally with magnetic flux trapped between alternative paths.*

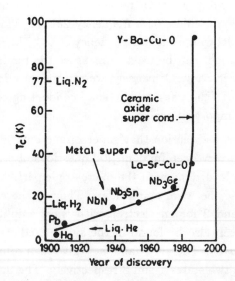

FIG. II.14 *The record of superconducting transition temperatures.*

is what Erwin Schrödinger, the founder of quantum mechanics after whom the Schrödinger equation is named, expressed rather picturesquely by asking if a cat can be found in a superposed state of being dead and alive! A superconductor provides us with a Schrödinger cat of sensible magnitude! This question remains unsettled.

---

12. SUPERCONDUCTOR COMES OUT OF THE COLD

Until about 1986, superconductivity belonged in the range of the liquid helium temperatures. This pre-occupation with low temperatures, and the age of the *low temperature superconductors* (LTSC) came to an abrupt end with the discovery of the ceramic superconductor (La-Ba-Cu-O, T_c about 35 K) by J.G. Bednorz and K.A. Müller in 1986. These ceramics were made by mixing oxides of lanthanum, barium, and copper and heating them to high temperatures so that they react to form the compound. With this the era of *high temperature superconductivity* (HTSC) had begun. After some initial skepticism, confirmations poured in from many laboratories around the world. Superconductors with still higher transition temperatures (T_c about 90 K to 125 K) were soon announced. Thus, substituting Y (yttrium) for La (lanthanum) could give $T_c = 90$ K. Reports of *unidentified superconducting objects* (USO'S) continue! Room-temperature superconductivity is awaited, though with some studied nonchalance. This is not the time or the place to tell the story of this breakthrough, except to affirm that the excitement generated by it among physicists, chemists, technologists, metallurgists and material scientists and the common man has no parallel in the recent and the not so recent history of science and technology. It also marks the culmination of a prolonged scientific endeavor (See Fig. II.14).

What is unusual about these high T_c superconductors? Are they really different from the conventional BCS superconductors? There are, of course, the obvious facts of high T_c (35 K–125 K) and the high critical field H_{c2} (100 tesla). But there is much more than that. The point is that normally one thinks of a superconductor as derived from a metal. The new superconductors seem more like derived from an insulator. These materials are oxides, in fact earthy ceramics, and are very poor conductors, almost insulators. But they are insulators with a difference. It is not that the bands of allowed one-electron states are completely filled. In fact the bands are half-filled. These are insulators because of the very large repulsion between the electrons that prevents them from occupying a state doubly — one with spin up and the other with the spin down as in normal metals. This single fact makes them behave abnormally even in the *normal* state above T_c. These materials hardly show any *isotope effect*. But *pairing* is not in doubt as confirmed by the flux quantization experiments. One strongly suspects that pairing may not be due to the phonons. It is believed that magnetism is involved in the superconductivity of these materials in an essential way. It is, however, too early to make a claim. There are just too many theories around. What one can definitely say is that the novel ideas generated by these superconductors will have a profoundly enriching effect on our thinking for years to come.

On the practical side, however, the possibilities are clearly enormous. High-T_c superconductors can do whatever the conventional superconductors do, and obviously do it much cheaper, for the simple reason that liquid helium costs a few dollars a liter while liquid nitrogen costs just a few cents a liter. You save on the cost of cooling. Besides, liquid nitrogen is much more efficient as a coolant.

Several applications come to the mind. Some are large scale applications involving high currents, high current densities and high magnetic fields as for power generation, transmission and energy storage. Others are small scale applications as in electronics involving low currents, but the current densities can still be very high. The zero electrical resistance and hence the absence of dissipation (heat loss) helps in two ways. It obviously makes the system more energy efficient and provides cheap magnetic flux. Less obvious, however, is the fact that it allows compact designs as we do not require large surface areas, cooling fins say, to remove heat as none is generated. This, for example, makes it possible to achieve highest packing density of electronic components by just making all the interconnects on a silicon chip superconducting — packing density of logic circuits comparable to that of human brain is realizable!

Compact and high-field superconducting magnets can be used in the giant particle accelerators to confine, store and direct the beams of charged particles. Just think of one such giant machine, the proposed six-billion dollar Superconducting Supercollider (SSC) having 10,000 such magnets! The cheap magnetic flux should come in very handy here. The same is true of fusion reactors (TOKAMAK) where the hot thermonuclear plasma has to be confined by a large toroidal magnetic field.

Large superconducting coils carrying persistent currents can be used for non-polluting energy storage. These Superconducting Magnetic Energy Storage (SMES) systems will have to be built underground for reasons of excessive magnetic stresses on the structure. Thus the magnetic energy stored in one cubic meter of space with a magnetic field of about 50 tesla can supply power at the rate of 1kW for a period of about ten days! The energy stored is proportional to the square of the magnetic field and, of course, to the volume. Magnetically levitated trains (mag-levs) are possible and already being contemplated to be operational by the year 2000. These ultrafast bullet trains (speeds of about 500 kmph) will be, for one thing, free from the 'click-click' of the conventional trains running on the 'permanent ways', and far safer too. One can also think of large amounts of dc (direct current) electrical power distribution using superconducting transmission lines, or better still underground cables, and cut the *copper-loss* that can easily amount to 5% of the power generated — in 1985 this had cost America $9 billion! Most of this could be saved using dc superconducting transmission lines. In all these cases, however, there is at present the technological problem of drawing wires of these highly non-ductile materials. But, encouragingly enough, 1 km long, 0.3 mm diameter fibers of these HTSC materials are already being made. These may be clad in copper and made into multifilament cables several meters long for actual use. The other limiting factor, almost as important as the critical temperature, is of a more fundamental nature — the critical current density J_c. High J_c is a *sine qua non* for most applications, but requires efficient pinning of the flux lines. Remember that for the HTSC, the lower critical field H_c is very low, about 10–100 gauss and, therefore, the flux lines enter the material rather easily. This is one of the most active areas of research in this field today. We are, however, close to having usable current densities normally used in copper conductors.

There are then the low-power applications. For example, sensors like SQUIDS can be made much cheaper with HTSC for mapping biomedical magnetic fields of the human brain (10^{-8} gauss) and the heart (10^{-5} gauss).

Even gravity waves, the weakest signal of all, may be detected with the help of these SQUIDS. Studies are under way for using very large Josephson junction arrays for faster, more compact fifth generation supercomputers with very large memory. Here the junction acts as a binary logic element that can be made to switch between the on-state (corresponding to zero junction voltage) and the off-state (corresponding to finite or normal junction voltage) in less than a picosecond (10^{-12} seconds) by passing a current in excess of the critical current. This is thousand times faster than the conventional switching devices based on transistors.

It may be another five years before these low power devices become commercially competitive. It may take longer for the large scale applications to become feasible. We may not be able to buy a spool of superconducting wire or ride a superfast mag-lev train for yet another decade. But there are no serious doubts about these things. In fact the possibilities are so diverse and so many that in any comprehensive planning for the future along these directions, we will do well to involve a science fiction writer. Just think of the curious toys that become possible with room temperature superconducting materials — the magnetically levitated gyros or the spinning tops, to name just two of them.

After all, room temperature superconductivity is a distinct possibility by the turn of the millennium. And in any case, in the foreseeable future when the coal and the oil will have nearly run out, man will have to turn to this *perpetual motion* — of the first kind.

Summary

Superconductivity is the complete disappearance of the electrical resistance of a material at and below a certain critical temperature which is characteristic of that material. The change of state occurring at the critical temperature is a thermodynamic phase transition. The supercondutor is, however, not merely a perfect conductor, it is also a perfect diamagnet: The magnetic flux is expelled from its bulk as the material is cooled in the presence of an applied magnetic field, through its transition temperature. This is the well known Meissner effect. This tendency to exclude magnetic flux makes it possible to trap the flux threading through a superconducting ring permanently, or to levitate a magnet above the superconductor. But, if the magnetic field exceeds a certain critical value, the superconducting state is destroyed and the material reverts to its normal resistive state. The flux expulsion is, however, not complete: The magnetic field does penetrate the superconductor up to a certain depth called the London penetration depth which is typically a few hundred Angstrom units. As the critical temperature is approached from below, the penetration depth tends to infinity (i.e. equals the size of the system), while the critical field tends to zero. All this is true for the so-called Type I, or the soft superconductors. For the Type II, or the hard superconductors on the other hand, the Meissner effect is observed up to a certain lower critical field, while the superconductivity is destroyed above a certain upper critical field. In between, there is the mixed phase where the magnetic flux enters bulk of the material in the form of flux tubes that have a normal core whose diameter is of the order of the coherence length that can vary from a few Angstroms to several thousand Angstroms. The latter is the smallest length scale over which the superconducting order may vary appreciably. When a supercurrent flows through the material it exerts a force on these flux tubes making them flow sideways, causing dissipation. It is necessary, therefore, to pin down these flux tubes to achieve high critical surrents that can flow without dissipation. At low enough temperature these flux tubes can order as a triangular flux lattice called the Abrikosov flux lattice.

Superconductivity is a purely quantum phenomenon on a macroscopic length scale. This strangely ordered electronic superfluid state is described by a complex order parameter whose phase leads to observable effects, such as the quantization (wholeness) of flux trapped in a superconducting ring, or a flux tube when measured in certain natural units, or the ability of the supercurrent to flow across an insulating layer, several Angstrom units thick, separating two superconductors. This is the famous Josephson junction effect.

The correct microscopic theory of superconductivity, the celebrated BCS theory named after J. Bardeen, L.N. Cooper and J.R. Schrieffer was proposed in 1957 almost half a century after the phenomenon was discovered by Kamerlingh Onnes in 1911. The basic idea is the formation of loosely bound electron pairs, the Cooper pairs, due to an indirect attraction induced by the polarization of the background lattice. The Cooper pairs behave roughly as bosons and undergo Bose-Einstein condensation at low enough temperature.

The condensate has the superfluid property.

Superconductivity occurs widely among elements, compounds and alloys. The transition temperatures are, however, abysmally low, typically a few degrees above the absolute zero of temperature. This necessitates the use of liquid helium as coolant which is both inefficient and expensive. The recent discovery by J.G. Bednorz and K.A. Mueller in 1986 of high temperature supercondutivity in certain oxides with transition temperatures now as high as 125 K, much above the boiling point of liquid nitrogen has changed all this. Liquid nitrogen can now be used as coolant which is much more efficient and costs much less.

Most applications depend on the Type II superconductors because of their higher critical temperatures, critical fields and critical currents. Superconducting magnets are already in use. Superconducting cables, energy storage devices, magnetically levitated trains and Josephson junction based supercomputers are becoming feasible. It seems that the revolution initiated by these novel high temperature superconductors may be comparable to that brought about by the transistor.

CHAPTER III

SYMMETRY OF NATURE

AND

NATURE OF SYMMETRY

CONTENTS

1. WHAT IS SYMMETRY
THAT WE SHOULD BE MINDFUL OF IT?

Our immediate sense of symmetry comes from looking at objects around us. It may well be that the idea of symmetry is very primitive and comes naturally to the human mind. Perhaps human mind can grasp it internally all by itself. But we shall leave these questions to the philosopher and to the artist. Instead, let us for a moment turn experimentalist and look at a sphere. Then we will be left in no doubt that we are in the presence of a perfect symmetry. We may view the sphere actively by turning it around everywhich way we like and find that it looks the same. We may view it passively by keeping the sphere fixed but shifting ourselves around it and find again that it looks just the same. It is this unchanging aspect of sameness against the changing viewpoint that symmetry is all about. But then we have to get sophisticated. We have to abstract the general idea of symmetry and make it free from this static and rather limited visual setting. This we must do and in doing so we will see more, and not less than what the artist can, for all his sensitivity and imagination, ever hope to see. Because, there is much more and subtle in the world of physics than meets the eye. However, we will continue to use the same ordinary word for it.

Symmetry suggests a sense of balance and proportion, of pattern and regularity, of harmony and beauty and finally of purity and perfection. These synonyms just about sum up all our subjective reactions to the symmetries that abound in Nature, with her myriads of inanimate objects and life forms — the celestial spheres of the sun, the moon and the planets, the hexagonal snowflake with its six-fold symmetry, the five-fold symmetry of the starfish and of many a wild flower, the bilateral symmetry of the butterfly with its outstretched wings and of the man in his poise (Fig. III.1). One even speaks of the fearful symmetry of the tiger. Examples will fill volumes. And as life imitates Art and Nature, we find something of it reflected in the art forms created by man — be it sculpture, architecture, painting, poetry or music. It is true though that in most of these cases the symmetry is only approximate. As a matter of fact the ancient Greeks used to intentionally and secretly

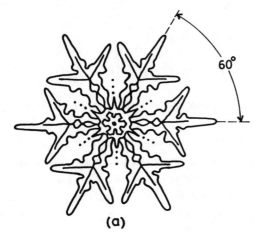

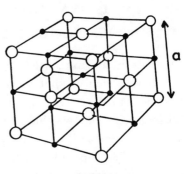

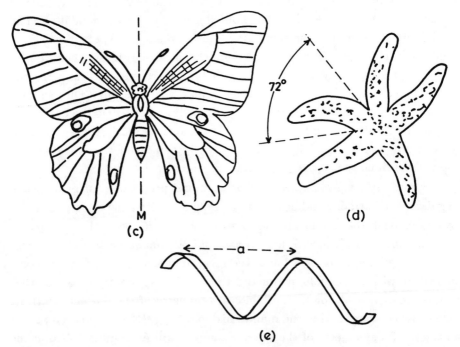

FIG. III.1 *(a) Snowlake with six-fold axis; (b) crystal of common salt; (c) butterfly with bilateral symmetry; (d) starfish with five-fold axis; (e) right-handed helix.*

introduce some degree of asymmetry in their otherwise symmetric designs. The fact remains, however, that the human mind is absolutely fascinated by Symmetry. In physics, the term symmetry takes on an objective meaning which is much deeper and far more precise, almost austere than what our vague feelings of it can command. Let us get acquainted with it.

Now, we can hardly do better than just repeat the definition of symmetry given by the great German mathematician Hermann Weyl — a thing is symmetrical if there is something you can do to it so that after you have finished doing it, it looks the same as it did before. This is an operational definition — it can decide. The 'thing' here is the *object* of interest. What you do to it is called the *Symmetry operation or transformation*. And 'looks the same' is yet another name for *invariance*. The 'look' itself is some discernible property of the object that remains invariant. Thus, there has to be an object with a discernible property that remains invariant under the action of the *'group'* of symmetry transformations. Now, the point of it all is that the object itself can be just about anything. It depends on our interest and on the level or the depth of our enquiry. At its simplest, the object may be a mere geometrical figure (a hexagon, a helix or a lattice), or the geometrical shape of a material body (a snowflake, a screw, or a crystal of common salt) (Fig. III.1). The symmetry operations involved here are purely geometric in nature — rotation by $360/6 = 60$ degrees or multiples of it about the sixfold axis of rotation, mirror reflection in the plane of the bilateral symmetry, translation in space by a repeat distance, or combinations of these (Fig. III.1). The object and its transform must be superposable if the symmetry is true. (This is obviously not so for a screw, or a helix. Although the screw is intrinsically identical with its mirror image, the two are not superposable. We will return to this interesting case later). But at its subtlest the object can be a mathematical entity, a (differential) equation expressing a physical law. Now, how do you rotate, reflect or translate an equation anyway? Well, we really don't do so literally. We perform these transformations passively on the independent variables, *i.e.*, the space-time coordinates occurring in the equation accompanied then by suitable transformations on the dependent variables. The invariance then is the invariance of the *form* of the equation under these symmetry transformations. More properly it is called covariance. Thus for instance an expression $x^2 + y^2 + z^2$ is invariant under any rotation of the Cartesian coordinate system (x, y, z) with its origin fixed at $x = 0$, $y = 0$ and $z = 0$. It just becomes $x'^2 + y'^2 + z'^2$, where the primed quantities are the coordinates of the same point, but with respect to the rotated (primed) coordinate system (x', y', z'). Similarly, the wave equation

$$\frac{\partial^2 \phi}{\partial x^2} + \frac{\partial^2 \phi}{\partial y^2} + \frac{\partial^2 \phi}{\partial z^2} - \frac{1}{c^2} \cdot \frac{\partial^2 \phi}{\partial t^2} = 0$$

keeps its form under the above symmetry transformation, and additionally under translation in space and in time. Just replace the unprimed quantities by the primed quantities. In particular $\phi(x, y, z, t)$ becomes $\phi'(x', y', z', t')$ and is numerically equal to it. If you take ϕ to be the pressure or the density, then the wave equation begins to describe a sound wave propagating in a medium such as air or water which is homogeneous (translationally invariant in space), isotropic (rotationally invariant) and unchanging in time (translationally invariant in time). In fact this equation has a much higher symmetry and it can describe the widest range of wave phenomena that occur in Nature. These symmetries of the medium (and the medium may well be vacuum as in the case of light) *almost uniquely* fix the form of this equation. Such is the restrictive power of symmetry.

One speaks of the symmetry of a particular law. Thus, we have the spherical symmetry of the Coulomb law of electrostatic attraction between a negatively charged electron and a positively charged nucleus of an atom. The Coulomb potential energy varies as the inverse of their distance apart, independent of the direction. The force on the electron, being gradient of potential, of course, varies as the inverse of the square of this distance and is directed radially inward. But for an atom embedded in a molecule or a solid, the potential law governing the motion of the electron has the symmetry of its environment which is necessarily lower than the spherical symmetry of the free atom. Much of the chemistry of molecules and the physics of solids depends on these environmental symmetries.

As we probe matter deeper, we uncover special laws that govern the goings-on at the nuclear and the subnuclear level — the domain of the elementary particles and the fundamental interactions between them. Here we encounter yet another kind of symmetry different from the space-time symmetries described above. These are the so called *local gauge symmetries* that seem to be at the very heart of the nature of things. It is already present in the interaction of light with charged particles where it was discovered first. But, of this more later.

A note of caution at this stage is in order. The symmetry of a law as expressed by a symmetric equation does not necessarily lead to symmetric phenomena resulting from it. The states of a system, the processes or the events represent the allowed solutions of the governing (differential) equation. But a particular solution gets selected by the initial conditions that can be imposed at will. These conditions need not have the symmetry of the system. And so it happens that the law of gravitational attraction between the earth and the sun is spherically symmetric, and yet the orbit of the earth round the sun is an ellipse — a foreshortened circle, with the sun at one of its foci. The same is true of a man-made satellite orbiting the earth. Its orbit depends on its height and the velocity at the time of its injection into orbit. A symmetry operation will not leave the particular orbit invariant but carry it into another, albeit allowed orbit. Thus, in general the particular solutions or the events or physical conditions themselves are not invariant. What is indeed invariant is the governing equation that fixes only the correlations between the successive events. The idea that the state of a system can have a symmetry lower than that of the governing law takes on a deep physical significance as we will see when we discuss the phenomenon of *spontaneous symmetry breaking*, which is the most symmetrical way of breaking the symmetry. In this we may catch a glimpse of the act of creation whereby Nature seems to have generated the observed diversity of fundamental laws as a result of a descent from the most symmetric, possibly a *'grand-unified'* law of interactions.

The all pervasive nature of symmetry is in itself a sufficiently strong reason for us to be mindful of it. But the most compelling reason of all is that symmetry is a great ordering principle and we can make it work for us. We will now demonstrate this power of it with the help of some simple and some not-so-simple examples.

To start with, symmetry simplifies things. Suppose you are asked to draw a butterfly with its out-stretched wings. Now, all that you really have to do is to draw only the left, or the right half of the butterfly, preferably on a tracing sheet. The other half is related to it by mirror reflection. It is more of the same. You can simply fold the sheet along the median line of bilateral symmetry and re-trace over your half-drawing. That is all. The reflection symmetry has halved your work, or very nearly so. In general, an n-fold symmetry divides your work by n. This is really a common trick and we should imagine that the makers of patterns use it all the time. This is, however, a trivial example. A highly non-trivial example of reduction of a problem by symmetry is provided by the case of a hydrogen atom. Here we have an electron bound to the nucleus (proton) by the attractive Coulomb potential which is spherically symmetric. In order to appreciate reasonably well the promised reduction of the problem we have to describe the atom properly. It is now well known that in the domain of the very small, and that is where the atoms belong, the proper theoretical framework is that of *Quantum Mechanics*, and not the *classical (Newtonian) mechanics* that describes our sensible world of middle dimensions so well (see Appendices A and B). Thus, we have to abandon the classical view of the hydrogen atom as a miniature solar system with sharply defined orbits for the planetary electron. We have, instead, an all pervasive waviness associated with the motion of the electron. We can picture the state of the electron as a fuzzy cloud around the nucleus, with the proviso that the density of the cloud at a point gives the probability (density) of finding the point-like electron at that point. (This replacement of the classical certainty of sharply determined orbits by the quantum uncertainty of dicey probabilities of being found somewhere is most disturbing. It was so to Einstein himself who was, ironically, one of the founders of this 'plutonic' republic of Quantum Mechanics, but never quite belonged there as a citizen. Quantum Mechanics is today the established *framework* theory for everything in the physical universe. Its predictions differ from those of classical mechanics and the difference gets more and more pronounced as we go deeper into the domain of the small). To get these probabilities one has to solve a certain wave equation, the *Schrödinger equation*, for the wave function ψ which is complex in general. The probability is then simply $|\psi|^2$, the square of its absolute magnitude. All that is important for our discussion is to note that ψ has both radial as well as angular dependence. The spherical symmetry of the Coulomb potential now helps us factor out the angular dependence and determine it completely without having to solve the Schrödinger equation. The spherical symmetry by itself determines the allowed values of the angular momentum $\ell(= 0, 1, 2....)$ and its component

$m (= -\ell, -\ell + 1, ..., \ell - 1, \ell)$ along a chosen direction in units of Planck's constant h divided by 2π. These are the labels, called *quantum numbers* that symmetry provides to specify completely the angular aspect of the state of the system. This is no mean reduction of the problem. In fact one can do better than this. In addition to the spherical symmetry, the Coulomb law has yet another 'dynamical' symmetry following from its form, namely that the force involves the square of the reciprocal of the distance, and not any other power such as the cube or the fourth power and so on. (This is, of course, a rather hidden dynamical symmetry and is for the preoccupied eyes of the mathematical physicists only). Properly treated, this symmetry solves the remaining radial problem too and provides yet another label, the principal quantum number n ($= 1, 2....$) that fixes the allowed electronic energies. That there is something special about the inverse square law that singles it out from among all possible central forces, can be seen from the following fact. Consider the motion of the earth around the sun, or better still the motion of a man made satellite around the earth. The orbits are elliptical as we know. But the real point which hardly ever gets emphasized is that the orbit closes upon itself! This will not be the case if you deviate ever so slightly from the inverse square law. For small deviations the orbit will still be close to being an ellipse but the ellipse will slowly precess or turn around the focus. The motion of perihelion (point of closest approach to sun) of the orbit of the planet Mercury around the sun may be viewed as due to small deviation from this dynamical symmetry of the inverse-square law caused by Einstein's general relativistic corrections to Newton's law of gravitation.

Now we turn to another aspect of this great ordering principle, namely, that symmetry classifies things. All classification is based on identification of a set of common characteristics. Thus we have the classification of the animal kingdom into vertebrates and invertebrates depending on the presence or the absence of the vertebral column. The *periodic table* of elements prepared by the great Russian chemist Mendeleyev is a classic example of classification. The most striking and rigorous example of classification by symmetry is the grouping of crystalline forms of solids. A crystal is a periodic arrangement of atoms in space. It can have spatial symmetries of discrete translation, discrete rotation and reflection and, of course, combinations of these. Symmetry considerations have led to the remarkable result that only a finite number of distinct groupings of these symmetry elements are possible. These are the celebrated 230 space groups of crystallography! Any of the nearly countless varieties of crystals, no matter how complex, must belong to one of these groups. We must hasten to add, however, that the crystals belonging to a given space group are certainly not identical, no more than all the vertebrates in the animal kingdom are identical. Finding the space group of a crystal is the first step towards understanding its molecular structure.

A much more profound example of symmetry-based classification in physics is the classification of identical particles as *fermions* (after the great Italian physicist Enrico Fermi) and as *bosons* (after the great Indian physicist S.N. Bose). Here the symmetry is with respect to permutation, or more simply reshuffling. Let us understand this. Consider a set of particles located arbitrarily in space. Let the particles be identical in all respects, *i.e.*, same mass, same charge and so on. You may think of a pack of cards, somewhat unusual in the sense that all the cards are alike — only queens of diamond, say. Now it is clear that any permutation of these identical particles (same as the reshuffles of the identical cards) will leave our system unchanged. After all, a permutation involves just pairwise interchanges and interchanging identical objects changes nothing. But not quite. The different permuted configurations are undoubtedly identical but they are distinguishable all the same. The reason for this is that nothing prevents you from keeping track of these identical particles as these are being moved around to their new permuted locations. This knowledge is sufficient to distinguish between the different permuted configurations even though the objects being permuted are identical. Thus the identical particles are distinguishable even if only by virtue of being initially located differently. You may wonder if this knowledge is of any consequence and if this distinction between identity and indistinguishability is not mere nit picking. Classically you are right. But as we have noted earlier, the correct framework for dealing with microscopic particles is quantum mechanics. And, most importantly, quantum mechanics does not allow sharply defined trajectories. It replaces them with an irreducible fuzziness. Therefore, even in principle we really cannot keep track of our identical particles in the process of permuting them as we did before. This idea of indistinguishability is brought home rather forcefully if you consider, *e.g.*, a pair of algebraic equations $x^2 + y^2 = 13$ and $x + y = 5$. These two equations are left invariant if we interchange x and y and hence permutation symmetric. Now, you can readily solve these two equations. You get either x = 3 and y = 2, or x = 2 and y = 3. Thus all you can say is that one of them equals 2 and the other equals 3, but which one is which you cannot say even in principle. So is the case with our identical particles. We can only say how many are at a given point of space (*i.e.*, the occupancy) but it is meaningless to ask which ones. This *indistinguishability* when treated properly leads to the great divide of identical particles into two classes — the fermions (*e.g.*, electrons, protons, neutrons, neutrinos, etc.) and bosons (photons, mesons, etc.). Identical fermions, electrons, say, exclude each other in that not more than one can occupy the same state. This is the Fermi-statistics. Kind of negative feed back at work. In contrast to this, any number of identical bosons, photons say, are allowed to occupy the same state. This is the Bose Statistics. In fact, bosons tend to clump together, a kind of positive feedback. What determines whether a given set of identical particles will be fermions or bosons requires deeper analysis of relativistic invariance. It is beyond our scope to go into that. But the result is simple. It turns out that a particle can have an intrinsic angular momentum called spin. You may roughly picture it as a spinning top much the same way as the earth spins about its own axis in addition to orbiting around the sun. The spin angular momentum is immutable (you cannot stop it spinning). It is quantized in multiples of $h/2\pi$, denoted by slashed $\hbar$. Now the rule is that particles with integral spin $(0, \hbar, 2\hbar, \cdots)$ are bosons and those with half odd-integral spin $(\hbar/2, 3\hbar/2, 5\hbar/2, \cdots)$ are fermions. This connection between spin and statistics has been one of the marvels of the symmetry principles in physics. The fact that two electrons (fermions with spin half) cannot simultaneously occupy the same point of space with their spins pointing in the same direction (*i.e.*, cannot be in the same state) is responsible for the stability of all matter, and for the fortunate circumstance that your hands do not go through the table which they might be resting on. For, in doing so the electrons in your hand must go through the electrons in the table which is clearly forbidden. The clumping tendency of photons (bosons with spin unity), on the other hand, makes it possible for any number of them to condense into a given state — the *Bose condensation*. This is what makes the laser beams so coherent (see Chapter I on Lasers). Similarly, superfluidity of ^{4}He (the isotope of helium with total spin zero), namely that it can flow through the finest capillaries without any viscosity, is due to the same Bose condensation of these atoms in the lowest energy state at low temperatures close to absolute zero. ^{3}He, the fermionic isotope, on the other hand, behaves differently even though chemically both isotopes are identical.

Elementary particle physics abounds in examples of order brought about by classification of the zoo of particles based on certain postulated, rather abstract and well concealed symmetries without knowledge of the details of the underlying laws (see Chapter VI).

Symmetry is also highly restrictive. It limits the possibilities allowed without detailed knowledge of the system. The classic example is the forbidden five-fold axis of rotational symmetry in a crystal. The only allowed ones are the two-fold, three-fold, the four-fold and the six-fold axes. The compatibility of the rotational and the translational symmetries rules out the five-fold axis as also the higher order axes of rotation. The five-fold axis is also conspicuous by its absence on the floor designs, or the tiling of a plane called tessellation. You see the square, the equilateral triangular and the regular hexagonal motifs, but never a regular repeating pentagonal pattern with the five-fold symmetry. However, individual molecules and other objects can and in fact do have the five-fold axis. Just think of a pentagram or the starfish. It is an interesting thought that the living organisms like the starfish may adopt the five-fold symmetry as a natural defence against the deadly 'capture' by the rigid crystalline formation.

The really restrictive power of symmetry in physics derives from the overriding conservation laws that it imposes — the conservation of energy, momentum, angular momentum and charge. We will return to this when we discuss this connection between invariance and conservation laws. Processes violating these are simply forbidden.

Symmetry is at its most powerful when it predicts. Let us illustrate it with an example from solid geometry. Suppose you are interested in regular convex polyhedra (poly=many, hedra=faces). A regular polyhedron is a volume bounded by plane faces which are identical regular polygons. A simple cube (the common dice) is one such polyhedron. It has six faces that are square (you can call it regular hexahedron). There are other regular polyhedra, namely the tetrahedron with four equilateral triangular faces, the octahedron with eight equilateral triangular faces, the dodecahedron with twelve regular pentagonal faces and finally the mysterious icosahedron with twenty equilateral triangular faces (Fig. III.2). These are the so called Platonic Solids contemplated by the Greek Pythagoreans. The question is if there are more. Well, the answer is a definite no. Symmetry forbids any other occurence. This is a restrictive aspect of symmetry. The predictive aspect is just the flip side of the coin. If there are intelligent inhabitants in some distant galaxy interested in these exotic dice-forms, we can predict that they will find just these five and no more.

But the real predictive power of symmetry is seen in particle physics. The basic idea is just this. Having identified or guessed the symmetry of the governing law, the processes, or the states or the particles related by the symmetry operations are all treated *at par*, *i.e.*, equally allowed and intrinsically the same. Thus, if you find one, the others, the missing ones are predicted. This is, for example, how the short-lived particle called Ω^- was predicted by Gell-Mann in 1962, and later confirmed in 1964 as the missing member of the family of 10 objects (resonances) predicted on the basis of a postulated symmetry $SU(3)$. This was a historic triumph of symmetry in physics.

There are two other aspects of symmetry of far reaching consequences. These are its unifying and creative powers. We will return to this point later.

There is an ingenious way the crystallographers use the power of symmetry constructively. Suppose you need to know the structure of a complex molecule. It may be a protein with some hundred thousand atoms, or a fragment of DNA. These are very very important but complex molecules. Proteins are the building blocks of cells and enzymes, while DNA (Deoxyribonucleic acid) carries the genetic information for making these proteins. Now you cannot use ordinary light to probe these. Its wavelength of several thousand Angstroms (1 Å = 10^{-8} cm) is much too large to reveal the finer molecular details on the scale of a few Angstroms. We must use X-rays with wavelength of about an Angstrom or so. If you shine X-rays on a sample containing these molecules, placed and oriented randomly, the scattered waves of X-rays will interfere randomly to produce a mere smudge on a photographic plate. If, however, you could somehow arrange the molecules periodically in space, that is to say if you could crystallize the substance, the waves scat-

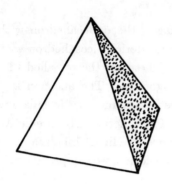

(a)

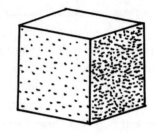

(b)

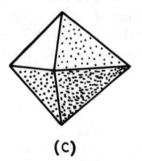

(c)

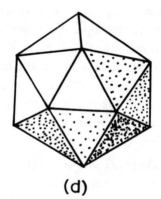

(d)

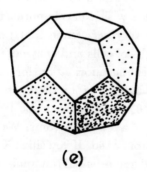

(e)

FIG. III.2 *Five Platonic Solids: (a) tetrahedron; (b) cube; (c) octahedron; (d) icosahedron; (e) dodecahedron.*

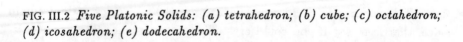

tered from the molecules will interfere constructively in certain well defined directions and thus produce a systematic pattern of bright sharp spots (the diffraction pattern) on the plate. This is like making Fourier series analysis of a periodic function. One can invert this to get at not only the periodic structure of the crystal lattice but also the structure of the molecules making it up! (One only hopes that the imposed crystalline arrangement has not done too much violence to the molecule whose structure we were interested in). This is why the crystallographers-turned-molecular biologists round the world are preoccupied with crystallizing these substances. At this point we should note that the crystalline order as a necessary condition for getting sharp X-ray spots has been called into question recently with the discovery of the so-called *quasicrystals* by D.Shechtman, I.Blech, D. Gratias and J.W. Cahn (1984). The first quasicrystal was an alloy, $Al_{14}Mn_{86}$, *i.e.*, 14 atomic per cent aluminum and 86 atomic percent of Manganese. Since then many more have been found. These materials show sharp spots like any other good crystal but the arrangement of spots has a five-fold symmetry which is of course, forbidden in the real space crystal lattice. The conclusion is that the conventional crystalline order is not necessary for sharp spots in x-ray diffraction. A two-dimensional quasicrystal is exemplified by the so-called *Penrose aperiodic tiling* of a plane with motifs of two rhombuses fitted as pieces of a jigsaw puzzle (Fig. III.3). The smaller and the larger rhombuses have angles 72 degrees and 108 degrees, and 36 degrees and 144 degrees respectively, and their areas and numbers are in the golden ratio $= (1 + \sqrt{5})/2$. This is the intellectual property of the Oxford mathematician Roger Penrose who constructed it for play. The tiling has no translational symmetry of the conventional crystals and yet would give a sharp diffraction pattern. It is now known that quasicrystals may be viewed as projection of ordinary-crystalline order from hypothetical higher dimensional spaces.

Our discussion of symmetry so far has been rather discursive. But, as we have remarked repeatedly, symmetry is a very precise concept. The proper language for a systematic study of symmetry is that of *group theory* which is a highly developed branch of mathematics. The basic idea is simplicity itself. Identify *all* the symmetry operations that leave a given object invariant. Call them $A, B, C, \cdots$. This is then an exhaustive list. It is clear from the very definition of symmetry that the successive applications of any two opera-

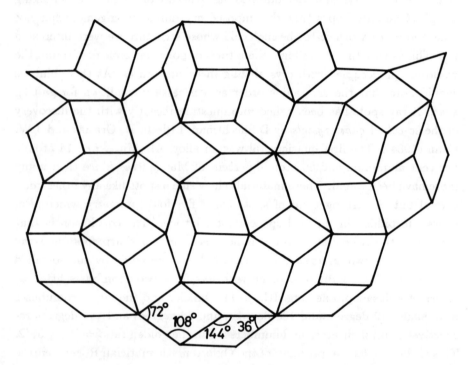

FIG. III.3 *Penrose aperiodic tiling of a plane.*

tions, first A and then B, say, will also leave the object invariant. Therefore, the combined operation must also be one of the symmetry elements we have listed exhaustively above. Let it be C. Then we can write $C = BA$. Mark the order of A and B in BA. It means A operates first, followed by B and the result is the same as C. This is a kind of multiplication, composition or successive operation, that gives the interlocking of the various symmetry operations. We say that the symmetry operations are closed under this multiplication. Next, we note that doing nothing at all to the object is also a symmetry operation because it trivially leaves it invariant. In fact it leaves it alone. We denote this trivial symmetry operation of 'doing nothing' by E (This is a fairly standard notation). Finally, we note that reversing a symmetry operation is also a symmetry operation — it restores *status quo ante*. Remember that the reverse of a clockwise rotation by an angle θ is an anticlockwise rotation by the same angle θ about the same axis. In obvious notation, we denote the reverse (more properly called inverse) of A by A^{-1}. It is now clear that $E = A^{-1}A$. (That is applying a symmetry operation followed by its inverse amounts to doing nothing). We are all set now. A set of elements having a law of multiplication (successive operation) under which the set is closed, with an identity (doing nothing) and where each element has a unique inverse is called a *group*. The symmetry operations then form a group. We are now compelled by the sheer logic of it. The inner structure of the symmetry group is given completely by enumerating the results of all pairwise multiplications, *e.g.*, $C = BA$. Constructing a multiplication table is like finger-printing the symmetry. Identical multiplication tables imply identical symmetry structures no matter how physically different the objects themselves may be. It is all very nice, but what can we do with all this, you may ask. Well, you can do a lot. An example will help illustrate the point.

Suppose the symmetry of the physical law in question turns out to be the symmetry of an equilateral triangle living on a plane (Fig. III.4). The symmetry operations are then E (identity), R_1 and R_2 (clockwise and anti clockwise rotations by 120 degrees, respectively, about the three-fold symmetry axis) and the three reflections m_1, m_2 and m_3 in the three mirror-lines (medians). One can readily construct the multiplication table — for example, $R_1 R_2 = E$, $m_1 m_2 = R_2$, $m_2 m_1 = R_1$, and so on. Remember that we have completely specified the symmetry structure of the physical law that governs

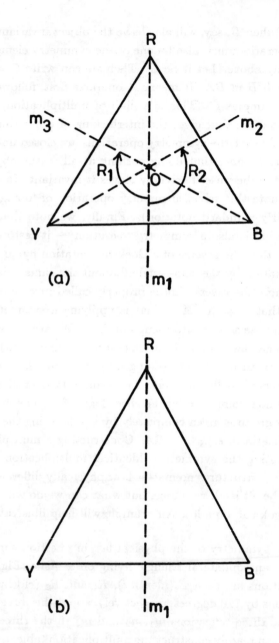

FIG. III.4 *Symmetry elements of (a) equilateral triangle; (b)isosceles triangle.*

our physical system. The latter may be a molecule with an atom literally at the center of an equilateral triangular environment formed by three other identical atoms. (Situations analogous to but more complicated than this are very common in chemistry, *e.g.*, an atom, or rather a doubly charged ion of copper at the center of a regular octahedron formed by the negatively charged atoms of oxygen in copper sulphate). Let us assume now that this system can exist in one of the three states, or *linear combinations of these*, which are permuted among themselves under the symmetry operations. Of course, we are assuming here that it is meaningful to speak of such linear superpositions. This is indeed the basic structure underlying quantum mechanics (see Appendix B). Thus we can identify the three vertices of our equilateral triangle with these three states. We provocatively label them as R (for red), Y (for yellow) and B (for blue). It is easily verified that our symmetry operations indeed permute them in all possible ways. (There are 6 ways in which 3 objects can be permuted and the number of elements in our symmetry group is also 6). While the symmetry operations do permute the three states among themselves, they do not mix them indiscriminately. Indeed, they split the possible linear combinations into two sets (called multiplets more properly) such that only members of the same multiplet mix among themselves. One multiplet, call it $W = R + Y + B$ has only one member (a singlet). The other has two members, $Z_1 = R - 2Y + B$ and $Z_2 = R - B$. Now Z_1 and Z_2 mix freely under our symmetry operations and, therefore, they are intrinsically the same — they differ according to our viewpoint only. Thus for instance they should have the same energy (or mass). We say that the multiplet is two-fold degenerate. Their energy, however, must be in general different from that of W with which they do not mix under symmetry. In this simple case we could write down this multiplet structure by mere inspection. In general one has to use the multiplication table in a systematic way. It is called the representation theory of groups. In our example W, and $\{Z_1, Z_2\}$ provide, respectively, one- and two-dimensional representations. We can go further and lower the symmetry to that of an isosceles triangle by pulling one of the vertices out (Fig. III.4b). Our symmetry group now will consist of only two elements $\{E, m_1\}$. It is a sub-group of the earlier larger group. The result is that the doublet is further split into two singlets. We now have three non-degenerate (unequal) levels.

This splitting or reduction of degenerate multiplets with the progressive lowering of symmetry is well known and well studied in chemistry and solid state physics, where the symmetry is mostly geometrical and known from structure. The situation is quite different in elementary particle physics where the symmetry is rather abstract and not directly accessible. Here symmetry takes on a creative role. This is made possible by the fact that a given group uniquely specifies the possible multiplet structures it can support. Thus one can postulate a symmetry and then work out the multiplet structures it implies and compare with the observed families of closely related particles. This is the idea underlying the unending quest for symmetries, *e.g.*, $SU(2)$, $SU(3)$ and so on. One is limited only by his ingenuity and insight. Thus $SU(3)$ (special unitary group of rotations in a three dimensional complex space) has a multiplet with 8 members (eight dimensional representation) and one with 10 members (10 dimensional representation) that fitted so well the observed families of 8 baryons and 10 hyperons — behold the 'unreasonable' effectiveness of symmetry in physics!

Finally, a remark on the group multiplication. Note, that in our example we had $m_1 m_2 = R_2$ and $m_2 m_1 = R_1$. Thus unlike ordinary multiplication of numbers the order in the applications of symmetry operations is important. We say that m_1 and m_2 do not commute. Such a symmetry group is said to be *non-abelian*. The important group of rotations in 3 dimensional space $SO(3)$ is non-abelian. The corresponding group of rotations in a plane is *abelian*. An amusing demonstration of this is the following. You fly out of the North Pole down the zero-degree longitude through Greenwich to the equator. You will be over the Atlantic, south of Ghana. This is a rotation by 90 degrees about the East-West axis. Now you turn and follow the equator to longitude 90 degrees East. You should be over the Indian ocean East of Sumatra. This amounts to rotation by 90 degrees about the North-South axis. Now, you perform these operations in the reverse order. Start out at the North Pole and turn by 90 degrees Eastward. But since you are right on the axis of rotation, you just stay put. Next fly down the zero degree longitude till you reach equator, and thus you end up over the Atlantic, South of Ghana, thousands of kilometers away from your earlier destination in the Indian ocean. It turns out that most of the symmetries in Physics are non-abelian, and that makes it richer. Abelian symmetry gives only non-degenerate one-dimensional or single-state multiplets.

---

2. SPACE-TIME SYMMETRIES: INVARIANCE AND THE GREAT CONSERVATION LAWS

Objects are located in space. They endure in time. This is true of all events and processes, of beings and becomings, that ultimately involve the elementary particles and their interactions that make up the world of physics. Admittedly, a highly reductionist view point but you can hardly fault it. It seems reasonable, therefore, that the study of symmetries of the objects and phenomena must be preceded by a proper study of the symmetries that this background space-time continuum may have. For obvious reasons we will call these the frame-work symmetries. These symmetries must be established as facts of experience, no matter how compelling *a priori* they may appear to be. To the best of our knowledge, then, the following symmetries are true.

Space is homogeneous. That is to say that the absolute position of an object is irrelevant. What it operationally means is that if we perform an experiment at a location and then repeat the same experiment somewhere else, in outer space, say, the results will be identical — translationally invariant in space. By the 'same experiment' we mean that all conditions relevant to the experiment must be reproduced exactly. Thus, if the change in earth's gravity in going out there is relevant, then the earth must be transported along with the apparatus. One may argue that this claim is then vacuous inasmuch as any discrepancy between the results of the two experiments can always be blamed on something that may have escaped our attention, to wit our altered position with respect to the distant stars! Now, this is perverse because it is possible to isolate our experiment far enough to any desired degree of accuracy by including larger and larger regions of space as part of our experimental set-up and, because one can assume reasonably that all effects are essentially local in nature. Ghosts are not admitted. In any case, there is nothing to suggest violation of this translational symmetry.

Next comes isotropy of space, or the irrelevance of absolute direction. Operationally, it means that if we perform a certain experiment and then rotate our entire setup to a new orientation and repeat the same experiment, the results will be identical. We can re-word all our earlier provisions and arguments in support of this. So far there is no empirical evidence in support of a preferred direction in space. Thus isotropy of space is a good symmetry.

There is an interesting connection between these two symmetries. Isotropy (relative to every point of space) implies homogeneity but not *vice versa*. This is readily proved. Let P_1 and P_2 be two points in space. Draw a sphere passing through P_1 and P_2, with center, say, at O. You can draw any number of such spheres. Now, viewed from O, P_1 and P_2 are related by isotropy and, therefore, are equivalent. You can repeat this process till you cover the entire space and thus establish homogeneity of space.

Next comes homogeneity of time, or time-translation symmetry. There is no absolute origin of time. If you perform an experiment now and repeat the same experiment at a later date, the results will be identical. Indeed, without these symmetries the universe will hardly be comprehensible. We should perhaps mention here that there is evidence that the universe is finite and that it had a beginning some 15 billion years ago — the *big bang*. We hope that we are at sufficient remove from these boundary and initial conditions to ignore these symmetry breaking effects here and now.

Finally, to these irrelevancies, namely those of absolute position, absolute direction and absolute time, we add the irrelevance of absolute rest, or of absolute uniform motion. Consider two unaccelerated platforms in uniform relative motion, that is to say that one platform is moving with a constant velocity as seen by an observer who is stationary on the other platform. Now, if we perform an experiment on one of these platforms and then repeat the same experiment on the other, the results should be identical. Thus no *local* experiment, i.e, without reference to the other platform will detect any effect that can distinguish between these two unaccelerated platforms — there is no absolute uniform motion. This equivalence of unaccelerated platforms is the great symmetry expressed by the *principle of relativity* and was a wonderful achievement of Galileo. Acceleration is, on the other hand, absolute and can be detected locally by an accelerometer — a mass attached to one end of a spring, the other end of which is fixed to the platform. (In all these discussions, we will ignore the presence of gravitation). This is quite consistent with our every day experience. We are hardly aware of the velocity with which the lift, by which we may be travelling, is moving except at the times of start and stop, *i.e.*, when there is acceleration or deceleration.

A platform is, more formally, a set of points at rest relative to one another. It is convenient to introduce a rectangular coordinate system (x, y, z) at rest with respect to these points. One may also assume a clock, an atomic clock, say, attached to every point of this set. The identical clocks may be synchronized by exchanging light signals. Thus, if A and B are two points and if t_1 is the time at which a light signal is sent out from point A, and if t_2 is the time at which the signal is received at and reflected by the point at B, and finally if t_3 is the time at which the signal is received back at the point A, then the clocks at A and B are synchronized if $t_2 - t_1 = t_3 - t_2$. Note that this is purely by symmetry and does *not* require knowledge of the speed of light. Thus an elementary event is completely located by giving spatial coordinates (x, y, z) and the time of its occurrence t, read out by the clock at the point (x, y, z). Such an unaccelerated platform equipped with the markers $(x, y, z; t)$ is called a Galilean frame of reference S, say. Another Galilean frame S', say, will have a primed space-time coordinate system $(x', y', z'; t')$. Now the relativistic invariance asserts that the laws expressed in terms of the primed and the unprimed space-time coordinates should have the same form. The question now is how the primed and the unprimed space-time coordinates of the same event are related. The relativity of motion encountered in every day life, also called *Galilean relativity*, would suggest the following answer. Time intervals are absolute. So are the space intervals. This means that, if $(x_1, y_1, z_1; t_1)$ and $(x_2, y_2, z_2; t_2)$ are the space-time coordinates of two events observed in a Galilean frame S, and $(x'_1, y'_1, z'_1; t'_1)$ and $(x'_2, y'_2, z'_2; t'_2)$ are those for the same two events but in another Galilean frame S', then the time interval $t_{12} = (t_1 - t_2) = t'_{12} = (t'_1 - t'_2)$ and the space interval squared $r_{12}^2 = (x_1 - x_2)^2 + (y_1 - y_2)^2 + (z_1 - z_2)^2 = (r'_{12})^2 = (x'_1 - x'_2)^2 + (y'_1 - y'_2)^2 + (z'_1 - z'_2)^2$.

This leads to the rules of vector addition of velocities and displacements well known from our high-school days. The symmetry operations here are the familiar translation and rotation (re-orientation) in space, 'boosting' to a relatively uniformly moving frame, and time translation. Galilean relativity is, however, based on our common experience with slow objects moving at small velocities, *e.g.*, the speed limit of about 100 kilometers per hour on national highways. Compare this with the speed of light, 1080 million kilometers per hour in vacuum. Can we extrapolate our tardy experience to such high velocities? Let us see. In Galilean relativity the speed of light in vacuum would depend on the relative velocity of the source of light and the observer. One can then, in principle, chase light and even outrun it. Or one can run just fast enough to keep pace, bringing light to a relative standstill. This is true, for instance, in the case of sound. But sound propagates only in a medium, *e.g.*, air. Light, however, can propagate in vacuum. Is vacuum too filled with an all pervasive medium — the 'aether' as was indeed thought for quite some time? This hypothetical medium, the aether, could then provide the preferred frame of reference at absolute rest, making thus the different Galilean frames moving relative to it in principle non-equivalent. Even the most careful laboratory measurements and astronomical observations have, however, failed to detect this aethereal medium. Einstein did not like this loss of symmetry anyway. The point is that light is an electromagnetic wave whose propagation in vacuum relative to a Galilean frame is described by a wave equation of the type we wrote down in the last Section. Notice that the speed of light 'c' occurs explicitly in this equation. The invariance (or rather covariance) of this equation with change from one Galilean frame to another then demands the invariance of the speed of light. Thus, we have the fundamental postulate of the absolute constancy of the speed of light for all Galilean frames of reference — the basis of *Einstein's special theory of relativity*. The changes from one frame of reference to the other are the symmetry operations that leave the speed of light unchanged. It is clear that for this to be so, the notion of *absolute* time interval t_{12} as separate from that of the absolute space interval r_{12} between the two events labelled 1 and 2 inherent in the Galilean relativity must be abandoned. Einstein's relativity replaces these two with a single *absolute* invariant interval s_{12} between the two events, given by $s_{12}^2 = r_{12}^2 - c^2 t_{12}^2$. The three-dimensional Euclidean space and time $(x, y, z; t)$ are replaced by a four-dimensional space-time (x, y, z, t) (the *Minkowski world*) that treats time t as just another co-ordinate to label the events, *at par* with space coordinates (x, y, z) (Fig. III.5).

An event is now located at a world-point (x, y, z, t) — the semicolon that set time apart from space has now been replaced by a common comma. The transformation from (x, y, z, t) to (x', y', z', t') is now the symmetry operation of displacement and rotation (Lorentz transformation) in this four dimensional world, keeping in mind, however, the technical point about the minus sign that occurs in $s_{12}^2 = (x_1 - x_2)^2 + (y_1 - y_2)^2 + (z_1 - z_2)^2 - c^2(t_1 - t_2)^2$.

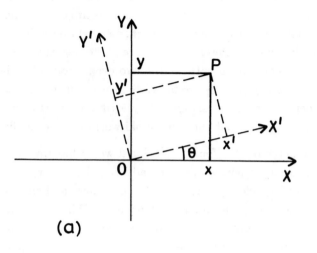

(a)

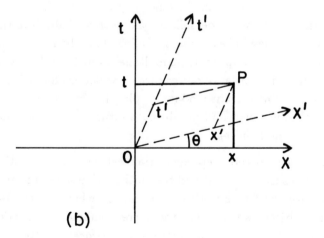

(b)

FIG. III.5 *Rotation in (a) ordinary space; (b) Minkowski space-time.*

One may mathematically absorb the minus sign by defining an imaginary time $\tau = it$, with $i = \sqrt{-1}$ and treat time formally completely at par with space. But it is actually better to leave it as such, as a gentle reminder that time is, after all, qualitatively different from space. The negative sign implies that the interval s_{12} can vanish without the two events coinciding in space-time, i.e., $s_{12} = 0$ but $r_{12} \neq 0$, $t_{12} \neq 0$. We can even have s_{12} negative — this would happen if two events separated by a time interval occur at the same spatial point. (We say that the Minkowski world has an indefinite metric).

The geometry of this four-dimensional world has important and interesting physical consequences. The well advertised popular effects — the variation of mass with velocity, the equivalence of mass and energy, the Lorentz contraction and time dilation, all belong here. The speed of light in vacuum is the limiting speed that cannot be exceeded. Our main concern here is, however, only the symmetry aspect of relativity — the great frame-work symmetry. Let us note one highly counterintuitive aspect of it because it has a deep significance for our discussion of invariance and conservation law later. Since it is only the interval s_{12} that remains invariant from the unprimed frame to the primed one, it is clear that we can have t_{12} zero but t'_{12} non-zero. That is to say that in the unprimed frame the two events are simultaneous, but in the primed frame they are not. This is the relativity of simultaneity that totally demolishes the notion of absolute time interval. (Incidentally, one may have an uneasy feeling that when simultaneous events in one Galilean frame appear non-simultaneous in the other, what happens to their chronological order of occurrence — which is older of the two. Well, relativity does allow certain amount of play in this game of courtesy, but there is an absolute past and an absolute future even here consistent with notions of cause and effect).

We now turn to the deep connection between these relativistic space-time symmetries (invariances) and the conservation laws. Inasmuch as these symmetries are the frame-work symmetries to which all the basic laws of physics are subject, we will call the corresponding conservation laws the Great Conservation laws. Consider a physical process written schematically as $x + y \rightarrow z + w$. A quantity is said to be conserved if its total value for the reactants $x + y$ is the same as its total value for the products $z + w$ of the process as observed in a given Galilean frame. Thus we speak of conservation of energy, linear momentum and of angular momentum. It turns out that the conservation of energy follows from the invariance with respect to translation in time. The conservation of linear momentum follows from the invariance with respect to translation in space (homogeneity of space). The conservation of angular momentum follows from the invariance with respect to rotation in space (isotropy of space). A proper discussion of conservation of these quantities (and even their definition in general) as a consequence of the invariances requires introduction of 'action' and 'action principle'. This is beyond our scope. The important point to note is that this connection between the invariance and the conservation law is not restricted to any specific dynamical laws such as Newton's laws of motion. The connection is purely kinematic. For the specific case of mechanical systems where, for instance, momentum is mass times velocity, one may derive conservation of linear momentum by applying Newton's three laws of motion. And so on for energy and angular momentum. But the connection is really much more general. After all, there are non-mechanical objects, light for instance, that also carry energy, momentum and angular momentum. We should note in passing that just as isotropy of space implies homogeneity, conservation of angular momentum implies conservation of linear momentum, but not vice versa.

Much of the restrictive and predictive power of these symmetries comes from the associated conservation laws. The striking example is radioactivity (β-decay) in which a neutron was thought to decay into an electron, a proton and something else. The electric charge is conserved (as required by another invariance called gauge invariance to be discussed later). However, a careful reckoning of energy and momentum of the system before and after the reaction led to an imbalance. Thus a new particle was suspected as a decay product that carries the missing energy and momentum. It was predicted to be neutral and to have zero rest mass. Also, recalling that the neutron, the electron and the proton all carry spin half (angular momentum $\hbar/2$), conservation of angular momentum required the then unknown particle to carry spin half. All this was confirmed happily later. This is the now well known but elusive elementary particle, the electronic anti-neutrino denoted by $\bar{\nu}_e$, and the corrected process reads $n \rightarrow e^- + p + \bar{\nu}_e$. These particles are now routinely and abundantly produced in laboratories, in nuclear reactors as well as accelerators.

The full power of these great frame-work symmetries is realized only when these are combined with the great frame-work theory — Quantum Mechanics (Appendix B). But this will take us very far afield. We will be content with just mentioning it. In addition to these continuous symmetries, there are discrete space-time symmetries too. One of them is the symmetry under space reflection, also called the mirror symmetry or parity. This produces enigmatic effects in the ordinary laboratory physics and chemistry as also in the extra-ordinary processes involving elementary particles. We will take this up next.

---

3. REFLECTION SYMMETRY

We have spoken of objects having bilateral symmetry, also called the left-right symmetry. A butterfly with outstretched wings or a maple leaf for example. When an object is reflected in a mirror, the left and the right sides of it get interchanged. Thus, an object having bilateral symmetry is by definition superposable on its mirror image. The mirror is just an optical device that enables us to visualize the result of reflection of objects in space through a plane. For these reasons the terms bilateral symmetry, left-right symmetry, mirror symmetry and the symmetry under space reflection are all used interchangeably. In physics, handedness is often referred to as *chirality*.

There is something that sets this symmetry apart from the rest that we have discussed so far. As noted above it is a discrete symmetry unlike the continuous symmetry of rotation or translation, say. Changes caused by continuous symmetry operations can be made arbitrarily small. Not so with discrete ones. You reflect or you don't. Nothing in between. Also, unlike these, it is a *non-performable* symmetry operation. Space reflection involves turning the object inside out laterally, an operation we can hardly perform continuously. But we can and we do visualize it by the optical trick of reflecting it in a mirror. Having visualized it so, nothing prevents us from making a physical copy of the image, using silly putty, say, which can then be tested for superposability on our object. This is the operational meaning of reflection symmetry as applied to shapes of material objects or geometrical figures. The non-performable nature of reflection symmetry conceals an important aspect of it that we will try to uncover now. Consider an arbitrarily shaped object and its reflection in a mirror. A rather handy example would be, well, your right hand itself. Its mirror image is constructed by translating every point on the hand to a point on the other side of the mirror, along the line perpendicular to the plane of the mirror and equidistant from it (Fig. III.6a).

Now, it is clear that this image (ideally your left hand) is not superposable on your right hand. Such an asymmetric object is called 'handed', and for very good reason. Ever tried your left-hand glove on your right hand? And yet nothing else is more like my right hand than its mirror image, that is my left hand. The reason that the two cannot be superposed is an inconvenient circumstance of life, namely, that the hand is a three-dimensional object and so is our physical world (space) in which cooped up we live. In a world of higher dimensions, the right hand could have been turned around by a temporary excursion into the extra dimensions, and thus superposed on the left hand. This can be demonstrated quite easily by an example taken from the world of lower dimensions — of a two-dimensional object, a flat-lander, living on a plane which is embedded in our familiar three dimension space. Thus, the symbol "Om" in Fig. III.6b can be superposed on its mirror image by simply folding the paper along the mirror line M. The act of folding involves a temporary lift or escape into the third dimension coming out of the plane of the paper. Science fiction is full of such excursions into the extra dimension — the "tesseract" in "A Wrinkle in Time" by Madeleine L'Engle is a fascinating case in point. These extra dimensions, somewhat curled up and rather inaccessible, are also the subject of serious thought by the physicists of our times. But we are digressing. All this suggests that an object and its mirror image are intrinsically the same. To emphasize this we call them a pair of *enantiomorphs* or antipodes — an impressive name for ordinary mirror images.

What is in a hand that makes it so 'handed'? To understand handedness more thoroughly we have to think of a screw — the common screw that we use to fasten things together, and without which much of our civilized world would simply come apart. The common screw is nothing but a helical ridge, called thread, cut into the surface of a cylinder, or a cone if it is a tapered screw. When the screw is turned as indicated by the circular arrow (Fig. III.6c), it advances (or recedes) along its axis as indicated by the linear

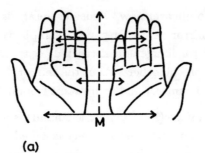

(a)

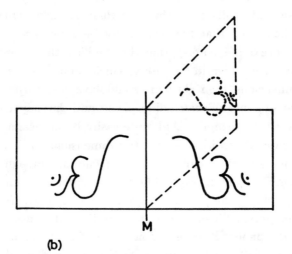

(b)

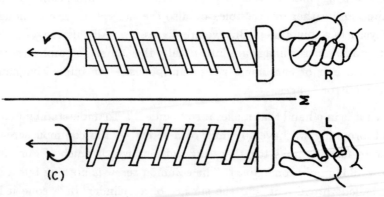

(C)

FIG. III.6 *(a) Mirror reflection; (b) continuous reflection via third dimension; (c) right-handed screw reflected as left-handed screw.*

arrow. This makes the screw a machine that converts a rotary motion about its axis into a translational motion along that axis. You may think of the Archimedes Screw that was used by the Egyptians to raise the waters of the Nile, and is still in use for similar purposes. It should be clear that there are two and only two classes of screws possible. These correspond to the two possible relations between the circular and the linear arrows. Let us give them names. Suppose that you clasp the screw in your right hand with your fingers pointing in the direction of the circular arrow while your thumb stays parallel to the axis of the screw. Now, if your thumb points in the direction of the linear arrow, then the screw is said to be right-handed. If, on the other hand, it points in the opposite direction, then the screw is said to be left-handed (Fig. III.6c). It is easy to see that the mirror image of a right-handed screw is a left handed screw. The two form a pair of enantiomorphs. That there are two classes of screws, *i.e.*, the two-ness of it is an absolute fact. But defining them as the left- and the right-handed screw is, of course, a matter of convention — a very useful convention though, which is followed uniformly all over the civilized world. This has been made possible by the intimate contacts we have had over centuries of togetherness, and not a little by our admirable practice of shaking hands. But a convention all the same. The non-triviality of this is brought home by the following thought evoking circumstance. Suppose we establish radio-contact with some advanced civilization in a galaxy far away. Such an eventuality can not be ruled out, thanks to the project SETI (Search for Extra-Terrestrial Intelligence) mounted by some serious minded people. Now we should have no difficulty convincing our distant correspondents that these are the two classes of screws possible. But try hard as we may we will not be able to explain to them what we mean by the right-handed screw. This is the famous problem of Ozma (named after the mythical prince Ozma in Lyman Frank Baum's classic "Wonderful Wizard of OZ"). The *Ozma problem*, suggested by Martin Gardner is a deep problem of communication theory, and its solution involves deeper understanding of symmetry in physics. We will return to it briefly later.

From the reflection symmetry of geometrical shapes let us now pass to the real question. Are the various laws of physics symmetric with respect to space reflection? To fix ideas consider a simple molecule CH_4, the molecule of methane (marsh gas found commonly in marshy lands). The molecule consists of a carbon atom surrounded by four equidistant hydrogen atoms arranged at the vertices of a regular tetrahedron (Fig. III.7).

The molecule is clearly reflection symmetric, *i.e.*, it is superposable on its image. We can break this reflection symmetry by replacing the four hydrogen atoms by four different atoms (or groups of atoms) X, Y, Z and W, say. Thus, for example if X=H, Y=CH$_3$, Z=C$_2$H$_5$ and W=OH, we get a molecule of butyl alcohol. Numerous other examples are possible. Chemists call such a molecule as having an *asymmetric carbon atom*, and the pair of enantiomers are called *stereoisomers*. Such a molecule is handed because it is no longer superposable on its mirror image. Thus for example, if you look down the XO direction, the atoms Y, Z and W will be seen as arranged either clockwise or anticlockwise. Now suppose we synthesize this molecule in the laboratory starting from the elements C, H, and O. The question is which one of the pair of stereoisomers will we get. The answer is simply this. If the laws governing the chemical reaction are symmetric with respect to space reflection, then the probabilities of getting the two stereoisomers are strictly equal. Therefore, at the end of the reaction we will get a mixture of the two in equal proportions. Chemists call this a *racemic* mixture. The mixture will have no net handedness. A law is said to be reflection symmetric, if a process or phenomenon and its mirror image are *equally* allowed by that law. Experimental evidence strongly suggests that the laws of physics that govern processes at low energies, like the chemical reactions, are indeed reflection symmetric. Thus the molecules of butyl alcohol in our example and the molecule of its mirror image will have the same physical and chemical properties, *e.g.*, the same boiling point, the same freezing point, the same density and, of course, the same molecular weight. Next we will demonstrate the predictive power of this symmetry of the physical law — we will predict optical activity. Consider again our handed molecule with the asymmetric carbon atom. A molecule of sugar is perhaps a more pleasing example. Sugar molecules are also handed but a bit more complex. We can hardly experiment on a single sugar molecule. So consider trillions of these identical sugar molecules — a solution of the sugar molecules in water, for example. The water molecules (H$_2$O) are mirror symmetric and, therefore, any handedness at all will be due only to the sugar molecules. We can and we will ignore

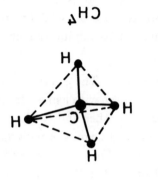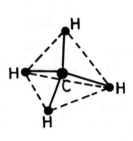

(a)

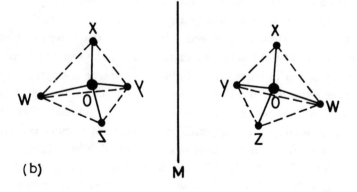

(b)

FIG. III.7 *Mirror reflection of molecules: (a) CH₄ with symmetric carbon atom; (b) OXYZW with asymmetric carbon atom, hence optically active.*

the solvent (water) completely in what follows. In the aqueous solution the sugar molecules are located and oriented randomly. This, however, does not neutralize or average out their handedness. After all, a chest full of left gloves can hardly be confused with a chest full of right gloves, no matter how randomly the gloves are placed. We will now send a beam of light, from a laser say, through our sugar solution. But first let us remind ourselves of some elementary facts about light.

Light is a transverse electromagnetic wave. The electric and the magnetic fields oscillate sinusoidally in time and space with a given frequency and wave-length. They are perpendicular to each other and to the direction of propagation of the wave (hence transverse). Light can be *circularly polarized.* Here the tip of the electric vector describes a helix with its axis along the direction of propagation. It may be left or right polarized according as the helix is left or right handed (Fig. III.8).

Light can also be *linearly (or plane) polarized* if the oscillating electric vector lies in a plane containing the direction of propagation. Finally, we note that a linearly (plane) polarized light may be viewed as a vector addition of the two oppositely circularly polarized light waves of the same frequency and wavelength. We are all set now. Let the beam of light passing through our sample of handed sugar solution be circularly polarized. The mirror image of this process will be an oppositely circularly polarized light passing through a sugar solution of opposite handedness. Given the reflection symmetry of the governing law, the two enantiomorphic processes must be equally and identically allowed. In particular the speed of light in the two cases must be exactly the same. But what if we keep our sugar solution the same and only reverse the sense of circular polarization of light. Well, this is not symmetry related to the earlier situation, and there is no sufficient reason to expect the speed of light to remain the same — it will in general be different. Thus the speed of light in a handed medium depends on the sense of circular polarization of the light! This effect can be made more spectacular by taking our light to be plane polarized. Recall that it may be viewed as a superposition of two oppositely circularly polarized light waves. Now that these two components must travel with different speeds, they will get out of phase as they traverse the handed medium. This results in twisting of the plane of polarization of the light relative to that of the incident light (Fig. III.9). This twisting or

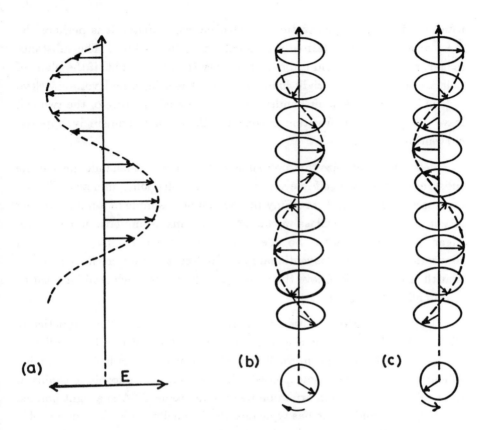

FIG. III.8 *(a) Plane polarized light; (b) left-circularly polarized light; (c) right-circularly polarized light.*

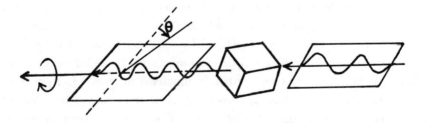

FIG. III.9 *Rotation of plane of polarization of light by optically active medium.*

rotation of plane of polarization is called *optical activity*. It is perhaps the most dramatic manifestation of handedness of the medium. The substance (sugar in our case) is said to be dextrorotary (dextro = right) if the plane of polarization twists as a right handed screw. It is said to be levorotary (levo = left) if it twists as a left-handed screw. A *racemic* mixture of the two will leave the plane of polarization unchanged. Somewhat, confusingly, opposite convention is also in use.

Let us re-emphasize that optical activity is due to the handedness of the substance. The law itself is even-handed, *i.e.*, reflection symmetric. This is expressed perhaps most forcefully by the famous example of a milk drinking kitten of the Looking Glass world. Milk contains asymmetric molecules of sugar, proteins and fats. So does, of course, the body of the cat. And conventional cats love conventional milk. Reflection symmetry now demands that the reflected kitten love the reflected milk just as much, and fare just as well in all respects.

The living world is, however, far from being racemic. Thus, practically all the 20 odd aminoacids that make up the proteins of the living cells are left handed. The proteins in the living cells, in turn, have a helical backbone which is almost always right handed. Each of the sugar phosphate chains in the double helix of the information bearing molecule DNA is a right handed (double) helix, and a man has typically 10^{11} kilometers of it. Left aminoacids are common and are assimilated by our body, but the right aminoacids are rare and filtered out by our kidneys. The nicotine commonly found in tobacco is known to be harmful but its reflected stereoisomer is rare and much less offensive. Same is true of other biochemicals such as the lactic acid found in milk or the table sugar (sucrose) found in sugarcane, etc. The sugar D-glucose is found throughout the animal kingdom but its mirror image L-sugar is unknown except by laboratory synthesis. There are indeed few exceptions to the rule that anything from lactic acid to the double helical DNA, having handedness, will occur biologically in only one form. Indeed if we let a colony of bacteria feed on a *racemic* (optically inactive) mixture of (L) and right (R) sugars, the bacteria feed preferentially on L- sugars, and then leave the mixture right-handed and optically active. The question now is, how do we understand this dominance of handedness, or shall we say high-handedness of the living world when the governing laws themselves are so just and even-handed? Well, not quite. But a highly plausible answer is something like this. The observed handedness of the living matter may be the result of a fantastic amplification of an initial chance asymmetry, ever so slight. This is made possible by the positive feedback inherent in the process of multiplication (reproduction) by self-replication that is all pervasive in the animate world. To see this clearly let us simplify things to the absurd limit and consider the first (single) helical strand of the DNA molecule ever formed. We know that it is potentially equally likely to be right- or left-handed. But once formed it has got to be just one of them. So let it be right-handed. Now, this single right-handed strand proliferates or muliplies by self-replication. It acts as a template and makes a copy of itself which is now necessarily right-handed. The process gets repeated over and over again. This is the positive feedback at work that may lead to the necessary amplification of an initial chance event over the aeons of chemical and biological evolution, and thus produce the handed life as we know it today. This is made all the more plausible by the observation that the inorganic world, by contrast, seems to be quite racemic. Consider the mineral quartz for example. It is one of the crystalline forms of silicon dioxide (SiO_2), the common silica sand. The basic unit here is SiO_2 which by itself is mirror symmetric, making quartz optically inactive when dissolved. But in a quartz crystal the units are arranged in the form of parallel helices which can be either left- or right-handed, making quartz optically active. In Nature both forms occur with equal frequency.

Does this not go against the laws of thermodynamics, the entropy principle, that makes states of equal energy equally probable? The left- and the right-handed strands are, of course, energetically equivalent, being related by reflection symmetry. Well, the point is that the thermodynamic statement is about a system in thermal equilibrium. But the living state is far from equilibrium. It is self-organized and maintained at the cost of 'freely' available energy that comes eventually from the sun. Once the cell is dead, the right-handed helices of the DNA molecule will begin to flip their handedness, and gradually tend to the racemic state as dictated by thermodynamics. Indeed, the rate of racemization can be and has been, used for the dating of dead cells older than 40,000 years, much better than the conventional dating based on the decay of ^{14}C, a radioactive isotope of carbon. Louis Pasteur regarded handedness as a sign of life. Racemization signalled death.

Are the fundamental laws of physics all strictly symmetric under space reflection? Is the antipodal world of the *Looking Glass* just as legal as our conventional world? We now know that the answer to this question is a definite *no*. There are fundamental processes such as the *β-decay* (*radioactivity*) controlled by the so-called *weak interaction* that break this reflection symmetry. There is a screw at the very heart of Nature. To see this we have to get more sophisticated. We have seen how to reflect geometrical figures and shapes of material objects. But how do we reflect magnetism?

Take a bar magnet with the poles N and S marked on its ends. Its mirror reflection will be just another bar magnet with the letters N and S laterally inverted (Fig. III.10). But this is a naive reflection of the body of the magnet. It hardly addresses the real question of how to reflect the magnetism of it. The magnetic field of the magnet may be regarded as due to an electric current circulating in a loop around the body of the magnet as indicated by the circular arrow (Fig. III.10). Remember Ampère's Law! (Incidentally, when Ernst Mach learnt of the sideways deflection of a compass needle when placed below and parallel to a current carrying wire, he was shocked out of his wits as he thought it to be violating the left-right symmetry. With our picture of the magnet, now we see that Mach's shock was a fake alarm as there is no such symmetry in this situation to start with). It should be clear now that when reflected, the sense of the circular arrow will reverse. And so will the polarity of the magnet. With this we are all set to describe the phenomenon that shook the world of physics — the fall of parity. *Parity* is yet another name for reflection symmetry.

Take an atom of cobalt, the isotope ^{60}Co to be precise. The nucleus of ^{60}Co has a spin. It is like a spinning top. This makes it a tiny magnet with the magnetic poles on the spin axis. As before this is equivalent to having a circulating current loop indicated by the circular arrow. We now apply a magnetic field. The nuclear magnet will align parallel to this field just as a compass needle aligns parallel to earth's magnetic field. The ^{60}Co is a radioactive nucleus. It decays by emitting, among other things, electrons, the β-rays, in all directions. The question is whether they come out equally in all directions, or there are some preferred directions. We can know this by placing detectors all around our sample and counting the number of electrons

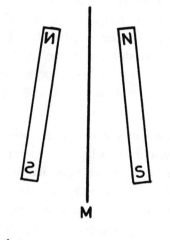

(a)

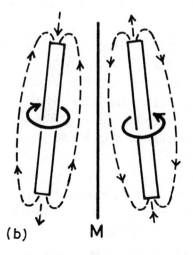

(b)

FIG. III.10 *Mirror (M) reflection of (a) body of a magnet; (b) magnetism.*

coming out in a given direction in a given interval of time. In particular let us compare the number of electrons shot out of the north pole (N) parallel to the field with the number shot out of the south pole (S) antiparallel to the field. If we perform this experiment as Madam C.S. Wu did in 1957, we will find that more electrons are shot out of the south pole than out of the north pole. Is this consistent with the reflection symmetry of the underlying law? To answer this all we need to do is to look at the process reflected in a mirror (Fig. III.11).

Everything looks the same except that the sense of the circular arrow and, therefore, the polarity, is reversed. Thus in the mirror world, more electrons would come out of the north pole antiparallel to the field than out of the south pole. The reflection symmetry is violated! (This *violation of parity* was predicted by the two Chinese-American physicists T.D. Lee and C.N. Yang in 1956 on theoretical grounds. It was confirmed by Madam Wu in 1957. The same year Lee and Yang won the Nobel Prize for Physics). In fact the nuclear spin of the ^{60}Co nucleus (the circular arrow) and the preferred direction of electron emission define a left-handed screw. Nature is weakly left-handed after all! This is more than a mere convention. One could perhaps use the ^{60}Co decay to communicate to our distant correspondent the meaning of the left and the right, and thus solve the *Ozma problem*. But not quite, as we will presently see.

We can go further and, as it were, pinpoint the screw by looking at the full β-decay reaction: ^{60}Co $\to^{60}$ Ni $+ e^- + \bar{\nu}_e$. The neutrino (or rather anti-neutrino $\bar{\nu}_e$), as we have noted earlier, is an elusive particle with zero rest mass. This particle has a zero charge and relativity requires it to move with the speed of light. Neutrino has spin one-half and this is important for us. For, relativity demands that the spin of this massless particle be either parallel or antiparallel to its velocity! Now the spin (the circular arrow) and the velocity (the linear arrow) form a screw or a helix that can be either right-handed or left-handed. In nature we find only left-handed neutrinos (and right-handed antineutrinos) (Fig. III.11). Here lies the screw at the heart of Nature! All reactions involving these handed objects violate parity.

Besides *parity* ($\mathcal{P}$), there are two other discrete symmetries — the *charge conjugation* ($\mathcal{C}$) and the *time reversal* ($\mathcal{T}$). The symmetry operation of *charge conjugation* ($\mathcal{C}$) replaces a particle with its antiparticle, denoted by an over-

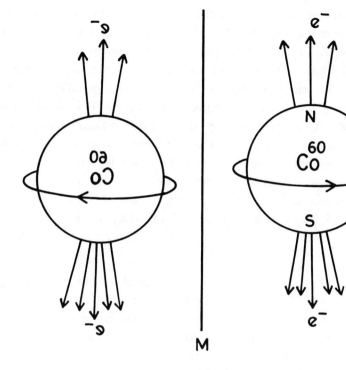

(a)

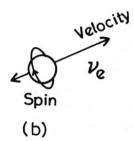

(b)

FIG. III.11 *(a) Parity violation in β-decay of cobalt-60 nucleus; (b) left-handed neutrino.*

head bar. A particle and its antiparticle have the same mass but equal and opposite electric charges, among other things. Thus we speak of anti-electron e^+ (commonly called positron), antiproton ($\bar{p}$), antineutron ($\bar{n}$), antineutrino ($\bar{\nu}$), and so on. Photon (γ) is its own antiparticle. Charge conjugation symmetry demands invariance of physical laws under the operation $\mathcal{C}$. Thus a reaction $X + Y \to Z + W$ and its conjugate $\overline{X} + \overline{Y} \to \overline{Z} + \overline{W}$ should proceed at the same rate. Antiworld is as allowed as our conventional world. And yet we see more electrons around than positrons, more protons than antiprotons and so on. The asymmetry seems to be of the cosmological origin, not fully understood at present. One thing should, however, be clear. We can hardly expect particles and antiparticles co-existing in close proximity. They would annihilate immediately producing a flash of radiation, $e.g.$, $e^- + e^+ \to \gamma + \gamma$. This positron annihilation is used in solid state physics to study electrons in metals.

Time reversal symmetry demands invariance of the law under the operation of time reversal ($\mathcal{T}$). Thus, if we take a movie of a process and then re-run the reel backwards, what we observe will be an equally allowed process. The reaction $X + Y \to Z + W$ is as legal as the time-reversed reaction $Z + W \to X + Y$. The time reversal operation ($\mathcal{T}$) requires reversing all velocities and spins in detail and interchanging past and future. Thus, at the level of elementary processes, there is no arrow of time. Microscopically, every process is reversible. (But how do we reconcile this microscopic reversibility with the all too common irreversibility of processes in complex systems — the irreversibility at the macroscopic level? What about ageing for instance? There is a thermodynamic arrow of time no doubt. The connection between the time reversal symmetry of the microscopic laws and the observed asymmetry of the complex processes has been and continues to be a subject of much debate. We will not pursue this matter here).

Like parity ($\mathcal{P}$), the time reversal ($\mathcal{T}$) and the charge conjugation ($\mathcal{C}$) symmetries are also approximate. There are subnuclear reactions in which $\mathcal{C}$ and $\mathcal{T}$ are individually violated. But amazingly, the combined action of these approximate symmetry operations (in any order) is an exact symmetry with no violation known. Thus, if in any process we replace all particles by their respective antiparticles, reflect the resulting process in a mirror, and then reverse all velocities and interchange past and future, we will get an equally allowed process. This celebrated $\mathcal{CPT}$ theorem expresses a deep symmetry of Nature. "All hell will break loose" if $\mathcal{CPT}$ invariance is ever found to be violated.

Finally, what about the Ozma problem? We now know why it is not sufficient just to ask our otherworldly correspondent to repeat the ^{60}Co experiment. For, he may belong to the antiworld (of antimatter). There will always be an ambiguity inasmuch as both a right-handed helix of matter and a left-handed helix of antimatter will interpret the results of the experiment equally well. We must somehow ascertain before-hand whether he is made of matter or antimatter. It turns out that this is in fact possible. There are subnuclear reactions that violate time-reversal symmetry and eventually provide us with a method of ascertaining the material versus antimaterial nature of the distant world. The details are much too complicated, but the happy ending is that the Ozma problem is solved in principle.

---

4. GAUGE SYMMETRY

We have been talking mostly about the geometric symmetries of space-time. These are the general framework symmetries without which the physical world will hardly be comprehensible. They seem so natural, almost *a priori*, that we take them for granted. Thus, the failure of symmetry under space reflection, even though a discrete and non-performable one, came as a great shock. Now we are approaching a symmetry of an entirely different kind — the *gauge symmetry*. It is special, it is abstract and it appeals only to a pre-occupied mind. For, here we are requesting invariance of the law that there be, with respect to transformations that are simply outrageous. And yet the experience of the last three decades points to these gauge symmetries as the basic dynamical principles on which the fundamental interactions (forces) of Nature are designed. The familiar *electromagnetic* fields that control much of chemistry and low energy physics, the *strong* interactions that hold neutrons and protons together in the nucleus, the *weak* interactions responsible for the radioactive decay of unstable nuclei, and possibly even the universal *gravitation* that holds the planets and the stars together, all seem to fit in with this general scheme as 'gauge-fields'.

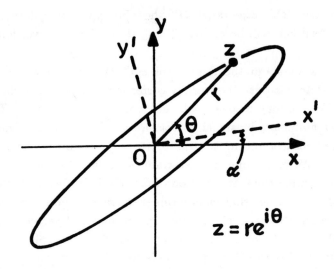

FIG. III.12 *Elliptical trajectory of a 2-dimensional harmonic oscillator in the complex plane representation.*

A proper understanding of gauge symmetry in physics requires a background knowledge of the framework theory, *quantum mechanics*, which is frankly outside the scope of this discussion (see, however, Appendix B). It is possible to get acquainted with the basic idea of gauge symmetry from an example that we know from our first year in college — the example of a simple harmonic oscillator (SHO). It will be a caricature, but real enough for our purpose. Let us get down to it without further apology.

Consider a particle performing a simple harmonic motion in a plane, *i.e.*, a two-dimensional SHO. What it means is that both its x and y coordinates oscillate sinusoidally with the same frequency. Thus the particle will in general describe an elliptical trajectory in the x, y-plane. It is convenient to combine the two motions along the x and the y axes into the motion of a single complex variable $z = x + iy$, where $i = \sqrt{-1}$ is the imaginary unity that keeps the real and the imaginary parts of z from getting scrambled up. The position of the particle in the x, y-plane is now labelled by a single complex variable z. This is the familiar Argand diagram for complex numbers (Fig. III.12).

The magnitude of z is $r = \sqrt{x^2 + y^2}$, which is the distance of the particle from the origin O, and the polar angle θ is its angular position, where $\tan \theta = y/x$. The SHO is described by the equation $d^2z/dt^2 + \omega^2 z = 0$. Here $2\pi/\omega$ is the time period of oscillation. We can even make ω time-dependent. As remarked before, the real and the imaginary parts of this equation do indeed describe the simple harmonic motions along the x and the y axes.

Now comes the crucial observation. We have reckoned the angle θ from the x-axis. But this is just a matter of convenience. The absolute origin of the angle is irrelevant. And with this irrelevancy comes the freedom of choice. We could, for example, rotate our x, y axes anticlockwise by an angle α to new axes x', y' and reckon θ from the new x'-axis. This trivially amounts to replacing θ by $\theta - \alpha$. We say that we have re-gauged θ. All we have to do is to multiply our equation by $e^{-i\alpha}$ and absorb this phase factor by redefining $z' = ze^{-i\alpha}$, and our equation reads the same in terms of z'. Nothing really has changed. We could do this, of course, because α was a constant, *i.e.*, time independent. This irrelevance of absolute θ and the associated invariance of our equation is what we call the *global gauge freedom* and invariance. Global because it was an overall shift of θ, fixed and the same for all time. Encouraged by this, we now become more demanding. We demand freedom of choosing α differently at different times. That is to say we demand invariance under time-dependent shift $\alpha(t)$. This is the *local gauge invariance*, *i.e.*, local in time. But with $\alpha(t)$ varying with time, the factor $e^{i\alpha}$ can no longer be absorbed by the redefinition of z because of the time derivative occurring in our equation. It will generate additional terms involving time derivatives of $\alpha(t)$. Our earlier invariance of the equation is obviously lost. The question is if we can regain it with as little and as reasonable or natural a modification as possible of our original equation. In other words, can we introduce something that will *compensate* for these additional terms. It comes as a pleasant surprise that the answer is *yes*. All we have to do is to replace the time-derivative d/dt occurring in our equation by $d/dt - iA(t)$ with the proviso that re-gauging θ locally as $\theta - \alpha(t)$ should be accompanied by a re-gauging of $A(t)$ as $A(t) - d\alpha/dt$. Here $A(t)$ is the compensatory, or the 'gauge' field. That is all. But what have we gained after all this, you may well ask. Let us see. The time-dependent shift $\alpha(t)$ amounts to rotating our reference frame with an angular velocity $d\alpha/dt$. Now we may recall from our high-school mechanics that such a rotation gives rise to 'fictitious forces', namely the centrifugal force and the Coriolis force acting on our particle. The centrifugal force is the radially outward directed force you feel while riding a merry-go-round. This is the force that makes the rotating earth bulge out at the equator. The Coriolis force is the force that makes you swerve sideways when you try to walk on a rotating platform. This is the force that deflects the winds and the ocean currents to the right (left) in the Northern (Southern) hemisphere due to Earth's rotation. After a little calculus our equation will show that the 'gauge field' $A(t)$ generates precisely these forces automatically. Thus, the requirement of local gauge invariance has created the right kind of forces acting on the particle in accord with experience. Is this not wonderful? This is the essence of the local gauge symmetry.

It is now believed that all the fundamental forces of nature, the *electromagnetic*, the *weak*, the *strong*, and the *gravitational*, are generated just this way. One has to simply identify the correct global symmetry (the irrelevancy) that is to be gauged locally. This is where all the ingenuity and the insight of the theorist lie. We have spoken of the irrelevance of the absolute origin of space and time, and the irrelevance of the absolute orientation in the Minkowski space-time. When these global symmetries are gauged locally, we get Einstein's *general theory of relativity* that replaces the old fashioned Newtonian gravitation acting in the old fashioned Euclidean space. Thus gravitation appears as a gauge field. The idea of local gauge invariance really comes in its own only when it is combined with the framework of quantum mechanics, with all its built-in redundancies, irrelevancies and unobservables. For instance, as we have remarked earlier, the absolute phase of the wave function ψ of an electron is irrelevant. It can be changed globally by an arbitrary constant. But when we gauge it locally, the compensating force turns out to be just the electromagnetic force that we know so well from our experience. It couples to (acts on) the charges and the currents as it should. In point of fact, should we replace the single independent variable t in our oscillator equation by the three space co-ordinates (x, y, z), generalize the gradients appropriately, and let $z(t)$ become $\psi(x, y, z)$, our equation will become the Schrödinger equation for a charged particle moving in a magnetic field represented by the gauge field $A(x, y, z)$, now become the so-called 'vector potential'.

When this gauge principle is applied to relativity-plus-quantum mechanics, it becomes the formidable gauge-field theory of physics today. The principle of local gauge invariance has become the guiding principle in our quest of fundamental understanding in the domain of the very small as well as the very large. Let us hasten to add that the same general principle appears again and again in our world of middle dimensions — the physics of *condensed matter*. So, next time you hear of the gauge invariance, it may well be the gauge theory of ordinary glass, or its magnetic cousin, the 'spin glass'.

---

5. SPONTANEOUS SYMMETRY BREAKING (SSB)

Finally, we come to discussing an idea which is as deep as the idea of symmetry itself, or perhaps even deeper. Its time came much later. But now it is seen as a physical principle that holds the key to unifying all the fundamental forces of Nature, the *electromagnetic*, the *weak*, the *strong* and possibly even the *gravitational*. This has been in one form or the other, the dream of physicists of all times. It is already partially realized now, and some say that the end is in sight. But, first, what is *spontaneous symmetry breaking*? Let us define it. A symmetry is said to be broken spontaneously if the symmetry of the state of the system is lower than (is sub-group of) the symmetry of the force law governing the system. Mark you, we do not break the symmetry of the law itself. We have already hinted at such a possibility — remember the elliptical orbit of the earth around the sun inspite of the spherical symmetry of the gravitational force of the sun! To fix the idea, let us consider two examples, the first a trivial one, taken from mechanics, and the second a highly non-trivial one taken from statistical mechanics where it all began.

Take a piece of wire. Bend it in a U-like shape and hold it vertically. Now slip a bead on the wire and let it slide freely on it. It is common knowledge that the bead will oscillate for a while and eventually settle down at the bottom of the U-wire (Fig. III.13), this being the state of lowest potential energy (equilibrium).

The gravitational potential energy measured from the bottom is proportional to the height. Thus, the U-wire is really a 'potential well' with a single potential minimum at the bottom. Notice that the potential is symmetrical about the vertical through this minimum. Thus the state of the system has

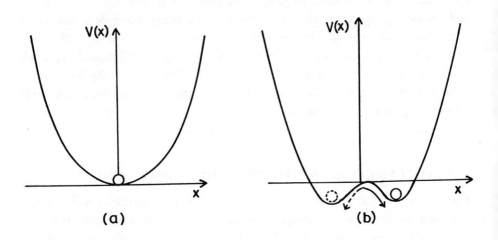

FIG. III.13 *Particle in a symmetrical potential well: (a) symmetric state;*
(b) spontaneously-broken-symmetry state.

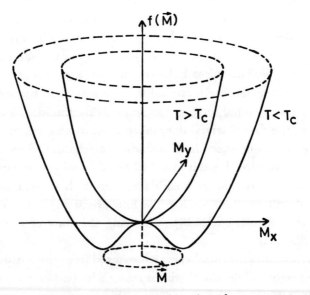

FIG. III.14 *Spontaneous symmetry breaking in a ferromagnet at T_c. $f(\mathbf{M})$*
denotes thermodynamic potential.

the same symmetry as the potential (the force law). Now, let us flatten the bottom part of our U-wire and finally make it convex upwards. We will now have two local minima of the potential located symmetrically about the mid-point which now becomes a local maximum. What should we expect now? The potential is still symmetrical about the vertical through the midpoint, but this is now a state of unstable equilibrium. A disturbance, ever so small, will tilt the balance in favor of one or the other of the two minima and the bead will roll down accordingly. Let it roll down to the right-side minimum. Now, the symmetry of this lopsided state is definitely lower than the symmetry of the potential which is still symmetrical about the vertical axis. This is *spontaneous symmetry breaking*. Broken symmetry agreed, but what is spontaneous about it, you may ask. After all we did need some disturbance to break it. Well, the point is this. The disturbance needed to break the symmetry of the state can be made arbitrarily small. Even the tiny thermal jiggling of molecules in the wire will do. The effect produced, namely the rolling down to one of the minima, is totally out of proportion to this tiny disturbance which could in principle be made almost zero. This is why it is called spontaneous. One is reminded of *Buridan's ass*. The ass was placed symmetrically between two identical stacks of hay. The ass was hungry but the very symmetry of the two options forbade him from making up his mind and, as the parable goes, he starved to death. But, of course, we know that the ass will eat — slightest bias, even auto-suggestion, will make him turn to one or the other of the two stacks of hay!

Now, we turn to the physically interesting example of a system of many interacting particles in thermal equilibrium. The branch of physics that deals with such systems is called *statistical mechanics* (see Appendix C). Most inanimate systems are of this kind. A good example is that of a ferromagnet. Take a piece of iron. For our purpose, we may regard the atoms of iron as tiny magnets, compass needles if you like. The origin of these tiny magnets, or the magnetic moments as they are called, lies in the spinning electrons. But this detail is not relevant for our discussion. These tiny magnets, shown as arrows in Fig. III.14, interact with each other. The interaction is due to the quantum mechanical 'exchange' of electrons because of their indistinguishability. But this is again a detail not important for our discussion. What is really important is that the interaction energy depends on the *angle between* these magnetic moments. Thus if we turn all the atomic magnetic moments by the same angle about the same axis, the energy of the system will remain the same. We say that the law governing the system is spherically symmetric. Furthermore, for a ferromagnet the energy is minimum when the magnetic moments are all paralled to each other. It is clear, therefore, that these atomic moments will tend to align parallel to each other. At high temperatures, however, the thermal agitation will make these moments point in different directions at random so that there is no net magnetization. The state of the system will be spherically symmetric — it has the same symmetry as the law of interaction. We call this disordered, high temperature symmetric phase the paramagnetic phase. As the sample is cooled sufficiently, however, the interaction energy favoring parallel alignment of the magnetic moments wins over the disrupting tendency of the thermal agitation. When this happens the magnetic moments align parallel to each other on the average and thus the system develops a net magnetization $\mathbf{M}$ which grows in magnitude with decreasing temperature. This low temperature ordered state is called the ferromagnetic phase. The temperature T_c at which the system makes the continuous transition from the high temperature disordered phase to the low temperature ordered phase is called the critical temperature, or the Curie temperature. The magnetization which is a measure of order is referred to as the *order parameter*. (The physics of continuous phase transition, often called the second order phase transition, at and about the critical temperature has been an extremely active area of research of our times. It is determined almost entirely by the symmetry and the dimensionality of the order parameter and is quite independent of the microscopic details of chemical composition, etc. This 'universality' of the 'critical behavior' is most fascinating but we must let it pass). The question now is what should be the direction of this net magnetization $\mathbf{M}$. Inasmuch as the energy depends only on the relative orientation of the atomic moments, all directions of $\mathbf{M}$ are equally probable statistically. And yet in a given realization some direction of $\mathbf{M}$ must get selected. This state is then symmetric only for rotations about this direction of $\mathbf{M}$ — it has only an axial symmetry which is a subgroup of the full spherical symmetry of the interaction law. The symmetry is thus *spontaneously broken*. For a large, in principle infinite, system an arbitrarily small magnetic field or anisotropy will fix the direction of $\mathbf{M}$. The connection with our mechanical example should be obvious. The order-parameter (magnetization $\mathbf{M}$) plays the role of displacement x. Instead of mechanical potential energy $V(x)$ which was to be minimized, here we have a thermo-dynamic potential (*free energy*) $f(\mathbf{M})$ which is to be minimized. The only detail that differs is that whereas the position x in our mechanical case was a scalar (one-dimensional), the order-parameter $\mathbf{M}$ is a vector. Thus, in the mechanical case the two equivalent minima were separated by a potential barrier in the broken symmetry phase, while in the ferromagnetic case all the equivalent minima (differing only by the direction of $\mathbf{M}$) are degenerate (*i.e.*, have the same free energy) and $\mathbf{M}$ can in principle *freely* gyrate among them. Perhaps a better mechanical analogue would have been the marble in a punted wine bottle. Incidentally, this freedom leads to the possibility of certain waves propagating in the broken symmetry phase whose frequency tends to zero as the wavelength tends to infinity (*i.e.*, they are massless). We call these Goldstone modes. For the magnetic case they are the spinwaves. All this plays an important role in the physics of phase transition in *condensed matter*. We emphasize that this is a highly cooperative phenomenon resulting from interaction among large number (infinite in principle) of particles, the spins in this case.

Now, how can all this possibly bring about unification of the fundamental forces of Nature? This is a highly technical and magnificent obsession of contemporary physics. We will try to give just the flavor of it in plain words. Any symmetry can be broken spontaneously. In particular and most importantly, it can be the *gauge symmetry*. It is the combination of gauge symmetry and *spontaneous symmetry breaking* that is central to unification. Consider the simplest case when the matter consists of charged particles (electrons). As we have already seen the global symmetry (namely, the irrelevance of absolute phase) when gauged locally, generates the electromagnetic field automatically, which in the quantum version is the photon. (It is inherent in this mechanism that the photon (the gauge field) be massless). Now in quantum theory, the interactions between particles are mediated by the exchange of quanta of some field (much the same way as exchange of a handball between two players will exert an effective force of repulsion between them. For attraction, let them exchange boomerangs). Thus, the photon mediates interaction between charges. It also follows from the general quantum principles, that the range of interaction be inversely proportional to the rest mass of the quanta exchanged. This is why the range of the electromagnetic interaction is infinite. This intimate 'genetic' connection between matter (electrons) and the gauge field (photons) leads us to expect an induced change in the character of the gauge field when the matter undergoes a phase transition in which the very global symmetry, whose local gauging generated the gauge field, undergoes spontaneous break down. In point of fact it would be very surprising if it were otherwise. We already see this effect in a superconductor in the laboratory (see Chapter II on Superconductivity). Here electrons undergo a transition in which the phase of this collective (macroscopic) wave function takes on a definite value breaking thus the global symmetry spontaneously. This induces the photon to acquire a non-zero mass, making it impossible for it to propagate very far into the superconductor. This explains the famous Meissner effect that magnetic field is expelled from a superconductor.

This was the simplest, but a most striking demonstration, of the change of character of a gauge field induced by the SSB of the matter (field). All we have to do now is to generalize to more complicated internal symmetries that can be imagined and indeed have been postulated. Thus, there may be several gauge fields generated by local gauging. They may be all symmetry related and thus of the same character. Now, if the matter undergoes SSB, the group of these gauge fields may be split into subgroups, and different subgroups may acquire different masses. Successive phase transitions (and the associated SSB's) may generate thus a gamut of fields with different characters. This generation of different gauge fields (fundamental forces) by the descent of symmetry due to SSB, from the single most symmetric initial entity is the dream of the *grand unified theory* (GUT). It has already been partially realized in the *unification* of the *electromagnetic* and the *weak* interaction by Glashow, Salam, and Weinberg for which they won the Nobel Prize in Physics.

One has the plausible scenario in mind that the universe began totally symmetric with a Big Bang some 15 billion years ago. As it expanded it cooled and underwent successive phase transitions. The associated SSB's led to the diversity of fields that survive at the present epoch. And what a diversity — if the strong interaction measures unity on a certain scale, the electromagnetic interaction will measure 10^{-2}, the weak interaction 10^{-5} and the gravitational interaction 10^{-34}! The strong interaction acts on hadrons (protons, neutrons, pions, etc., or their postulated building blocks, the quarks) but not on leptons (electrons, neutrinos, etc.) and is short-ranged. The electromagnetic interaction acts on all charged particles and has infinite range. The weak interaction involves neutrinos and has a still shorter-range. The gravitational interaction is the weakest of all, but acts universally on everything and has infinite range. And yet all may have a common origin. This reminds one of the concept of the Nirguna Brahma of the ancient Hindus — formless, featureless, totally symmetric pure existence, from which all the diversity originated, shall we say, by spontaneous symmetry breaking.

This brings us to the end of our exploration of symmetry. We have seen its power. Obviously, symmetry cannot answer all the *why*'s but it does reduce them to a fewer *why*'s. To the philosophical question of why Nature is so symmetric, we can perhaps answer thus. Symmetry is, in the ultimate analysis, absence of bias. It is an expression of justice. There is a principle of insufficient reason against asymmetry. A sphere is admitted. But a deviation from sphericity must bide our question.

Galileo had spoken of the great Book of Nature. We should perhaps add that the first and the last Chapters of this Book are on *symmetry* and its *spontaneous breakdown*, respectively.

Summary

Symmetry means invariance of an object with respect to a set of operations called symmetry operations to be performed on it. The object may be the geometrical form of body such as a crystal of common salt and the set of operations may the geometrical operations of translation along a direction, rotation about an axis or reflection in a plane. The symmetry operations may be continuous or discrete, physically performable or non-performable. More importantly, the object may be a law of nature itself expressed mathematically by a certain equation. Symmetry then means the invariance, or rather covariance, of the form of the equation under the mathematical transformations corresponding to the symmetry operations that may not be geometrical in nature. Symmetry is a powerful physical principle that helps us not only simplify calculations and classify and unify diverse objects, it also restricts the possibilities in the absence of complete knowledge of the physical world. It creates new physics when a symmetry is requested on intuitive grounds.

There is a branch of mathematics called *group theory* that provides the proper and powerful language for dealing with symmetry. Symmetry has played a fundamental role in quantum physics, particularly in the domain of high energy physics where its predictive power has been fully vindicated. The rather abstract idea of *gauge symmetry* is one of the profoundest concept produced by the human mind. Any symmetry, however compelling aesthetically it may be, must be established experimentally. Thus, *parity* signifying the left-right symmetry between an object or a process and its mirror image turned out to be false in certain fundamental processes involving neutrinos. Symmetry can also be broken spontaneously when there is a phase transition. The idea of spontaneous symmetry breaking has played a decisive role in our understanding of phase transitions in general and in the context of the early universe in particular. The search for deeper, hidden symmetries of Nature continues.

CHAPTER IV

CHAOS:

CHANCE OUT OF NECESSITY

CONTENTS

1. INTRODUCTION: CHAOS LIMITS PREDICTION

Physics is ultimately the study of change — of becoming. Changes are determined by the laws of physics that be. For example, we have Newton's three laws of motion. The laws themselves are, of course, believed to be changeless. The necessary connection of events implied by the deterministic nature of the physical laws leaves us no chance of freedom except for that of the choice of initial conditions, that is the initial positions and velocities of all the elementary subunits that make up our system. Once this set of initial data is entered, the future course of events, or the process, is uniquely determined in detail and, therefore, predictable in principle for all times, indeed just as much as the known past is! Thus, for instance the trajectories of all the 10^{18} molecules that belong in each cubic centimeter of the air in your room, suffering some 2×10^{27} collisions each passing second, are in principle no less predictable than those of an oscillating pendulum or an orbiting planet — only much more complex.

You may recall that in order to specify the initial conditions for a single particle, taken to be a point-like object, we need to enter a set of three numbers, its Cartesian coordinates x, y, z, say, to fix its position. We say that the particle has three dynamical degrees of freedom. Another set of three numbers is required in order to fix the corresponding components of its velocity. For N particles these add up to $6N$ independent data entries. The state of the system can then be conveniently represented as a point in a $6N$-dimensional abstract space called *phase space*. The motion of the whole system then corresponds to the trajectory of this single representative phase point in this phase space. (The nature of phase space, or more generally speaking *state space*, of course, depends on the context. Thus, in the case of a chemical reaction we may be concerned with the concentrations x, y, z, say, of three reacting chemical species whose rate of change (kinetics) depends only on these concentrations. In that case the state space will be a three dimensional one). Deterministic dynamics or kinetics implies that there is a unique trajectory through any given phase or state point, and it is calculable in principle. In our example of the air in the room, $N = 10^{18}$! Behold the tyranny of large numbers. The complexity here is due to our having to store large amounts of input information and to solve as many equations of motion — the computational complexity of information processing. We will do well to remember here that a molecular dynamicist of today with free access to the fastest supercomputer available, capable of a billion floating point operations per second, can barely simulate the digitized motion of some 10^4 particles, and that too only approximately. But these practical limitations are besides the point. In principle the motion is calculable exactly and hence predictable — or so a perverse purist would claim.

It is true that we speak of chance and probability in physics, in statistical physics to wit, where we have the Maxwell-Boltzmann distribution of velocities of molecules of a gas, the Gaussian distribution of errors in a measurement, or the random walk of a Brownian particle (a speck of pollen floating in water for example) and various averages of sorts. But these merely reflect an incompleteness of our knowledge of the details. The randomness of the *Brownian motion* is due to its myriads of collisions with the water molecules that remain hidden from our sphere of reckoning. In point of fact even if we could calculate everything, we wouldn't know what to do with this fine-grained information. After all our sensors respond only to some coarse-grained averages such as pressure or density that require a highly reduced information set. It is from our incompleteness of detailed information as also, and not a little, from our lack of interest in such fine details, that there emerges the convenient concept of chance and of probability. But strictly speaking, everything can in principle be accounted right. There is truly no game of chance: the roll of the dice, the toss of the coin or the fall of the roulette ball, can all be predicted exactly but for the complexity of computation and our ignorance of the initial conditions. This absolute determinism was expressed most forcefully by the the 19th century French mathematician Pierre Simon de Laplace. Even the whole everchanging universe can be reduced to a mere unfolding of some initial conditions, unknown as they may be to us, under the constant aspect of the deterministic laws of physics — *sub specie aeternitatis*.

But this turns out not to be the case. The Laplacian determinism, with its perverse reductionism, is now known to be seriously in error for two very different reasons. The first, that we mention only for the sake of completeness, has to do with the fact the correct framework theory for the physical happenings is not the classical (Newtonian) mechanics but Quantum Mechanics (See Appendix B). There is an Uncertainty Principle here that limits the accuracy with which we may determine the position and the velocity (momentum) of a particle simultaneously. Try to determine one with greater precision and the other gets fuzzier. This reciprocal latitude of fixation of the position and the velocity allows only probabilistic forecast for the future, *even in principle*. Quantum uncertainty, however, dominates only in the domain of the very small, *i.e.*, on the microscopic scale of atoms and molecules. On the larger scales of the 'world of middle dimensions' of common experience and interest, the deterministic classical mechanics is valid for all practical purposes. We will from now on ignore the quantum uncertainty. We should be cautioned though, that the possibility of a fantastic amplification of these quantum uncertainties to a macroscopic scale cannot be ruled out.

The macroscopic uncertainty, or rather the unpredictability, that we are going to talk about now, emerges in an entirely different and rather subtle manner out of the very deterministic nature of the classical laws. When this happens we say that we have *chaos*, or rather deterministic chaos to distinguish it from the thermal disorder (or 'molecular chaos' of Brownian motion).

1.1 The Butterfly Effect

But how can a system be deterministic and yet have chaos in it. Is there no contradiction in terms? Well, the answer is *no*. The clue to a proper understanding of deterministic chaos lies in the idea of *Sensitive Dependence on Initial Conditions*. Let us understand this first. As we have repeatedly said before, the deterministic laws simply demand that a given set of initial conditions lead to a unique and calculable state of the system at any future instant of time. It is implicitly understood here, however, that the initial conditions are to be given to infinite precision, *i.e.*, to infinite number of decimal places if you like. But this is an ideal that is frankly unattainable. Errors are ubiquitous. What if the initial conditions are known only approximately? Well, it is again implicitly assumed that the approximately known initial conditions should enable us to make approximate predictions for all times — approximate *in the same proportion*. That is to say that while the errors do propagate as the system evolves, they do not grow with the passage of time. Thus, as we progressively refine our initial data to higher degrees of accuracy, we should get more and more refined final-state predictions too. We then say that our deterministic system has a predictable behavior. Indeed, operationally this is the only sense in which prediction acquires a well defined meaning. In terms of our state space picture, this means that if we started off our system from two neighboring state points, the trajectories shall stay close by for *all* future times. Such a system is said to be well behaved, or regular. Now, the point is that deterministic laws do *not* guarantee this regularity. What then if the initial errors actually grow with time — that too exponentially? In our phase space picture then, any two trajectories that started off at some neighboring points initially, will begin to diverge so that the line joinging them will get stretched exponentially (as $e^{\lambda t}$, say) with the passage of time. Here λ measures the rapidity of divergence or convergence according as it is positive or negative.It is called the *Lyapunov exponent.* The initial instant of time can, of course, be taken to be any time along the trajectory. This is precisely what we mean by the sensitive dependence on initial conditions. It makes the flow in the phase space complex, almost random. For, then the approximately known initial conditions do not give the distant future states with comparable approximation. The system will lack error tolerance making long-time prediction impossible. This is often referred to picturesquely as the Butterfly Effect: Flap of a butterfly's wings in Brazil may set off a tornado in Texas. Some sensitivity! When this happens, we say that the dynamical system has developed chaos even though the governing law remains strictly deterministic. We might aptly say that chaos obeys the letter of the law, but not the spirit of it.

There is yet another way of expressing the sensitive dependence on initial conditions, without comparing neighboring trajectories. After all a given dynamical evolution is a one-shot affair and it should be possible to express this characteristic sensitivity in terms of that single one-shot trajectory *per se*. It is just this: One cannot write down the solution of the dynamical equations in a closed, smooth (analytic) form valid for all times. For, this would mean that the state variable at any time must be a smooth function of the initial conditions, which negates the sensitive dependence. This means that the evolution equation (*i.e.*, the *algorithm* for change) must be solved step by step all the way to the final time and the calculated values of the state variable catalogued. There is no short cut. There is a kind of algorithmic complexity in this.

Now, what causes this sensitive dependence on initial conditions? Does chaos require a fine tuning of several parameters, or does it persist over a whole range of the parameter values? Is chaos robust? How common is chaos? Is every chaotic system chaotic in its own way or there is universality — within a class may be? How do we characterize chaos? Can a simple system with a small number of degrees of freedom have chaos, or do we need a complex system with a large, almost infinite number of degrees of freedom? Is the claim of distinction between a chaotic and a statistically random system a mere nit picking, or one of physical consequence? These are some questions that we will address in the following sections in a somewhat intuitive fashion. Collectively, these problems are studied under the forbidding heading of 'Dynamical Systems'. Deep results have been obtained in this field over the last two decades or so. Some questions still remain unanswered or only partially answered. But a physical picture of chaos has emerged which is already fairly complete. It is likely to improve greatly with our acquaintance with some examples of chaos.

1.2 Chaos Is Common

One may carry the impression that the *sensitive dependence on initial conditions* necessary for chaos must require a fine tuning of control parameters which can happen only accidentally in nature. This would make chaos an oddity that can rarely occur in our sensible world and hence an object only of mathematical curiosity. This notion is, however, erroneous. Chaos is a robust phenomenon that persists over a wide range of values of the control parameters. The sensitive dependence on initial conditions does not wash easily. Chaos is, therefore, not rare. Indeed, the contrapositive is true. Nature is full of it. Almost any real dynamical system, driven hard enough, turns chaotic.

The most celebrated example of chaos is, of course, the *fluid dynamical turbulence* — the last unsolved problem of classical physics. Great physicists of this century have bemoaned, 'Why turbulence'? We see turbulence in jets and wakes, in water flowing through pipes and past obstacles. As the flow rate (the control parameter) increases beyond a critical value, and this

is the meaning of the phrase 'driven hard enough', the smooth laminar flow
becomes unstable. It develops waviness and quickly turns into a complex,
almost random pattern of flow, live with swirling eddies of all sizes as we move
downstream. The flow pattern is *aperiodic* — it never repeats itself. This
is fully developed turbulence. We can demonstrate this easily by injecting a
marker dye in the flow tube, that makes the flow pattern visible to the eye
(See Fig. IV.1a).

At the onset of turbulence the otherwise fine and straight thread of
marker dye undulates wildly and quickly disrupts into a complex ramified
pattern, down to length scales too fine for the eye to discern, making the
water appear uniformly colored as we move downstream — as if through
thorough mixing. The threshold condition for the onset of this instability
can be conveniently expressed in terms of a dimensionless control parameter
called the *Reynolds number* (R), after the great fluid dynamicist Osborne
Reynolds. The Reynolds number $R = UL\rho/\eta$, where U is the typical flow
velocity, L the tube diameter, ρ the fluid density and η the fluid viscosity
(treacliness). Turbulence sets in for R greater than a critical value R_c which
is about 2000 for water, but can be as high as 10^5 if the flow is increased very
gently and the pipe is very smooth. (This is something like the superheating
of water beyond its boiling point). It is clear that the critical Reynolds
number R_c is not a universal number. It depends on the geometry of the
problem, whether the tube cross section is circular or elliptical for instance.
But importantly, the sequence of flow patterns, or the route to turbulence
turns out to be universal — within a class. (This is very reminiscent of
the universality class of a second order phase transition. Thus, different
ferromagnets have different Curie (critical) temperatures. But the behavior
of magnetization near the critical point is universal within a wide class. These
are deep problems).

Another example of chaos that has played a decisive role in the experimental and theoretical study of turbulence-as-chaos is the *Couette-Taylor
instability* problem. Here the fluid is contained in the annular space between
two long co-axial circular cylinders, one of which, usually the inner one, is
made to spin. At low spin rates the fluid is dragged viscously around by
the inner cylinder and there is a time independent laminar flow in the form
of rotating circular cylindrical sheets. Beyond a critical spin rate, however,

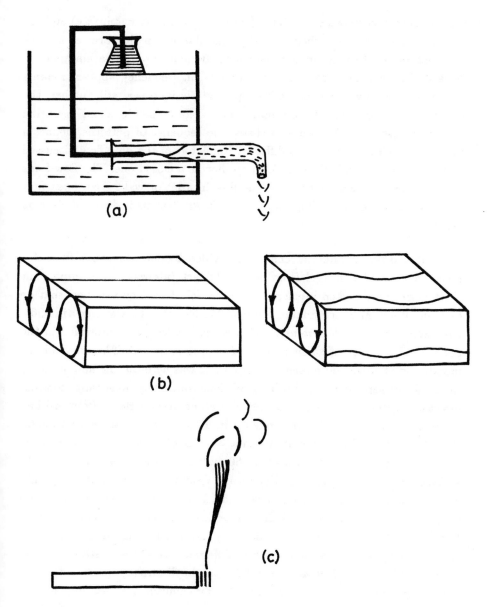

FIG. IV.1 *Examples of instabilities leading to chaos: (a) turbulence in pipe flow marked by a dye; (b) convective instability showing cylindrical rolls and their wobbling; (c) rising smoke column developing whorls.*

this laminar flow becomes unstable (the Couette-Taylor instability) and the circular cylindrical flow sheets develop undulations in the form of a stack of doughnuts or tori. A single independent frequency of oscillation appears. At a still higher spin rate another independent frequency appears making the system doubly periodic. As the spin rate is increased still further, the pattern suddenly jumps to a chaotic one. Again the threshold condition is non-universal but the route to chaos, the sequence of patterns, defines a universality class. This seems different from that of the pipe flow discussed above. We should note that the Couette flow is a closed one: The flow feeds back on itself. In contrast to this the flow in a jet or wake is an open flow where the turbulence develops as we move down stream. It does not close on itself.

Yet another example of chaos is provided by the convective flow of a fluid mass which is heated from below. In fact this was the example that E. N. Lorenz studied so intensively as a model of his 'toy weather' that he had simulated on his computer, a Royal McBee, at MIT way back in 1963. His paper entitled *'Deterministic Non-periodic Flow'* was published in the *Journal of Atmospheric Sciences, vol.20, pp.130–41, (1963)*, that laid the foundation of the modern science of *chaos*. It was here that he had observed the sensitive dependence on initial conditions that makes long-range weather forecast impossible. The great French mathematician Henri Poincaré had hinted at such a sensitive dependence in 1903! (It is a sobering thought that Lorenz did all his simulations on a Royal McBee, a vacuum-tube based computer capable of doing just about 60 multiplications a second — a snail's speed compared to the modern Cybers and Crays capable of hundreds of millions of floating point operations a second). Returning now to our convection cell, nothing much happens if the temperature difference is small enough. Heat is simply conducted to the cooler top surface through molecular diffusion. At a higher critical temperature difference, this steady state becomes unstable (the *Rayleigh-Bénard instability*) and cylindrical rolls of convection develop in our cell, assumed rectangular for convenience (see Fig. IV.1b). This happens when the forces of buoyancy overcome the damping effects of viscous dissipation and thermal diffusion. Heated fluid at the bottom expands, and thus made lighter, ascends along the center to the cooler top layers, delivers heat there and then moves out and down the sides completing the roll. On further heating beyond a threshold, this pattern becomes unstable and the rolling cylinders begin to wobble along the length. More complex flow patterns appear successively, cascading to a point of accumulation at which the flow becomes turbulent. The route to chaos (turbulence) is the so-called *period-doubling* bifurcation that we will discuss later. Incidentally, the control parameter in this case is the dimensionless *Rayleigh number* $R_a = g\rho\alpha h^3 \Delta T / \eta\kappa$, where g = gravity, κ = thermal conductivity, α = coefficient of thermal expansion h = height of cell and ΔT the temperature difference. The first instability sets in when R_a exceeds a critical number that depends on the geometry, *e.g.*, the aspect ratio of the cell. Convectional instability is common in Nature. On a hot summer day one cannot fail to observe the pattern of clouds imprinted on the sky by the rising convectional currents.

There is an amusing example of turbulence that you can readily observe if you are a smoker, or have a smoker friend willing to oblige. The column of smoke rises from the lighted tip straight upto a height and then suddenly disrupts into irregular whorls (see Fig. IV.1c). Here again the hot smoke rises faster and faster through the cooler and so denser air until it exceeds the critical Reynolds number and turns turbulent. It is really no different from an open jet.

The examples of chaos cited above are all from fluid dynamics — turbulence. But chaos occurs in the most unexpected places. It is revealed in the ECG traces of patients with arrhythmiac hearts and the EEG traces of patients with epileptic seizures. It is suspected in the apparently random recurrence of certain epidemics like measles. It lurks in the macroeconomic fluctuations of commodity and stock prices. It is seen in chemical reactions and in the populations of species competing for limited resources in a given region. The irregular pattern of reversals of Earth's magnetic field is suspected to be due to the chaotic geodynamo. Chaos is, of course, well known in the non-linear electrical oscillators — the classic Van der Pol oscillator and lasers where chaos masquerades as noise. Chaos is implicated in the orbits of stars around the galactic center. The list is endless. But most of these examples suggest that chaos requires an interplay of large number of degrees of freedom. Is this really so?

1.3 Can Small Be Chaotic?

Small here means smallness of the number of degrees of freedom that the dynamical system may have. A number as small as three, say. Can such a small system exhibit the apparent randomness that we associate with chaos? Chaos, as we have seen in the case of turbulence, means an aperiodic flow with a long-range unpredictability about it. The flow pattern must not repeat itself at the very least. Can a system with a small, indeed *finite*, number of degrees of freedom be so 'infinitely inventive' of the flow patterns as to go on surprising us forever without exhausting its stock of possibilities and without having to repeat itself? That is the question. And the answer that has emerged slowly but surely over the last two decades or so is that it can, provided there is *non-linearity* in it. This is a necessary condition. Dissipation will help — it is there in all real systems anyway. This view disagrees sharply with the ideas about turbulence that have prevailed since the time of the great Soviet physicist L.D. Landau. Landau's paradigm for turbulence visualized a sequence of *infinite* number of competing oscillations with incommensurate frequencies, emerging one at a time as the control parameter crossed the successive thresholds of instabilities. So it is 'infinite' versus 'finite'. To appreciate this fully let us examine some of the examples of chaos mentioned above more carefully.

The examples of chaos we have described so far are mostly about turbulence in fluids. It is clear that the fluid flow is described completely by Newton's laws of motion. Of course, they have to specialize, to the case of the fluid continuum where they take on the form of the macroscopic equation of fluid dynamics — the Navier-Stokes equation. This equation simply expresses the law that mass times acceleration of an element of fluid equals the net force acting on it due to the difference of pressures fore and aft, and the viscous drag on the sides. (Viscosity, of course, implies dissipation of energy that will cause the motion to die down, unless kept alive by an external driving agency, the rotating inner cylinder in the Couette flow for instance). How can such a deterministic system exhibit the apparent randomness of turbulence. The idea that goes back to Landau is this. The fluid has an infinitely many degrees of freedom. How? Well, at each point of space the fluid can move up and down, right and left, and forward and backward. That is three degrees of freedom. There are infinitely many points in any finite volume and that adds up to infinite number of degrees of freedom.

It is often convenient to combine these local point-wise degrees of free-dom distributed in space into extended oscillations or waves, called modes, of different wavelengths and talk in terms of these infinite number of modes instead. It is just as a wave on the surface of water is made up of movements of the elements of water all over the surface. Conversely, the movement of the fluid at any given point can be re-constituted from the superposition of these waves. The two descriptions are completely equivalent and are related by what mathematicians call (spatial) Fourier Transform. Likewise, a flow at a point fluctuating in time can be regarded as a superposition of periodic oscillations of different frequencies through a temporal Fourier transform — giving a frequency spectrum. The strength of the different Fourier compo-nents gives the power spectrum of the flow.

This then is the infinite number of oscillations or modes of the fluid that Landau invoked to describe turbulence. As the frequencies are incom-mensurate, the flow is only quasi-periodic, never quite repeating itself. Two frequencies are said to be incommensurate if their ratio is irrational, *i.e.*, it cannot be expressed as a ratio of two whole numbers. A fully developed tur-bulence has infinitely many such frequencies making the flow aperiodic for all practical purposes. The power spectrum is then a continuous one with no dominant sharp frequencies in it. This is what chaos is. This aperiodicity can be roughly appreciated with the help of a simple example. Imagine a system of large number, 26 say, of simple pendulums of periods 0.5, 0.6, 0.7, $\cdots$, 2.8, 2.9, 3.0 seconds and let them go simultaneously from the right of their equilibrium positions. Some thought and arithmetic will show that it will take 7385 years before the system repeats its initial state! And that when the frequencies (or the periods) *are still commensurate* and finite in number. Incommensurate frequencies will push this recurrence time to eternity.

This scenario for turbulence, and for chaos in general, seems eminently appealing but has really never been put to test. No one has seen this route to turbulence via the successive emergence of these infinite number of mode frequencies. Perhaps it applies to open flows — jets and wakes. On the contrary, there are known examples of turbulence that refute it. The Couette flow as discussed above is a case in point. So is the case with the convective flow which was modelled by Edward Lorenz by a finite number, in fact three of interacting macroscopic degrees of freedom. It showed chaos in no uncertain terms. And the route to this convective chaos was studied experimentally in an ingenious experiment on a liquid helium cell by Albert Libchaber, and was found to be of the period-doubling kind, quite different from that of Landau, as we shall see later. In any case there are a number of computer models with small number of degrees of freedom that show chaos. Question is what makes them chaotic. Answer lies in non-linearity. The Navier-Stokes equations are non-linear.

Non-linearity is a mathematical way of saying that the different dynamical degrees of freedom act on each other and on themselves so that a given degree of freedom evolves not in a fixed environment but in an environment that itself changes with time. It is as if the rules of the game keep changing depending on the present state of the system. A simple example will illustrate what we mean. $y = ax$ is a linear relation. Changes in y are proportional to those in x, the constant of proportionality being 'a'. In this game the effect is proportional to the cause and the controlling parameter 'a' is fixed. Now consider $y = ax^2$. Here the effect is proportional to the square of the cause. We may rewrite this in the old form as $y = (ax)x$ and say that the parameter controlling the game, namely (ax) itself depends on x. This is non-linearity or non-linear feedback if you like. In point of fact interactions that are linear in state variables are no interaction at all. One can re-define new state variables as linear combinations of the old ones such that these new state variables do not interact at all. This is called normal mode analysis. It is this non-linearity that makes for the complex behavior — in particular it can generate the sensitive dependence on the initial conditions. It can amplify small changes. One does not need infinite number of degrees of freedom.

But you may turn around and say that the fluid dynamical system does have an infinite number of degrees of freedom anyway. Well, this is where dissipation (friction) comes handy. It turns out that all but a few of these get damped out rather quickly by this friction, and the system settles down to the few remaining, macroscopic degrees of freedom that really define its state space. As we will see later, we need a minimum of three to have chaos in a continuous flow.

A final worry now. A dynamical system with a small number of degrees of freedom will have a low dimensional state space. Moreover, the degrees of freedom are expected to have a finite range — we don't expect the velocities to be infinite for example. Thus the phase point will be confined to a finite region of the low dimensional state space. This raises a geometrical question of packing. How does the phase trajectory, confined to a finite region of the low dimensional state space, wind around forever without intersecting itself or closing on itself? There is an ingenious geometrical construction that illustrates how this is accomplished. We know that two trajectories from neighboring points diverge, stretching the line joining them exponentially. But this cannot go on indefinitely because of the finiteness of the range. The trajectories must fold back, and may approach each other only to diverge again (Fig. IV.2).

This will go on repeating. This process of stretching and folding is analogous to a baker's transformation. He rolls the dough to stretch it out and then folds it, and then repeats the process again and again. We can drop a blob of ink on the dough to simulate a dust of phase points. The process of stretching and folding will generate a highly interleaved structure. In fact after a mere two dozens or so of iterations we will have 2^{24} layers and thus a fine structure down to atomic scales. The inky spot would have spread out throughout the dough coloring it apparently uniformly, suggesting thorough mixing. Yet, actually it is finely structured. The neighboring points on the inky spot would have diverged out and become totally uncorrelated after a few rounds of baker's transormation. Indeed, stretching with folding is a highly non-linear process that generates the above sensitivity. Baker's transformation seems a general algorithm for chaotic evolution.

Our discussion so far has centered on some deep philosophical questions about and a general qualitative understanding of chaos. We will now turn to a simple but real, in fact household experiment with chaos that should be refreshing.

We are going to experiment with a leaking faucet! We will be following here the idea of Robert Shaw of the University of California, Santa Cruz.

---

2. LESSON OF THE LEAKING FAUCET

Leaky faucets are known universally. They are perhaps best known as a very effective form of torture as many an insomniac wait attentively for the next

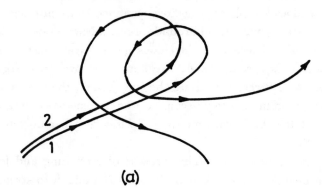

(a)

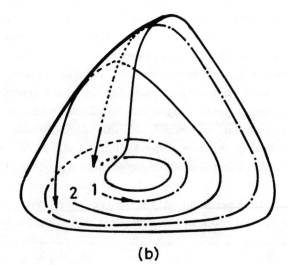

(b)

FIG. IV.2 *Sensitive dependence on initial conditions: (a) divergence and,
(b) divergence-cum-folding-back of neighboring trajectories.*

drop to fall. It is, however, less well known that there is a universality to their pattern of dripping as the flow rate is turned on gradually and sufficiently. There is some pretty physics about it. It also happens to be rather easy to experiment with. All you need is a leaky faucet, preferably the one without a wire mesh, and a timing device to monitor the time intervals between successive drops as the flow rate (our control parameter) is gradually increased. Unfortunately, the time intervals can get rather short, on the order of a fraction of a second, down to milliseconds and even microseconds, just when interesting things begin to happen. We will, therefore, have to employ detectors other than our unaided eyes and ears. We could, for instance, use a microphone to pick up the sound signals of the falling drops, or better still, arrange to have the falling drop interrupt a beam of laser (or ordinary) light to be detected by a photodiode. The data can then be acquired by and stored on a PC, which is now become part of the science kit in most high schools. At low enough rate of flow, the faucet leaks with a monotonous periodicity — drip, drip, drip, $\cdots$. The successive drops fall at equal intervals, T_0 say (Fig. IV.3).

This dripping pattern persists up to a fairly high threshold flow rate. The time interval T_0, of course, gets shorter and shorter. At and beyond this threshold, however, this pattern becomes unstable and a new pattern emerges — pitter-patter-pitter-patter-$\cdots$ (Fig. IV.4). The intervals between successive drops become unequal now. We have a short interval T_1 alternating with a long interval T_2, generating thereby a sequence $T_1, T_2, T_1, T_2,$ $\cdots$. We say that the period has doubled! The original single period T_0 has bifurcated into a pair of unequal periods T_1, T_2. The repeat 'motif' now consists of a pair of two successive periods, one long and the other short. Note that the *period-doubling* refers to this two-ness, and not to the absolute value of the time interval, which, if anything, gets only shorter as the flow is turned on progressively. This new pattern in its turn persists up to still higher threshold, and then becomes unstable. There is a period doubling again: Each of the intervals T_1 and T_2 bifurcates into two unequal intervals, leading to the pattern $T_3, T_4, T_5, T_6, T_3, T_4, T_5, T_6, \cdots$. Let us summarize our findings so far. We started with period 1 $(= 2^0)$, which bifurcated to period 2 $(= 2^1)$ and which in turn again bifurcated to period 4 $(= 2^2)$. The trend continues. At the nth bifurcation, we get the period 2^n. It turns out

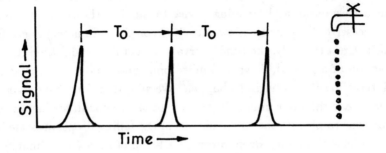

FIG. IV.3 *Leaking faucet dripping with single period T_0 at low flow rate.*

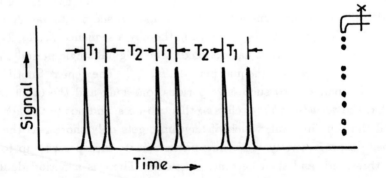

FIG. IV.4 *Leaking faucet dripping with doubled period (T_1, T_2) at a higher flow rate.*

that the successive bifurcations come on faster and faster and they pile up at a point of accumulation as $n \to \infty$ (infinity). This is the critical value of our control parameter. At this point, the period becomes 2^∞, which is infinite. The dripping pattern never repeats itself. It has become aperiodic. It is as if the drops fall to the beats of the 'infinitely inventive drummer'. This is *chaos*. We have just discovered a period-doubling route to chaos!

The chaos is robust and it has a universality about it that we will discuss later. Thus, you can change the faucet; you can replace water with beer if you like; you can add some surfactants to reduce its surface tension or whatever. The period-doubling sequence will still be there, and many numbers characterising the approach to chaos will remain unchanged. This is reminiscent of the universality familiar from the critical behavior of second-order phase transitions.

What happens if we push our control parameter beyond this critical value. Well, the aperiodic chaos persists but there are now finer structures such as the appearance of narrow windows of periodicities, that we cannot pause to discuss here.

It is amazing that a simple looking system such as a leaking faucet should reveal such complexity, or conversely, that chaos should hide such simplicity. We will now turn to a more quantitative description of chaos.

---

3. A MODEL FOR CHAOS

Now that we have experimented with chaos, let us make some simple model of it that we can play around with, and hopefully, solve it in the process. Solving here requires no more than elementary operations of addition, multiplication and raising to power (exponentiation), done repeatedly (that is iteratively), all of which can be entered easily on a hand-held calculator. In fact this is precisely what was done by Mitchell J. Feigenbaum of the Los Alamos National Laboratory in early seventies using his HP-65 hand-held calculator. And that made the study of chaos such a refreshing exercise with numbers, and, of course, led to many a breakthrough.

3.1 The Logistic Map

The model we are going to study is the so-called *logistic map* or equation: $X_{n+1} = bX_n(1 - X_n)$. It is an algorithm, that is a rule for the growth of a quantity 'X' controlled by a growth factor 'b'. Given the value X_n of this quantity at the nth instant, or go-round, all you have to do is to substitute this value in the right-hand side expression, evaluate it and voila! You have the value of 'X' at the $(n + 1)$th go-round, namely X_{n+1}. You can iterate this process and thus generate the value of 'X' at any future instant, starting from an initial value X_0, called the seed. This is all the mathematics that there is to it. Simple and yet it holds in it the whole complexity of chaos. But let us first examine how such a *logistic equation* may arise in a real physical situation.

Consider a savings bank account with the compound interest facility. So, you deposit an amount X_0 initially and a year later it becomes $X_1 = bX_0$. The growth factor 'b' is related to the rate of compound interest. Thus, for an interest rate of 10%, the control parameter $b = 1.1$. Two years after the initial deposit we will have $X_2 = bX_1 = b^2X_0$ and so on. In general we have $X_{n+1} = bX_n$. This linear rule, or algorithm, gives an unlimited growth, in fact an exponential growth. Of course, if 'b' were less than unity, our algorithm would lead to total extinction too and your deposit would dwindle ultimately to nought. (This could happen if X represented the *real* value of your money and the rate of inflation exceeded the rate of interest). The quantity X could equally well denote the value of your stocks or shares in a stock or share market. A much more revealing example for our purpose, however, will be that X_n represents the population of a community at the nth round of annual head count, or census. It may be the population of fish, or gypsy moths, or the Japanese beetles or even that of cancer cells. This is a matter of great interest to population biologists, and was studied extensively by Robert May at the Princeton University, using the logistic equation. But why this logistic equation? Let us see.

The rate of growth of population is obviously controlled by the natural birth and death rates. The growth factor 'b' depends on their difference. Thus, if the birth rate exceeds the death rate, then 'b' exceeds unity and we have the classical Malthusian scenario of unlimited exponential growth. If the inequality is reversed, then 'b' will be less than unity and we face total extinction. So, very aptly, Robert May called 'b' the 'boom and bustiness' parameter. The real life is, however, different. Communities live in a finite ecosystem competing for the common resource (food) which is limited. They often have the predator-prey relationship. (They may also be self-limiting because of moral constraints or ritualistic cannibalism). Now, if the population of a community grows too large, it faces death by starvation and the growth rate declines automatically. Thus there is a logistic limit to growth. (In such a case it is convenient to express the population 'X' as a fraction of its maximum possible value. Then X will lie between zero and unity). This slowing down of the growth rate is a kind of negative feedback. We can easily simulate it by replacing our growth factor 'b' by $b(1 - X)$. Clearly then the effective growth rate declines with increase in the population 'X'. The result is that our Malthusian growth scenario, represented by the linear rule $X_{n+1} = bX_n$ gets modified to the logistically limited growth scenario $X_{n+1} = bX_n(1 - X_n)$ which is non-linear. Most real systems are non-linear. This makes all the difference.

3.2 Iteration of Map

Having thus convinced ourselves of the reasonableness of the *logistic map* as a model for self-limiting growth, we have set the stage for action — on our programmable calculator. Select some value of the control parameter 'b'. Enter then a seed value X_0 as input on the right hand side of the logistic equation, evaluate the expression $bX_0(1 - X_0)$ and just output it as X_1. To get X_2, we have to use the just evaluated X_1, as the new input and out comes X_2. The procedure can be iterated n times to get X_n at the nth round, and so on for the whole sequence. Now repeat this numerical exercise with another value of the seed X_0, and look for any change in the pattern of the sequence of numbers that come out. (Note here that our X's by definition lie between zero and one. It is readily verified that this constrains the control parameter 'b' to lie between zero and four). Is the pattern periodic? Does the period change with 'b'? Or else, is there a pattern at all? These are the questions to be answered. We are zeroing on chaos.

All this can be viewed live on the screen of your PC with the help of a few lines of statements in BASIC. But, for a clear understanding of what is really going on, it is best to resort to the following graphical construction. Just plot X_{n+1} (vertically) against X_n (horizontally) using them as the Cartesian coordinates. This is our phase space if you like: A finite phase space, a unit square. On this plot our logistic function $bX_n(1 - X_n)$ against X_n is nothing but a parabola standing on the unit horizontal base (Fig. IV.5).

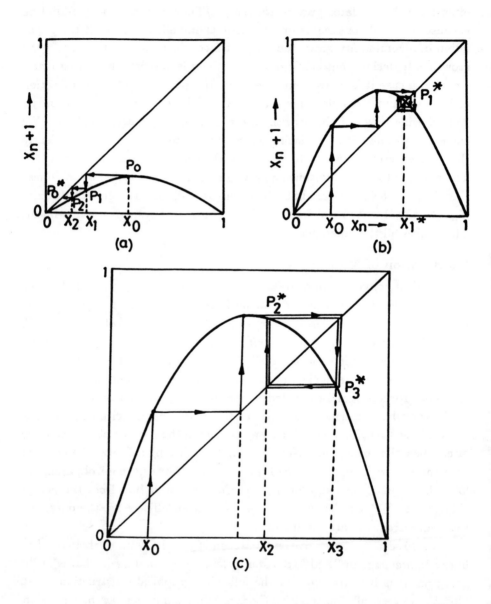

FIG. IV.5 *Graphical construction of logistic map: (a) stable fixed point* P_0^* *at origin; (b) stable fixed point* P_1^* *away from origin; (c) period-two attractor* P_2^*, P_3^*.

Now start with the input seed X_0 marked on the horizontal axis. To get X_1, just move vertically up (or down as the case may be) to the parabola at the point P_1 on it. It is clear that P_1 has the coordinates (X_0, X_1). In order to get X_2 now, we have to use X_1 as the new input on the horizontal axis and repeat the above procedure. It is much more convenient, however, to draw a 45° straight line, diagonally across the phase space on which $X_n = X_{n+1}$. With this, to get X_2 from X_1, all we have to do is to move horizontally from the point P_1 across upto the diagonal and then move vertically meeting the parabola at a point P_2, say. Clearly, P_2 will have the coordinates (X_1, X_2). The procedure can now be iterated to generate the whole sequence $X_0, X_1, \cdots X_n, X_{n+1}, \cdots$, ad infinitum (Fig. IV.5).

3.3 The Period Doubling Bifurcation

With the graphical construction in hand, we can at once make several observations by mere inspection. For sufficiently small values of the control parameter 'b', the parabola lies entirely below the diagonal line. It is readily seen that starting with any seed value X_0 whatsoever, we quickly cascade down and finally converge to the origin (Fig. IV.5a). This makes the origin an *attractor*. Let us denote this point by P_0^* and the corresponding X value by X_0^* (=0, of course). It is clear that on further iteration $X_0^* \to X_0^*$. Hence it is a Fixed Point, in fact a stable fixed point. That is, starting with any X_0, we end up there.

This situation persists as we increase our control parameter 'b' until the diagonal is just tangential to the parabola at the origin. This threshold value, b_1 say, of the control parameter can be obtained straightforwardly by equating the two slopes at the origin. We get $b_0 = 1$. For larger values of 'b', the parabola gets steeper and intersects the diagonal once again. This alters the situation completely. For now, starting with any seed X_0, other than zero, we converge not to P_0^* but to P_1^* (Fig. IV.5b). The corresponding stable fixed point value is readily found to be $X_1^* = 1 - 1/b$. The earlier fixed point X_0^* has become unstable now — it is a repeller rather than an attractor. It is readily appreciated that the question of *stability* under iteration is related to the slope of the parabola at the point of intersection with the diagonal. A slope steeper than $+1$ or -1 means 'unstable', otherwise stable. This slope-stability relation is a standard result in the so-called linear stability analysis. But graphically, it is obvious by construction.

As we hike up the control parameter further, things become really interesting when the control parameter exceeds a second threshold b_1 (=3). The fixed point X_1^* becomes unstable! Starting with a seed value X_0, the sequence $X_0, X_1, \cdots, X_n, X_{n+1}, \cdots$ fails to converge to any fixed point at all. Instead it settles down to a *2-point periodic* cycle $X_2^* \to X_3^* \to X_2^* \to X_3^* \to \cdots$. We call X_2^* and X_3^* elements of the cycle (Fig. IV.5c). We say that the earlier fixed point X_1^* has *bifurcated* to the alternate pair (X_2^*, X_3^*) that forms an attractor of period 2, or a 2-cycle. This is the famous *period doubling*. The alternating sequence $X_2^* \to X_3^* \to X_2^* \to X_3^* \to \cdots$ is reminiscent of the pair of alternating long and short intervals of the dripping faucet discussed in the last section. The general trend should be clear now. At the next threshold value b_2, say of the control parameter b, the 2-cycle becomes unstable and we have instead a *4-point periodic* cycle $X_3^* \to X_4^* \to X_5^* \to X_6^* \to X_3^* \to X_4^* \to X_5^* \to X_6^* \to \cdots$. So we have a 2^2-cycle now. It is simply that each element of the 2-cycle has bifurcated to two elements (See bifurcation plot in Fig. IV.6). And so on to 2^3-cycle, and in general to 2^k-cycle at the kth threshold b_k. It turns out that successive thresholds of instability come on faster and faster and converge to a point of accumulation b_∞ as $k \to \infty$, where we have a 2^∞-cycle: the period becomes infinite! The pattern never repeats itself. It has become aperiodic. This is the onset of *chaos*, and the entire sequence is known as the *Feigenbaum scenario* of period-doubling bifurcation route to chaos. The critical value of the control paramter for the onset of chaos turns out to be $b_\infty = 3.569$, the *Feigenbaum constant*.

3.4 Universality

The approach to criticality is subtle and interesting. The ratio of the successive intervals between the threshold values of the control parameter, the so-called bifurcation ratio $(b_k - b_{k-1})/(b_{k+1} - b_k)$ tends to a limiting number $\delta = 4.669\cdots$ as k tends to infinity. The entire sequence of events leading up to chaos seems to simulate the dripping patterns of the leaking faucet surprisingly well. Could this be a mere coincidence? Well, it could have been suspected to be so but for the fact that the behavior observed for the logistic equation actually turns out to be universal within a whole class. As was shown by Feigenbaum, we can replace the parabolic (quadratic) map by any other single-hump map, but with a smooth maximum, without changing the period doubling bifurcation sequence or the associated limiting ratio δ

and other critical scaling exponents characterising the onset of chaos. On the other hand, replacing the parabola with a 'tent' having a sharp triangular apex at the maximum point is a different matter. It will be a different class. But $X_{n+1} = r\sin(\pi X_n)$ will belong to the same class as the logistic map. This is something familiar from the modern theory of second-order phase transitions due to Kenneth G. Wilson. It is this universality of the critical behavior that makes an algorithm as simple looking as the logistic map capture the essential physics of diverse systems close to a crisis. After all it is the crisis that brings out the intrinsic character common to a whole class of *individuals*.

3.5 Fully Developed Chaos

What happens if we push our control parameter b beyond b_∞? The behavior continues to be generally chaotic except for windows of periodic oscillations for narrowly tuned values of the parameter. There is a phenomenon of *intermittency* where apparently periodic behavior appears in the midst of chaos. But these are structures too fine for us to discern here (Fig. IV.6).

It is, however, very revealing to consider the extreme case where b takes on its maximum possible value, *i.e.*, $b = 4$. For then, we can simplify matters tremendously by a simple change of variable. Let $X_n = \sin^2(2\pi\Theta_n)$, where Θ_n is an angle in radians. Remember that the trigonometric function $\sin(2\pi\Theta)$ is periodic in its argument with a period 1 ($\pi = 3.1415\cdots$ is the ratio of circumference to diameter of a circle). Our logistic equation written in terms of Θ's reads $\Theta_{n+1} = 2\Theta_n$ (mod 1). This simply means that to obtain Θ_{n+1} from Θ_n, multiply Θ_n by 2 and drop the integral part of it so that the remainder lies in the interval 0 and 1. This is the meaning of the notation 'mod 1'. The remainder is then Θ_{n+1}. Thus, for example, if $\Theta_n = 0.823$, then $\Theta_{n+1} = 2\Theta_n$ (mod 1) = 0.646, and $\Theta_{n+2} = 2\Theta_{n+1}$ (mod 1) = 0.292, and so on. Thus, starting from a seed value Θ_0, say, we can generate the entire sequence $\Theta_0, \Theta_1, \cdots, \Theta_n, \Theta_{n+1}, \cdots$. You see now that the scale factor 2 in the relation $\Theta_{n+1} = 2\Theta_n$ (mod 1) will 'stretch' the interval between any two neighboring seed values on iteration. The condition (mod 1), however, *folds* back the values as soon as they try to get out of the unit interval. This is the non-linear feedback. The above stretching and folding operations are precisely the ones we had discussed as baker's transformation in the previous Section. This is what gives the *sensitive dependence on initial conditions*. It

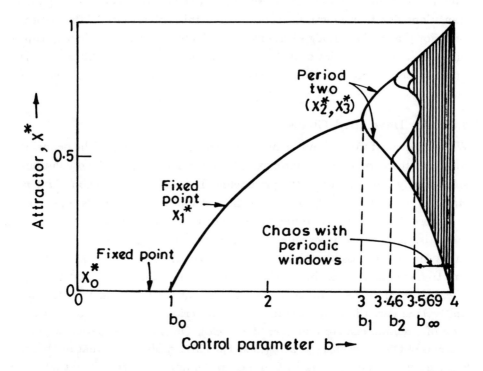

FIG. IV.6 *Bifurcation plot of attractors X^* against control parameter 'b' for period-doubling route to chaos — after Feigenbaum (schematic).*

can be shown that under this stretching and folding, the Θ_n values hop over the entire unit interval at random, eventually covering it densely. The entire unit interval is the *attractor* — *strange attractor* as we will see later. This is the fully developed *chaos!* Similar things happen in the range $b_\infty < b < 4$ also. However, here the Θ_n values hop randomly on a subset of the unit interval whose *dimension* is less than unity as we will see later: It is a *fractal*.

The *logistic map* is a kind of 'Feigenbaum Laboratory' in which we can do numerical experiments with chaos and learn a great deal. One wonders, however, how such a discrete map can approximate reality when all real processes happen in continuous time. We address this question next.

3.6 Poincaré Sections: from Continuous Flows to Discrete Maps

A discrete map of the kind we have discussed above can be constructed for a real dynamical system quite naturally by taking the so-called *Poincaré section* of the continuous flow, or the trajectories, in the phase space of the system.

In Fig. IV.7 we have shown such a Poincaré section for a dynamical system having a three-dimensional phase space. The *surface of section* has been taken to be a plane perpendicular to the z-axis. But other choices are possible. Here, instead of watching the trajectory continuously in time, we simply record the sequence of the points $P_0, P_1, \cdots$ of intersection of the trajectory with the chosen plane as the trajectory crosses it in a given direction — of the positive z-axis, say. In effect we have replaced the continuous phase flow in time described by a *differential* equation, by a discrete mapping relating the (x, y)-coordinates of the successive points P_n, P_{n+1} of intersection — an algebraic, *difference* equation. We may have, for example $X_{n+1} = 1 - cX_n^2 + Y_n$ and $Y_{n+1} = bX_n$, where 'c' and 'b' are adjustable parameters, *e.g.*, we may have $c = 1.4$ and $b = 0.3$. This two-dimensional discrete map, called the *Hénon map* after the French astronomer Michel Hénon, describes the Poincaré section of a *chaotic attractor* in a three dimensional phase space, of which more later. The Poincaré section of a three-dimensional flow will be in general a two-dimensional surface. But if the dissipation is large and the areas contract rapidly, then the section will consist of points distributed along a curve. In other words, we have now a one-dimensional map $X_t \to X_{t+1}$. This is often called the *first return map*.

Taking Poincaré sections is a very revealing technique for studying, or

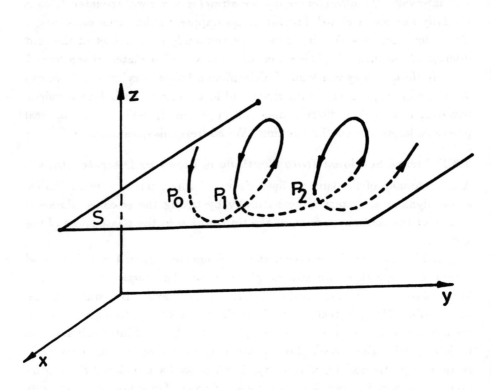

FIG. IV.7 *Poincaré section* $(P_0, P_1, \cdots)$ *of a phase space trajectory by the plane S.*

rather visualizing the geometrical forms, such as orbits in the phase space (the *phase portraits*). It certainly reduces the dimension by one. And yet it retains the all important qualitative, global features of the dynamics, *e.g.*, the divergence or convergence of neighboring trajectories, expansion or contraction of phase-space volume elements, the periodicity or aperiodicity of orbits and many other signatures of dynamics. Also, the algebraic difference equations are easier to handle than the original differential equations. One can, for example easily implement the Hénon map on a programmable calculator. Poincaré section is particularly useful for the study of attractors in higher dimensions. We will look at these attractors next.

---

4. STRANGE ATTRACTORS AND ROUTES TO CHAOS

Attractors are geometric forms in the phase space to which the phase trajectories of the dynamical system converge, or are attracted and on which they eventually settle down, quite independently of the initial conditions. The idea of an attractor is quite simple but very powerful for a global qualitative understanding of the motion, both regular and irregular without having to solve the equations of motion which is seldom possible. It is best illustrated through examples.

4.1 Stable Fixed Point

The simplest attractor is a *fixed point*, or rather a *stable fixed point*. Consider the phase portrait of a damped linear oscillator, a pendulum with friction for example. The phase space is two-dimensional comprising the velocity (momentum) and the position coordinates. Because of damping the phase point spirals in onto the origin and rests there (Fig. IV.8a).

This spiral fall to the origin corresponds to oscillations dying down to a dead beat due to dissipation. The origin is thus a fixed point. It is obviously stable — displaced from it, the system returns to it eventually. Also, all trajectories, no matter where they began, are attracted towards it — hence the name attractor. It is also readily seen, that any element of phase space will contract in its extension as it is attracted towards the fixed point. As there is contraction along both the independent directions in the 2-dimensional phase space, both the Lyapunov exponents are negative. This 'contraction'

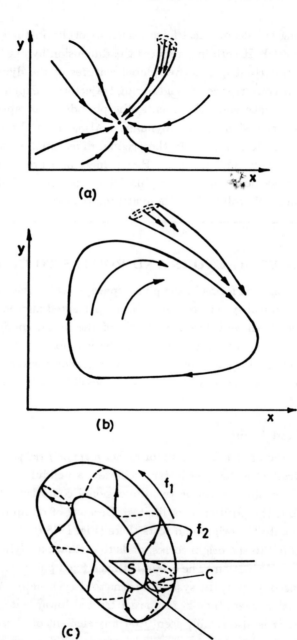

FIG. IV.8 *(a) Stable fixed point; (b) limit cycle; (c) biperiodic torus.*

is the meaning of the term 'dissipative flow' in the phase-space language. Most interesting flows in Nature are dissipative and have to be maintained by external driving forces, such as stirring, heating, pumping, kicking, etc. We call them open systems.

4.2 Limit Cycle

Next comes the *limit cycle*. It is a closed loop in the phase space to which the trajectories converge eventually (Fig. IV.8b). The limit cycle corresponds to a stable oscillation. Here one Lyapunov exponent is zero (along the loop) and the rest are negative. Its Poincaré section is just a point. Again the flow is dissipative. For a two-dimensional phase space, the limit cycle is the only attractor possible, aside from the fixed point. This is a direct consequence of the condition that the phase trajectories cannot have self-intersection. To have anything more complicated, the trajectory must escape in a third dimension. (Recall that from a given point a unique trajectory must pass). The limit cycle is at the heart of the most simply periodic processes in Nature — the beating of the heart, the 'circadian rhythms' of period 23 to 25 hours in humans and animals, the cyclic fluctuation of populations of competing species in an ecosystem, oscillating chemical reactions like the Beluzov-Zhabotinsky reaction, marked by colour changes every minute or so, etc. The limit cycle thus is a natural clock.

4.3 The Biperiodic Torus

The next most complicated attractor has the geometric form of a doughnut or anchor ring — the *torus* (Fig. IV.8c). The trajectories converge on the surface of the torus, winding in small circuits around the axis of the torus (at a frequency f_1) while orbiting in large circles along the axis (at a frequency f_2). Here two of the Lyapunov exponents are zero and the rest are negative. This corresponds physically to a compound or biperiodic oscillation resulting from the superposition of two independent motions. If the frequencies f_1 and f_2 are commensurate, that is if the ratio f_1/f_2 can be expressed as a ratio of two integers, then the trajectory on the torus closes on itself — frequency locking. The motion is actually periodic then. A Poincaré section of the trajectory will be a finite set of points traversed by the successive go-rounds. If, on the other hand, the frequencies f_1 and f_2 are incommensurate, the trajectory will cover the torus densely and the Poincaré section will be a continuous closed curve. The motion will be quasi-periodic. Such a bi-periodic torus attractor is known to show up in the Couette flow. But the simplest example will be two coupled oscillators of different frequencies. Incidentally, if the frequencies are close enough and the coupling is non-linear, the oscillators may get 'locked' into a common frequency mode. This is called 'frequency entrainment', noted first by the great 17th century Dutch physicist Christiaan Huygens. He was surprised that two proximate church lamps should oscillate at exactly equal frequencies.

4.4 The Strange Attractor

The zero-dimensional *fixed point*, the one-dimensional *limit cycle*, and the 2-dimensional *torus* are all examples of low dimensional attractors that characterize dissipative flows which are regular, that is stable and predictable to any degree of accuracy. These flows are essentially periodic. There is, however, an entirely different type of low-dimensional attractor that characterizes flows which are irregular, that is unstable and unpredictable in the long term. Such an attractor is a phase-space non-filling set of points or orbits to which all trajectories from the outside converge, but on which neighboring trajectories diverge. Thus at least one Lyapunov exponent has to be positive. It is *'strange'* in respect of the geometry of its form as well as in terms of its manner of traversal of this finite region of phase space. It is called *strange attractor* and is the engine that drives *chaos*. It was discovered jointly by the Belgian mathematician David Ruelle and the Dutch mathematician Floris Takens around 1971.

Strange attractor is the answer to the question we had posed earlier — how can a trajectory remain confined forever to a finite region of phase space without intersecting itself, as it must not, and without closing on itself, as again it must not, for that would mean a periodic motion. To make matters worse, the system happens to be dissipative — eventually the phase-space volume must contract to zero. You see the conflict! Just think of a one-meter long strand of the contortionist DNA packed within a cell measuring one millionth of a meter across, and you will get some idea of the actual problem which is infinitely worse. The conflict of demands for zero volume and eternal self-avoidance is resolved by making the attractor into a *fractal* — a geometric form with fractional dimension lower than the phase space in which it is embedded. Fractal is an ingenious way of having surface without volume if you like. We will now illustrate all this with two celebrated examples of strange attractor: The *Hénon attractor* of the discrete two-dimensional *Hénon map* for its simplicity, and the *Lorenz attractor* of the three dimensional convective flow for its complexity. The latter is also historically the first known example of strange attractor.

4.5 The Hénon Attractor

As already mentioned in Subsec. 3.6, the Hénon map, discovered by Hénon in 1976, is given by $X_{n+1} = 1 - cX_n^2 + Y_n$ and $Y_{n+1} = bX_n$. Thus, starting with any point in the (X, Y)-plane, one can generate the entire attractor by iterating the map a large number (usually 10^5–10^6) of times. It can be shown that the map is dissipative for the magnitude of b less than unity, *i.e.*, for $|b| < 1$. (It is conservative (phase volume preserving) for $|b| = 1$). For $b = 0$, the dissipation is so large that the map contracts to a one-dimensional quadratic map of the kind we have discussed earlier in Section 3. For a convenient value of $b = 0.3$ (moderate dissipation), we can now study the map as function of the control parameter 'c'. The map follows a period doubling route to chaos. For $c = 1.4$ we have a strange attractor, shown schematically in Fig. IV.9.

Two crucial features are to be noted here. One has to do with the geometry and the other with the motion. First, the attractor has a self-similar microstructure: on magnification an element of it reveals details similar to the whole. This self-similar geometry leads to a fractional dimensionality for the attractor. It is a fractal. The second crucial feature is the manner in which the fractal is traversed by the phase point. The pattern of points in Fig. IV.9 tends to guide the eye along certain lines. This is, however, merely the 'closure tendency' of the human eye. Actually, the successive points on these lines are *not* traversed successively by the phase point. In fact the phase point hops almost randomly over the entire attractor. This is, of course, best seen on your screen as the points appear iteratively, one by one, at totally unexpected places, but eventually a pattern of points is generated that looks like that in Fig. IV.9.

4.6 The Lorenz Attractor

We have already discussed convection in a mass of fluid heated from below and the chaos that results when the temperature difference exceeds a criti-

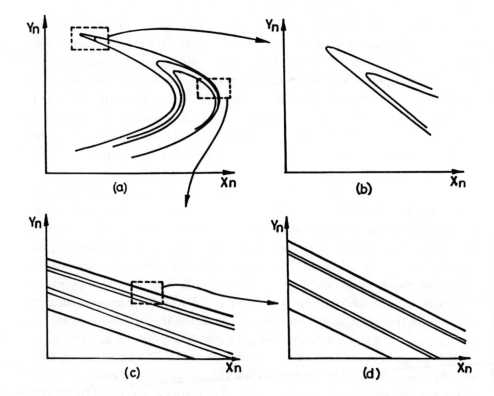

FIG. IV.9 *The Hénon attractor and its fractal structure: (a), (b), and (c) reveal self-similarity under magnification.*

cal value. Of course, such a system has infinitely many degrees of freedom but, all except three get damped out and the chaotic regime is just a three-dimensional phase space (X, Y, Z). Roughly speaking X measures the rate of convective overturning, Y the horizontal temperature variation and Z the vertical temperature variation. Just to give a flavor of things, we write down the equations obeyed by these variables:

$$\dot{X} = \sigma(Y - X)$$
$$\dot{Y} = \gamma X - Y - XZ$$
$$\dot{Z} = XY - bZ.$$

These are the equations that Lorenz studied in 1963 as his model for weather. Overhead dot denotes rate of change with time. Here, σ (sigma), γ (gamma) and b are parameters defining the conditions of the experimeter. Thus, b is a geometry factor (depends on certain aspect ratio). Sigma (called the *Prandtl number*) measures the ratio of viscosity to thermal conductivity-times-density, and γ is the control parameter — the Rayleigh number normalized in a certain way. In our experiment, as you increase the temperature difference from zero, you increase γ from zero. For γ between zero and one, we have the steady diffusion of heat. For γ greater than 1 but less than about 24.74, steady convective rolls develop. Eventually, for γ greater than about 24.74 we have motion on a strange attractor giving chaos (the Lorenz attractor).

A strange attractor in three dimensions is hard to visualize. One could either resort to making Poincaré sections, or just look at the projection on a plane (see Fig. IV.10).

One can readily see that the trajectory winds around an unstable fixed point C, spirals outward and after reaching out to a critical distance from C, crosses over to wind around the other unstable fixed point C'. The number of circuits the trajectory makes around any one of the unstable fixed points before it switches to the other, generates a sequence which is as random as the tossing of the coin.

It turns out that the trajectories are actually confined to a sheet of small but finite thickness, within which is embedded the entire complex geometry of the attractor: It is a fractal.

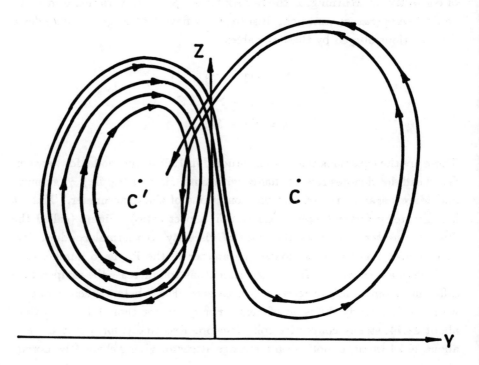

FIG. IV.10 *The Lorenz (strange) attractor projected on YZ-plane. C and C' are two unstable fixed points. The strange attractor depicts chaos (schematic).*

The Lorenz system turns out to describe other systems as well, for example the irregular spiking of laser outputs. The irregular reversal of Earth's magnetic field may be linked to the random jumping of the trajectory between C and C'.

4.7 Routes to Chaos

We have seen that chaos is described by strange attractors. In most of the physically interesting cases these attractors are the low-dimensional ones. We will now summarize two commonly observed routes to chaos described by such low-dimensional strange attractors.

The first route is that of an infinite sequence of period-doubling bifurcations. We begin with a stable fixed point (stationary point) that becomes unstable and bifurcates to a limit cycle (periodic orbit) which undergoes a period-doubling sequence leading to a pile-up, that is an accumulation point (aperiodic orbit) — onset of chaos. This is the Feigenbaum scenario that we have already come across in the case of the leaky faucet. It is also the route followed by the convective turbulence (Lorenz attractor). This route has already been shown schematically as the bifurcation plot for the logistic map in Section 3. In Fig. IV.11a we have exhibited this route in terms of the projection of the phase trajectory in a plane.

There is another route to chaos that is qualitatively different. Here we start with a stable fixed point (stationary point) that becomes unstable and bifurcates to a limit cycle (a periodic orbit). The limit cycle in turn becomes unstable and bifurcates to a biperiodic torus (doubly periodic orbit) with incommensurate frequencies, i.e. quasi-periodic. Finally, from the biperiodic torus we jump to chaos — without any intervening sequence of appearances of new frequencies. This is the sequence leading to turbulence observed in the Couette flow discussed in Section 1. Here the fluid is confined to the annular space between two coaxial cylinders one of which is made to spin. In Fig. IV.11b we have shown schematically this route to chaos.

It is important to appreciate a qualitative difference between these two routes to chaos. In the first route (period-doubling route), it is simply the period of the limit cycle that changes at bifurcation. No new states are being generated. In the second route, the bifurcation from the limit cycle to the torus is a topological change involving new states.

There are other routes to chaos possible. But there is no experimental

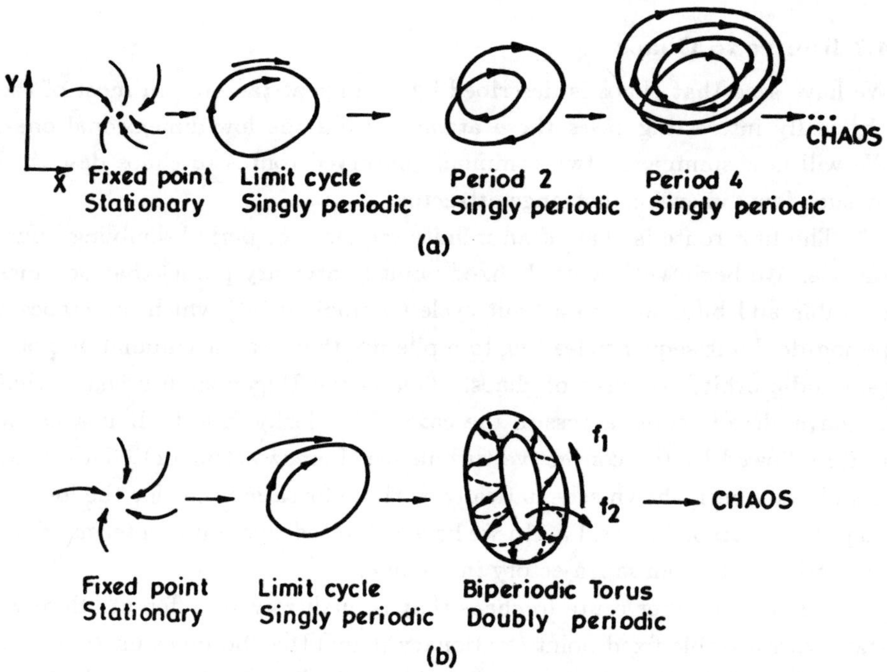

FIG. IV.11 *Routes to chaos involving (a) infinite number of period-doubling bifurcations; (b) finite number of topological transitions.*

evidence for the Landau scenario involving the appearance of incommensurate frequencies, one at a time, leading eventually to a confusion that we call chaos.

---

5. FRACTALS AND STRANGE ATTRACTORS

Fractals are geometrical objects with shapes that are irregular on all length scales no matter how small: they are self-similar. To understand them, let us re-examine some familiar geometrical notions.

We know from experience that our familiar space is three dimensional. That is to say that we need to specify three numbers to locate a point in it. These could, for example, be the latitude, the longitude and the altitude of the point giving respectively, its north-south, east-west, and the up-down positions. We could, of course, employ a Cartesian coordinate system and locate the point by the triple (x, y, z). Any other system of coordinates will equally suffice. The essential point is that we need *three* numbers — more will be redundant and less won't do. Similarly, a surface is two-dimensional, a line one-dimensional and a point has dimension zero.

These less-dimensional spaces, like the two-dimensional surface, the one-dimensional line, or the zero-dimensional point, may be viewed conveniently as embedded in our higher three-dimensional space. All these ideas are well known from the time of Euclid and underlie our geometrical reckonings. So intimate in this idea of dimension that we tend to forget that it has an empirical basis. Thus, it seems obvious that the dimension has to be a whole number. Indeed, the way we have defined the dimension above it will be so. (We call it the *topological dimension*). There is, however, another aspect of dimensionality that measures the spatial content or the capacity. We can ask, for instance, how much space does a given geometrical object occupy. Thus we speak of the volume of a three-dimensional object. It scales as the cube of its linear size — if the object is stretched out in all directions by a factor of 2, say, the volume will become $2^3 = 8$ times. Hence it is three dimensional. Similarly, we speak of the area of a surface that scales as the square of its linear size, and the length of a line that scales as the first power of the linear size. As for the point, its content, whatever it may be, stays constant — it scales as the zeroth power of the linear size, we may say. In all these cases, the *'capacity dimension'* is again an integer and agrees with the topological dimension.

But this need not always be the case. To see this let us consider the measurement operation in some detail. Consider a segment of a straight line first. To measure its length we can use a divider as any geometer or surveyor would. The divider can be opened out so that the two pointed ends of it are some convenient distance apart. This becomes our scale of length. Let us denote it by ϵ. Now all that we have to do is to walk along the line with the divider marking off equal divisions of size ϵ. Let $N(\epsilon)$ be the number of steps (divisions) required to cover the entire length of the line. (There will be some rounding off error at the end point which can be neglected if ϵ is sufficiently small so that $N(\epsilon) \gg 1$). Then the length of the segment is $L = \epsilon N(\epsilon)$. Let us now repeat the entire operation but with ϵ replaced by $\epsilon/2$, that is with the distance between the divider end points halved. The number of steps will now be denoted by $N(\epsilon/2)$. Now we must have $L = \epsilon N(\epsilon) = (\epsilon/2)N(\epsilon/2)$. This is because we have covered exactly the same set of points along the straight line skipping nothing and adding nothing. Thus, halving the measuring step length simply doubles the number of steps required. This can be restated generally as the proportionality or scaling relation $N(\epsilon) \sim \epsilon^{-1}$, that is the number of steps $N(\epsilon)$ required is inversely proportional to the size of the linear step length for small enough ϵ. This seems obvious enough. The '1' occuring in the exponent gives the *capacity dimension* (D) of the straight line. In a more sophisticated language, $D = -(d\ln N(\epsilon)/d\ln \epsilon)$ as $\epsilon \to 0$. In our example this indeed gives $D = 1$. Here ln stands for the natural logarithm. Thus, finding the capacity dimension is reduced to mere counting. It is indeed sometimes referred to as the box counting dimension.

So far so good. But what if we have a curved line. For in this case as we walk along the line with our divider, we will be measuring the chord length of the polygon rather than the true arc length of the curved line as can readily be seen from Fig. IV.12a.

Are we not cutting corners here? Well, no. The point is this. If the curve is a *smooth* one, it will be straight locally. Thus, as we make our step length ϵ smaller and smaller, the chords will tend towards the arcs, and the polygon will approximate the curve better and better, until it eventually hugs the curve as $\epsilon \to 0$. (This is the significance of the proviso $\epsilon \to 0$ in our

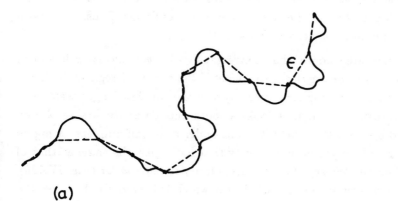

(a)

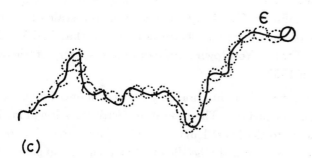

FIG. IV.12 *(a) Covering a line with scale length ϵ; (b) self-similarity of a fractal curve; (c) covering a line with ϵ discs or balls.*

logarithmic formula). In the end, therefore, there is no distinction between the case of the straight line and that of a curved but *smooth* line: in both the cases, we will get $D = 1$. The smooth curve is rectifiable! Thus, all smooth curves have the same capacity dimension $D = 1$.

Curves that are smooth, that is straight locally (*viz.*, have a well defined tangent) are said to be regular. Now, what if the curve is irregular — that is to say that it has a zig-zag structure down to the smallest length scale. You take a small segment of it that looks smooth enough at the low resolution of the naked eye and then view it again under a magnifying lens of higher resolution. It will reveal an irregularity that looks just the same as that of the original curve. We say that the irregularity is *self-similar* (Fig. IV.12*b*). It has no characteristic length scale below which it is smooth. It is like the infinite 'matryoshka' with a doll within a doll within a doll, ad infinitum, all exactly alike but for the size. How do you find the dimension and the length of such an irregular curve? Intuition is no guide now. But we can get back to our divider and walk the curve as before. The polygon traversed by us will now never quite hug the irregular curve. But we still have our ϵ and $N(\epsilon)$, and we can hopefully find the scaling $N(\epsilon) \sim \epsilon^{-D}$ as $\epsilon \to 0$. The exponent D is then the capacity dimension of the curve by our definition. However, D is now not necessarily an integer. It will in general be a fraction! The zig-zag curve is then called a *fractal* and D its *fractal dimension*. For example, we may survey the coastline of England and expect to find that its fractal dimension is about 1.2. What about the length of this fractal coast line? Well, we still have $L = \epsilon N(\epsilon)$. Thus, as $\epsilon \to 0$, the length of the coastline tends to infinity! The length depends on the scale of resolution. It was the study of such irregular shapes and forms that had led Benoit Mandelbrot of the IBM at Yorktown Heights to the discovery of fractals and fractal geometry in 1960.

A fractal line is something intermediate between a one-dimensional line and a two-dimensional surface. The fractal line with its self-similar irregularity structure down to the finest length scale has an amusing but revealing aspect to it. If we were to run the needle of a gramophone on such a fractal track, it will produce a symphony that is independent of the speed of the needle!

Same ideas can now be extended to an irregular (fractal) surface. In this case, of course, we will employ elementary squares or plaquettes measuring ϵ on the side, to cover the surface. If $N(\epsilon)$ is the number of plquettes required to cover it, then we expect $N(\epsilon) \sim \epsilon^{-D}$, where D is the fractal dimension. In all these examples, our geometrical objects, the lines and the surfaces, are assumed to be embedded in the background Euclidean space of dimension 3, which is our familiar space. But we can always imagine our objects to be embedded in a background Euclidean space of a higher dimension denoted by d. We will call this the *embedding dimension*. It is important to realize that the fractal dimension of the geometrical object is its intrinsic property. It does not depend on the space in which it may happen to be embedded. Of course, the embedding space must have a Euclidean dimension greater than the fractal dimension of the object embedded in it. Thus for example, a zig-zag curve embedded in our 3 dimensional Euclideon space may be 'covered' with 3-dimensional ϵ-balls or ϵ-boxes instead of 1-dimensional ϵ-segments (Fig. IV.12c). The fractal dimension will work out to be the same.

And now we are ready for an important generalization. We will ask for the fractal dimension of a *set of points* embedded in a Euclidean space of dimension d. We need this generalization for our study of chaos. Because here the backgound Euclidean space is the state space that can have any integral dimension. In this state space is embedded the set of points, or more like a dust of points, that forms the attractor whose fractal dimension (D) we are after. The dimension d of the state space is the embedding dimension. Our method remains the same. We cover the set of points with d-dimensional balls of radius ϵ and then fit the scaling relation $N(\epsilon) \sim \epsilon^{-D}$ or the equivalent logarithmic formula discussed earlier. All this can be done rather quickly on a PC. Before we go on to the fractal attractors, let us get acquainted with some of the fascinating fractals that have become classic now. We will actually construct them.

5.1 The Koch Snowflake

Take an equilateral triangle. Now divide each side in three equal segments and erect an equilateral triangle on the middle third. We get a 12 sided Star of David. The construction is to be iterated ad infinitum. Smaller and smaller triangles keep sprouting on the sides and finally we have a zig-zag contour that is irregular on all length scales in a self-similar fashion, resembling a snowflake (Fig. IV.13).

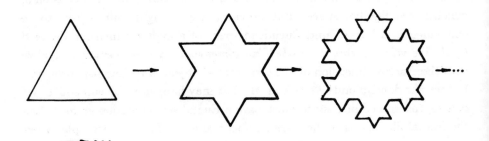

FIG. IV.13 *Koch snowflake. Construction algorithm for this fractal of dimension 1.26.*

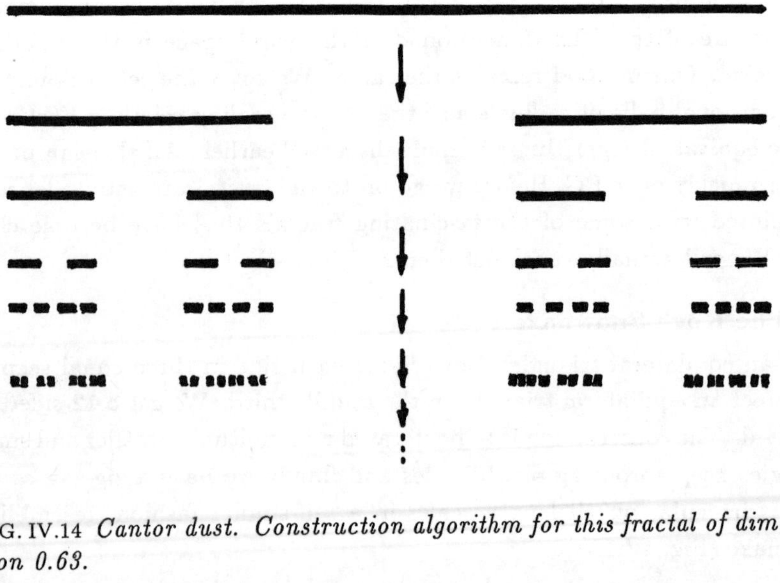

FIG. IV.14 *Cantor dust. Construction algorithm for this fractal of dimension 0.63.*

To find its fractal dimension, all we have to note is that at each iteration the scale of length gets divided by 3 and the number of steps required to cover the perimeter increase by a factor of 4. This gives the fractal dimension $D = -\ln 4/\ln(1/3) = 1.26$. It is interesting to note that while the actual length of the perimeter scales as $(4/3)^n$ to infinity as expected, the area enclosed by the figure remains finite, close to that of the original triangle.

5.2 The Cantor Dust

Take a segment of a straight line of unit length, say. Trisect it in three equal segments and omit the middle third, retaining the two outer segments (along with their end points). Now iterate this construction by trisecting these two segments and again omitting the middle thirds, and so on ad infinitum. Eventually, you would be left with a sparse dust, called the *Cantor set*, of uncountably infinite points (Fig. IV.14). What is its capacity dimension? Well, in each iteration, the length of the segments gets divided by 3 while the number of such segments is multiplied by 2. Hence the fractal dimension $D = -\ln 2/\ln(1/3) = 0.630$. Of course the total length of the set at the nth iteration is $(2/3)^n$ which tends to zero as $n \to \infty$.

Aside from these and many other fascinating examples of fractal sets which are just mathematical constructs, we do have fractal objects occurring in nature. We have already mentioned coastlines naturally. Other examples would be clouds, mountain sides, foams and sponges, lightning discharges, galactic structures, aggregates of different sorts, human and animal lungs, etc. Indeed, a fractal geometry would be Nature's strategy for enhancing the surface to volume ratio. This has relevance to the problem of 'missing mass' in astrophysics. The mass distribution may have fractal structure.

The fractal dimension D is a purely geometrical measure of the capacity of a set. Thus in 'box counting', all the boxes, or rather the neighborhoods, are weighted equally. In the context of chaos, however, fractal is a set of points in the state space on which the 'strange attractor' lives. The different neighborhoods (boxes) of the fractal are visited by the phase point with different frequency in the course of its 'strange' aperiodic motion: the fractal set is physically inhomogeneous. The fractal dimension D as defined above fails to capture this physical fact. It is possible to define yet another fractal dimension that weights different neighbourhoods according as how frequently they are visited. In fact a whole spectrum of fractal dimensions are defined in what is called *multifractal analysis* of the fractal set. This is a highly technical matter and we will not pursue it further here. Suffice to say that the multifractal analysis combines geometry with physics and thus reveals more information.

---

6. RECONSTRUCTION OF THE STRANGE ATTRACTOR FROM A MEASURED SIGNAL: THE INVERSE PROBLEM

A dynamical system may have a number of degrees of freedom. This whole number is the dimension 'd' of its phase space. The dynamics will eventually settle down on an attractor. If the dynamics is chaotic, it will be a strange attractor (a fractal) of fractional dimension D, embedded in the phase space of (embedding) dimension $d > D$. Both d and D are important specifications of the nature of the system and its strange attractor. Now, in general these two dimensions are not given to us. All we have is an experimentally observed, seemingly random signal representing the measurement of a single component of the d-dimensional state vector in continuous time. It could, for instance, be the temperature in a convection cell, or a component of fluid velocity of a turbulent flow at a chosen point. This can be measured, for example, by the laser Doppler interferometry, in which the frequency of a laser light scattered by the moving liquid at a point is Doppler shifted. The shift is proportional to the instantaneous velocity of the fluid at that point. The question is, can we reconstruct these important attributes such as D and d from such a single-component measurement in time? It was proved by Floris Takens that we can indeed retrieve some such vital information about the system from this stream of single-component data.

The philosophy underlying this claim is quite simple. In a dynamical system with interacting degrees of freedom, where everything affects everything all the time, it is conceivable, indeed highly plausible, that a measurement on any one of them should bear the marks of all others, however implicitly. The real question now is how to recover this information. This is a rather technical matter. The technique is, however, very straightforward to apply and is worth knowing about.

Let us take $x(t)$ to be the measured signal. The first step is to discretize time. For this, pick a convenient time interval τ and form the sequence $x(t)$,

$x(t + \tau)$, $x(t + 2\tau)$, $\cdots$. This is known as a *time series*. Next we are going to construct vectors out this series. First, construct the two-dimensional vectors $(x(t), x(t + \tau))$, $(x(t + \tau), x(t + 2\tau))$, $(x(t + 2\tau), x(t + 3\tau))$, $\cdots$. Each vector has two Cartesian components, for example $x(t)$ and $x(t + \tau)$, as any two-dimensional vector should. Plot these vectors in a two-dimensional phase space, so as to generate a dust of points. Now, count how many neighbours a point has, on the average, lying within an ϵ radius of it. Let the number be $C(\epsilon)$. For small ϵ, we will find that $C(\epsilon)$ scales as $C(\epsilon) \sim \epsilon^{\nu}$. Then ν is said to be the *correlation dimension* of our set. We will use a subscript and call it ν_2 because we had constructed it from a 2-dimensional phase space. Now repeat this procedure, but this time with the 3-dimensional vectors constructed out of our time-series, *i.e.*, $(x(t), x(t+\tau), x(t+2\tau))$, $(x(t+\tau), x(t+2\tau), x(t+3\tau))$, $\cdots$. Each vector has three Cartesian components. We will, therefore, get the correlation-dimension ν_3 this time. And so on. Thus we generate the numbers ν_1, ν_2, ν_3, $\cdots$. We will find that ν_n will saturate at a limiting value, a fraction generally, as n becomes sufficiently large. That limiting value is then the Correlation Dimension of the Attractor of our system. It is usually denoted by D_2 and is less than or equal to the fractal dimension D of the attractor. Also, the value of the integer n, beyond which ν_n levels off, is the minimum embedding dimension of our attractor. That is, it is the minimum number of degrees of freedom our dynamical system must have to contain this attractor.

This technique, due to P. Grassberger and I. Procaccia, seems to work well with low dimensional attractors. It has been used extensively to characterise chaos in all sorts of systems including human cortex. Thus the EEG *time series* for various brain-wave rhythms, *e.g.*, with eyes closed, with eyes open, with and without epileptic seizure have been analysed with startling results. The epileptic brain is less chaotic (*i.e.*, it has lower fractal dimension) than the normally active brain!

The time-series analysis and reconstruction of attractor gives other important information as well. It can distinguish deterministic chaos from the truly random noise. Thus, for example, the correlation dimension D_2 is infinite for a random noise, but finite and greater than 2 for chaos. (For a limit cycle it is 1 and for a torus (quasiperiodic) it is 2). Thus, we know, for example, that the noisy intensity output of a laser can be due to a low dimensional strange attractor in its dynamics.

---

7. CONCLUDING REMARKS: HARVESTING CHAOS

That a simple looking dynamical system having a small number of degrees of freedom and obeying deterministic laws can show chaos is a surprise of classical physics. It has made chaos a paradigm of complexity — one that comes not from the statistical law of large numbers, but from the highly fine-structured self-similar organization in time and space. There is no randomness put in by hand. Chaos is due to the *sensitive dependence* on *initial conditions* that comes from the stretching and folding of orbits. Mathematically, this is represented by the all important non-linear feedback. Thus, chaos can amplify noise, but it can also masquerade as 'noise' all by itself. It is robust and occurs widely. It has classes of universality.

Theory of chaos rests on a highly developed mathematical infrastructure that goes by the forbidding name of *non-linear dynamical systems*. But high-speed computing has brought it within the capability of a high school student. The contemporaneous ideas of fractal geometry have provided the right language and concepts to describe chaos.

There are still many open questions. For instance, we have discussed chaos for open dissipative systems only. There are closed dynamical systems that are conservative (non-dissipative). A celebrated example is that of a three-body problem in astronomy, the problem of three planets gravitationally attracting each other. These are non-integrable systems. Then there is the problem of two non-linearly coupled oscillators for example. These can be treated quantum mechanically. Does chaos survive quantization? What is the signature of quantum chaos? There are other rather technical questions. But it can be safely claimed, that for many dissipative systems, the theoretical framework of chaos is complete. It is now time to harvest chaos.

Undoubtedly, the most outstanding application of our knowledge of chaos is to understand fluid dynamical turbulence — the last unsolved problem of classical physics. For several types of closed flow, routes to turbulence are now understood. The problem of turbulence in open flows such as wakes and jets is, however, still unsettled. Do we have here after all the Landau scenario of confusion of incommensurate frequencies, appearing one at a time?

Next comes the problem of short and intermediate range weather forecast that started it all from the time of early work of E.N. Lorenz. Chaos, of course, rules out long-range forecast. But can we extend our present 3-4 day forecasts to a 10-12 day forecast? One must estimate the time over which memory of the initial data is lost due to folding back of the phase-space trajectories. This ultimately limits the range of prediction.

Reconstruction of the strange-attractor, if any, is at the heart of an apparently random behavior is a powerful diagnostic tool that is finding application in diverse fields of study including neurology, psychology, epidemiology, ecology, sociology, macro-economics and what have you.

There is yet another way in which we can make chaos really work for us. If we want to explore a large portion of phase space efficiently, it may be best to use a chaotic trajectory. Nature seems to have made use of this chaotic access to, or exploration of possibilities in the vital matters of evolution and in the working of our immune system. Chaos in our neural network is also an important subject for study.

There are other highly speculative aspects of chaos. Chaos is physically intermediate between the 'thermal bath' and the macroscopically ordered motion. What happens when a chaotic system is coupled to a 'thermal bath'? Can we make 'intelligent heat engines'? What is the thermodynamic cost of driving a system chaotic? Or an even more speculative question such as whether chaos is ultimately responsible for the well known quantum uncertainties. We do not know.

It has been claimed that the scientific revolution, or rather the shift of paradigm caused by 'chaos' is comparable to that caused by quantum mechanics. To us this comparison does not seem very apt. In any case, judging by their respective impacts, the claim is overstated. But what can definitely be claimed is that chaos has been a surprise of classical physics. It has the subtlety of number theory. You can certainly have fun with chaos if you have a Personal Computer, or even a hand-held calculator.

Summary

A dynamical system is said to be chaotic if its long time behavior cannot not be computed and, therefore, cannot be predicted exactly even though the laws governing its motion are deterministic. Hence the name deterministic chaos. A common example is that of turbulence which develops in a fluid flowing through a pipe as the flow velocity exceeds a certain threshold value. The remarkable thing is that even a simple system, such as two coupled pendulums, having a small number of degrees of freedom can show chaos under appropriate conditions. The chaotic behavior results from the system's sensitive dependence on the initial condiditions. The latter amplifies uncontrollably even the smallest uncertainty, or error unavoidably present in the initial data as noise. This sensitivity derives essentially from the non-linear feedback, and often dissipation inherent in the dynamical equations. The dynamical evolution may be described by continuous differential equations or by discrete algebraic algorithms, and can be depicted graphically as the trajectory of a point in a phase space of appropriate dimensions. The representative point may converge to a stable fixed point, or a limit cycle, or a torus, and so on. These limiting sets of points in the phase space are called *attractors* and signify respectively a steady state, a singly periodic motion, or a doubly periodic motion, and so on. The chaos corresponds to a *strange attractor* which is a region of phase space having a fractional dimension (a fractal) and its points are visited aperiodically.

There are several routes to chaos. We have, for example, the period doubling route of Feigenbaum in which the dynamical system doubles its period at successive threshold values (bifurcation points) of the control parameter, until at a critical point the period becomes infinite and the motion turns chaotic. These different routes define different universality classes of common behavior.

Chaos has become a paradigm for complex behavior. Many apparently random systems are actually chaotic. Powerful techniques have been developed to reconstruct the strange attractor from the monitoring of a single fluctuating variable. This helps us make limited predictions and diagnostics for diverse problems, e.g., the weather forecast and turbulence, epileptic seizures, populations of competing species, macroeconomic fluctuations, recurrence of epidemics, etc.

CHAPTER V

BRIGHT STARS AND BLACK HOLES

CONTENTS

1. WATCHERS OF THE SKIES

Up until about 1945, astronomers relied almost exclusively on land-based optical telescopes for their observations of the skies. But light, the kind of signals these instruments can detect, forms a very small part of the electromagnetic spectrum (Fig. I.2, Chapter I). Electromagnetic signals coming from space are very likely to interact (Fig. V.1) in one way or another with the atoms and molecules — oxygen, nitrogen or other scarcer elements — that make up the Earth's relatively thin atmosphere and disappear in high altitudes. Only photons with energies† in the interval $1.7 - 3\,\mathrm{eV}$, around $10^{-3}\,\mathrm{eV}$, and also in the interval $10^{-8} - 10^{-4}\,\mathrm{eV}$ can reach the Earth unobstructed. These parts of the spectrum are the windows of transparency of the Earth's atmosphere: the *visible window*, the *microwave window*, and the *radio window*.

Following the second World War, astronomers vigorously exploited advances made in radio and space technologies to extend their observations beyond the optical band and to explore other regions of the electromagnetic spectrum, each a window with a different view of the Universe.

Radioastronomy was developed first. Radio waves are sometimes of thermal origin, produced when a free electron loses its energy by radiation in the electric field of an ion (a mechanism called the *bremsstrahlung*, see Fig. V.2). More often, it is a nonthermal process (the *synchrotron radiation*) in which an electron is accelerated along a curved trajectory by some cosmic magnetic field, causing it to emit photons. The detection of these radio sources revealed the widespread presence of relativistic *plasma* (that is, an electrically

---

† The photon's energy, E_γ, is related to its wavelength, λ, by the equation: $E_\gamma = hc/\lambda$. It is conventional to use h for the Planck constant and c for the speed of light. They have the following numerical values: $h = 4 \times 10^{-15}\,\mathrm{eV\,s}$ and $c = 3 \times 10^{10}\,\mathrm{cm\,s^{-1}}$. Wavelength is then expressed in centimeters (cm) and energy in electron-volts (eV). This energy unit is related to the more familiar cgs unit by $1\,\mathrm{eV} = 1.602 \times 10^{-12}\,\mathrm{erg}$. In the following both units of energy will be used interchangeably.

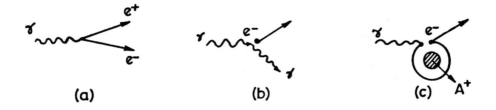

FIG. V.1 *Interactions of photons with atoms. (a) Pair production: a photon with energy $E_\gamma > 1 MeV$ may annihilate into a pair of electron and positron; (b) Compton scattering: a photon with energy between 50 keV and 1 MeV may scatter from an atomic electron to which it transfers some of its energy; (c) Photoelectric ionization: a photon with energy below 50 keV may self-destruct upon knocking out an electron from an atom which then becomes charged or ionized.*

neutral mixture of energetic electrons and ions) in galaxies and intergalactic space, which in turn has led to the discoveries of the actual origins of that plasma — *pulsars, quasi-stellar radio sources*, or *quasars*, and, more generally, active *galactic nuclei*. Another major contribution of this branch of astronomy is the detection of the *microwave background radiation* by Penzias and Wilson in 1965, a discovery of greatest importance for astrophysics (see Chapter VII). This radiation has a thermal spectrum corresponding to a temperature of 2.7 K, and represents the cooled remnant of a hot radiation emitted in the early phases of the formation of the Universe.

X-ray astronomy is the study of cosmic objects under extreme physical conditions — matter subjected to disruptive processes or to temperatures of hundreds of millions of degrees. These objects are powerful sources of radiation with energies ranging from a few hundred electron volts to a few hundreds of thousands of electron volts per photon. Many of such sources emit nonthermal radiation produced when electrons are accelerated to ultra-relativistic velocities by strong fields. The corona of the Sun is an example of a place hot enough to emit such radiation. The first strong X-ray source discovered beyond the solar system was Scorpius X-1, which probably derives its emission from accretion of matter by a *neutron star* in a close binary, or double-star, system. The first dedicated X-ray satellite Uhuru, launched in 1971, detected two other binary systems, Hercules X-1 and Centaurus X-3;

FIG. V.2 *(a) Bremsstrahlung (braking radiation): electromagnetic energy emitted by a rapidly moving charged particle when it is slowed down by an electric field; (b) Synchrotron radiation: radiation emitted by relativistic charged particles moving in circular or helical orbits in a magnetic field.*

it was followed in 1978 by the Einstein X-ray Observatory, which sent back to Earth high-resolution images of many other X-ray sources.

Gamma rays found in space are generated in nonthermal processes or, at higher energies (hundreds of MeV), in the decay of certain unstable particles (*e.g.,* neutral π mesons), which are themselves products of violent collisions between cosmic-ray particles.† The detection of this radiation has resulted in the mapping of a diffuse gamma-ray background along the Milky Way and the identification of several discrete sources. Some of these — the Crab nebula, Cygnus X-3, and Hercules X-1 — emit ultra-high energy (10^{12} eV) gamma-rays. Astronomers believe this high-energy radiation might contain clues to the way matter and energy are transformed in the hearts of stars and galaxies, and hope that the Gamma Ray Observatory, launched in 1991, would give them further insight into these phenomena.

Yet, despite prodigious advances made in the study of electromagnetic radiation at non-optical wavelengths, supplemented even by the study of non-electromagnetic signals (*e.g.,* neutrinos and cosmic rays), *optical astronomy,* with the support of sophisticated detectors and powerful computers, continues to play a central role in astrophysics. The reason is that stars contain most of the *visible matter* of the Universe and they emit most of their radia-

---

† *Cosmic-ray particles* are particles and nuclei originating from highly evolved stars boosted to extremely high speeds by star (supernova) explosions, or from interstellar material accelerated by various fluid motions. Active galactic nuclei are also copious sources of relativistic particles.

tion in the visible band. Narrow as it is compared to the whole spectrum, this band can yield a tremendous amount of information on the stellar contents and, accordingly, on the major component of the Universe. For this very same reason, there is no better way to start our exploration of the Universe around us than by a study of the stars, their birth, life, and death.

Summary

The Earth's atmosphere is transparent to visible light, radiowaves, microwaves and partially to infrared light. Thus, observations of the sky from the ground give us an incomplete and perhaps biased view of the Universe. Radio and space technologies have been adopted by astrophysicists to extend their study of cosmic electromagnetic radiation to all parts of the spectrum. As a result, our understanding of the Universe has gained both in detail and scope. Yet, optical astronomy continues to play a crucial role in all astrophysics because it is capable of gathering a vast amount of information on the most important component of the Universe, the stars.

---

2. GATHERING CLOUDS

On a clear night, if you look up at the sky you will likely see a broad band of faint distant stars stretching from horizon to horizon which the ancient Greeks likened to a river of milk and accordingly called a '*galaxy*'. This band is an edge-on view of our galaxy, which we simply refer to as the Galaxy, or the Milky Way. If we could observe our galaxy far away from outside, it would appear as a gigantic thin disk with a central bulge. The disk is about* 30000 pc across, 200 to 400 pc thick, packed mostly with stars traveling in nearly circular orbits about the galactic center with very small up-and-down motions. The central bulge contains stars moving with small circular speeds in random directions; this accounts for their roughly spherical distribution. Above and below the disk, a spherical swarm of globular clusters with large

---

* Astronomers often express distances in units of *parsec* (pc) or *light year* (ly). 1 light year = 9.46×10^{17} cm, 1 parsec = 3.26 ly = 3.08×10^{18} cm. More on the definitions of these units and the measurements of distances in astronomy can be found in Chapter VII.

random velocities forms a halo concentrated toward the center of the Galaxy. If we could look at our galaxy face-on, as we can with external systems such as the galaxy NGC4622, we would see in the plane of the galactic disk a beautiful spiral pattern composed of several spiral arms outlined by very brilliant stars. These bright stars, strung along the arms like 'beads on a string', contrast sharply with the very dark inside edges of the arms and the faint, more diffuse patches throughout the disk. These irregular formations suggest the presence of matter in the space between the stars.

All matter between the stars in our galaxy, collectively called the *interstellar medium*, has a mass of several billion solar units,† or a few percent of the mass of all the *visible matter* of the Galaxy. The interstellar material is a mixture of gas (of atoms and molecules) and dust; although this dust amounts to only about one percent of the gas mass in the Galaxy, it plays an important role in the dynamics of matter between the stars and affects appreciably the way we see distant stars. Gas and dust gather in large formations of clouds, of which astronomers recognize several types by their different physical properties: dark clouds of dust, molecular clouds, neutral (HI) and ionized (HII) hydrogen regions, planetary nebulae, and supernova remnants. The last two types are specifically connected with star deaths and will be considered later on in this Chapter. In the present section, we will discuss the general nature of the interstellar medium and its role in the formation of stars.

2.1 Basic Thermal Properties of Matter and Radiation

It was only in 1904 that Johannes Hartmann showed conclusively that the space between stars was not a complete void, as many had thought until then, but was rather filled with rarefied matter. This discovery, like many others about outer space, was made by analyzing the incoming light. Before

---

† It is sometimes convenient to express certain astronomical measures in *solar units*. The most commonly used are the following:

Mass of the Sun	$1\,M_\odot = 2 \times 10^{33}\,\mathrm{g}$
Radius of the Sun	$1\,R_\odot = 6.96 \times 10^{10}\,\mathrm{cm}$
Luminosity of the Sun	$1\,L_\odot = 3.90 \times 10^{33}\,\mathrm{erg\,s^{-1}}$
Distance Earth-Sun	$1\,\mathrm{Astronomical\,unit(Au)} = 1.50 \times 10^{13}\,\mathrm{cm}$

we describe this observation, let us review a few basic results of the radiation theory (for more details see Appendix C).

Consider a box with perfectly reflecting walls containing a large collection of free, nonrelativistic particles in random motion. We call such a collection an *ideal gas*. After a sufficiently long time, the gas will most probably have the equilibrium distribution. If we cut a small hole in a wall and count the number of particles that come out at different energies in a small energy interval and if we repeat our little experiment with different other holes, we will find that the number of particles per unit velocity volume varies exponentially with kinetic energy E as $N(E) = C \exp(-E/kT)$, where C is a normalization constant determined by the number of particles in the system, and k is a universal constant independent of the nature of the particles, called the Boltzmann constant.‡ The parameter T is called the *temperature* of the system. It is clear that the particle count decreases as E increases at a rate determined by T: it drops by a factor $e = 2.71$ for each energy increase of kT (see Fig. C.1 of Appendix C). If the total number of particles is the same in two given gas samples at different temperatures, there are more high-energy particles and less low-energy particles at the higher temperature than at the lower temperature.

The above relation for $N(E)$ is known as the *Maxwell-Boltzmann distribution* of a classical ideal gas in thermal equilibrium. It describes the equilibrium situation that a dilute gas prepared in an arbitrary initial state will almost certainly reach after a sufficiently long time which depends on the collision rate of the gas molecules. It yields two useful results. First, it implies that the *average kinetic energy* of a particle in the system is given by $\frac{3}{2}kT$, which can also be taken to define the *kinetic temperature*, or often simply the temperature of the gas. (For a system in thermal equilibrium, the parameters called 'temperature' appearing in various relations have the same values, but it is not necessarily the case for a thermally unstable system). A simple yet important fact is manifest from the gas distribution law: there will always be some particles with very high kinetic energies which can initiate reactions that one might not expect from the average energy value

---

‡ Two basic physical constants used in thermal physics are the Boltzmann constant k and the Stefan-Boltzmann constant σ whose numerical values are: $k = 0.86 \times 10^{-4}$ eV/K, and $\sigma = 5.67 \times 10^{-5}$ erg cm^{-2} s^{-1} K^{-4}.

alone. Another direct consequence of the Maxwell-Boltzmann law is the well-known *pressure-temperature relation* for an ideal gas in thermal equilibrium: $P = nkT$, where P denotes the pressure and n the particle number density.

The Maxwell-Boltzmann relation applies to *classical, distinguishable* particles. To take into account *quantum effects*, that is, the statistics of *indistinguishable* particles, it must be modified. The new relations are called the *Fermi-Dirac distribution* for half-integral spin particles, or fermions, and the *Bose-Einstein distribution* for integral spin particles, or bosons. Photons are massless bosons of spin one; their distribution law is given by $1/[\exp(E/kT) - 1]$ where the photon energy, E, is related to its wavelength λ by $E = hc/\lambda$. When properly weighted by a λ-dependent factor, this function gives the spectral distribution of wavelengths, known as *Planck's law of thermal radiation*, or also the black body radiation (see Fig. C.3 of Appendix C). This result means that any system with uniform temperature T and in thermodynamic equilibrium with its own radiation field (*i.e.*, sufficiently opaque) will emit photons having the Planck spectral distribution of wavelengths. This distribution curve has a single maximum. The wavelength at this peak, denoted λ_{max}, is given by *Wien's displacement law*: $\lambda_{max}T = 0.29\,\mathrm{cm\,K}$. Another important consequence is the *Stefan-Boltzmann law*, which gives the *energy flux* (emission rate per second per unit area) leaving the surface of a thermal radiator as $f = \sigma T^4$, where σ is the Stefan-Boltzmann constant. Note the very steep rise in the rate of radiation from the surface of an opaque body with increasing temperature. If we assume an ordinary star to be a sphere of radius R that radiates like a black body, then its *luminosity*, or power, is given by the energy flux over the whole surface area of the star, that is, $L = 4\pi R^2 f$. Remember: with T known, Wien's law gives the star's color; with both T and R known, Stefan-Boltzmann's law gives the star's luminosity.

We emphasize that the validity of Planck's law only requires the emitting body to be sufficiently opaque and to have a uniform temperature. It does not matter what the body is made of, nor how specifically the radiation is produced. In fact, the way black body radiation is produced depends on the nature of the system (for example, solid-state or gas) and on the energy available. In the case of interest, the normal stars, the main contribution to thermal radiation is the bremsstrahlung of electrons (*i.e.*, loss of energy by radiation through interaction with an electric field) and its inverse process. As the electrons move in the (gas) medium, they get pushed and pulled by the electrostatic field produced by the ions and emit a photon at each deflection. The energy of the photon produced at each event is precisely the kinetic energy lost by the interacting electron at this event but need not be the same at any other. Since the kinetic energy of electrons has a random distribution, the photon distribution over energy (or wavelength) is a continuum. Inversely, a free electron can also absorb a photon, increasing its own energy; to conserve momentum, there must be a heavy nucleus nearby to absorb momentum but little energy. Equilibrium is attained when photons are emitted and absorbed in equal numbers; it is preserved in stars because the photons are effectively trapped in the stellar interior. Thus, a star can be considered as an ideal radiator. Observations confirm this. The measured spectral distribution of sunlight has a peak at about $\lambda_{max} = 5000$ Å (an Ångstrom equals 1×10^{-8} cm). For the main part, from the ultraviolet to the infrared, it agrees with the spectral distribution of a black body at temperature 5800 K. This number then defines the *surface temperature* of the Sun, or the *effective temperature* it would have if it were a perfect black body radiating at that temperature.

2.2 Spectral Analysis

The general applicability of the radiation law, which makes it so useful for many purposes, is also its weakness. The radiation law tells us nothing, for example, about the chemical composition of a star. For this kind of detailed information, we have to look elsewhere, at the *line absorption spectrum* formed in the more diffuse upper layers of the star's atmosphere. The mechanism that yields absorption spectra is as follows. As thermal photons rise up from the star's deeper layers with a continuous distribution of wavelengths, those with the right energies will be absorbed by atoms in the upper layers. The excited atoms then de-excite either by collisions or by radiative transitions. In the first case, the surplus energy is converted into kinetic energies for the particles involved in the collisions; no radiation is emitted. Since the upper layers of the atmosphere are cooler than the lower layers, there are statistically more excitations followed by collisional deexcitations than collisional excitations followed by radiative deexcitations. Thus, there is a net loss of photons in the outgoing flux at the line energy. In the second case, a

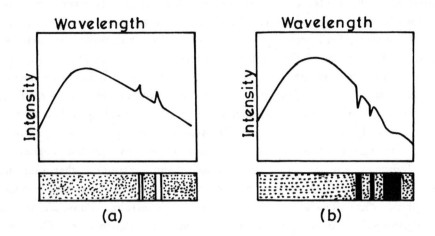

FIG. V.3 *(a) Emission lines and (b) Absorption lines on a background continuum.*

photon is emitted directly into space. It has no preferred directions of travel and is equally likely to take the original direction as any others. Because the hot surface of the star is surrounded by a dark absorbing space, there are more photons originally traveling toward us being scattered out of our line of sight than the other way around. Again, there is a net loss of line photons traveling in our direction. In both cases, photons are removed from the underlying continuum radiation field at the transition energy, leading to the appearance of dark lines, or *absorption lines*, in the stellar spectrum (Fig. V.3).

Each species of atom has a unique inner structure governed by its own dynamics. This uniqueness is reflected in a distinctive, characteristic pattern of energy levels representing the quantum states of the atom. Radiative transitions between these levels give rise to line spectra which act like a signature of the atom, and observations of absorption lines can yield clues to the chemical composition of the stellar atmosphere. Absorption spectra can also yield information about the star surface temperature. For example, if there exist dark lines identifiable with ionized hydrogen, it means that the star outer layers must be hot enough to produce a large number of particles with kinetic energies high enough to ionize hydrogen. Since the hydrogen ionization energy is 13.6 eV, the temperature at which the ionization is spectrally detectable is about 15000 K. Similarly, the presence of dark lines of ionized helium in a stellar spectrum implies that the star surface temperature must be 28000 K or more.

2.3 The Discovery of Interstellar Medium

Hartmann discovered, quite by accident, the existence of interstellar gas while investigating the spectrum of a star called δ-Orionis, the lighter member of a double-star system. As a star in a binary system moves along its orbit, it executes a periodic motion, alternately approaching us and receding from us (assuming its orbital plane is not precisely perpendicular to our line of sight), causing the wavelengths of its spectral lines to vary periodically, first in one direction, then in the other. This is the well-known *Doppler effect*: when a source of radiation with wavelength λ is moving toward an observer with velocity v, the wavelength measured will decrease by $\lambda v/c$, where c is the velocity of light; if the source is moving away, the wavelength will increase by the same amount. What Hartmann found was that, among the absorption lines that underwent periodic Doppler shifts and therefore must come from the atmosphere of δ-Orionis, there were stationary lines whose wavelengths remained unchanged. He identified those narrow, dark lines as belonging to once-ionized calcium and concluded that the 'stationary lines' did not come from the orbiting star itself, but arose from absorption of the star light by ionized calcium in an interstellar cloud lying between the binary stars and Earth.

Further studies show that the interstellar lines are always much narrower than the lines of the underlying spectrum and each is often split into several components lying close together. For an absorption line to be narrow, all the factors that could contribute to its broadening (thermal motion, rotation of the medium, etc.) must be very small. A detailed analysis of interstellar line widths indicates the temperature of the absorbing medium to be of the order of 100 K. Further, the splitting of the line into closely-spaced components means that the intervening cloud formation is divided into several lumps moving more or less together, each with a slightly different Doppler velocity. In short, interstellar gas must be relatively diffuse and cold (100 K), and must exist as 'clouds' moving more or less as a unit at small relative velocities (10 km/s). Astronomers refer to this structure as the *diffuse clouds*. In the

TABLE 1

Main components of interstellar medium

---

(1) **Gas** *of atoms and molecules, either neutral or ionized. Hydrogen is the predominant element, but molecules containing up to 13 atoms can also be found. The gas is aggregated in large formations in varying concentrations and temperatures. Dark clouds (e.g.,the Giant Molecular Clouds) are among the coolest (10 to 100 K). Hot ionized gas (10000 K) is observed around young stars and hot dying stars (ejection by planetary nebulae or supernovae).*

(2) **Dust** *is present throughout the interstellar gas. It is highly opaque in the optical band but transparent in the infrared. It protects newly formed molecules and acts as a moderating agent for infrared-emitting protostars. It is also a depository of material ejected by old stars.*

(3) **Cosmic-ray particles** *are highly energetic particles and nuclei produced by active galactic nuclei, supernovae, and supernova remnants. In our Galaxy, they are confined by the galactic magnetic field.*

---

years following Hartmann's discovery, absorption lines in visible spectra due to heavy elements, such as sodium, potassium, titanium, and iron, and simple molecules, such as CN, CH, and CH^+ (ionized molecule), were identified. Why are heavy elements observed, and not other, lighter elements? Where do they come from? And where do the interstellar clouds come from? Why are they broken up in lumps? How are they related to stars and other objects in the Galaxy? The first question will be discussed presently; we will take up the other questions later on.

• *Optical and ultraviolet observations of interstellar medium*

In interstellar space, both matter and radiation are very dilute, and processes whereby particles might become excited, involving either absorption of radiation or collisions between particles, take place very rarely. Practically all atoms and molecules must be found in their ground states. Thus, for instance, when an atom happens to absorb a photon, it is very likely in its lowest energy level. The resulting atomic transition usually involves the lowest frequency and shows up as the most prominent line in the spectrum, called *the resonance line.* Some atoms, like calcium and sodium, have a structure such that their resonance lines are in the *visible* part of the spectrum. Ionized calcium was detected first in interstellar space not because of its abundance (in fact, as it turns out, it is a million times less abundant than hydrogen), but because its resonance line happens to be in the *visible*, at 3968 Å. But for most other elements the resonance lines lie in the *ultraviolet.* For example, the resonance line of hydrogen (the Lyman-α line) occurs at wavelength 1216 Å, while that of helium at an even shorter wavelength, 586 Å. As we already know, all radiation at wavelengths shorter than 2900 Å is absorbed by the Earth's atmosphere and cannot be seen by a ground-based observer (see Sec. 1). It is only rather recently, with the use of balloons, rockets, and satellites, that scientists have been able to obtain stellar spectra in the ultraviolet and get a more accurate picture of the space between the stars. They discover the presence of neutral atoms, ions, as well as molecules. They further find that the chemical composition of the interstellar gas reflects fairly closely that of the solar and stellar atmospheres, that is, dominated by hydrogen and helium, with only traces of heavier elements.

But observations in the ultraviolet, while very useful, have the drawback in allowing only a sampling of gas clouds lying relatively close to the solar sytem. The ultraviolet radiation, with its short wavelengths, cannot penetrate great distances of interstellar gas and dust. Its detection gives us a picture of a rather limited region around us. What lies beyond, in between the visible stars? In particular, how can we detect that most important element in the Universe, hydrogen?

• *The 21-cm radiation*

Fortunately, neutral hydrogen emits a very special radiation, at wavelength of 21 cm, right in the middle of the radio frequency window. Because of its long wavelength, this radiation can travel great distances in space without being absorbed by the intervening particles, and thus gives us the best clue we have to the distribution of hydrogen in most parts of the Galaxy. The detection of the 21-cm line plays a crucial role in the study of the interstellar medium.

The mechanism that gives rise to the 21-cm transition in hydrogen is as follows. Besides the usual electrostatic force between the electron and the proton that together make up the hydrogen atom, there exists a weak magnetic

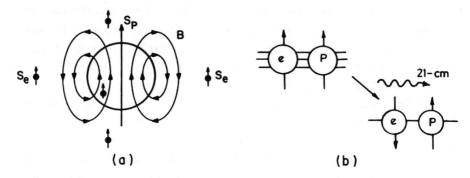

FIG. V.4 *Spin flip hydrogen radiation. (a) Electron in the magnetic field B of a proton. (b) A 21-cm photon is emitted when the atomic hydrogen makes a transition from a state of parallel electron and proton spins to the state of antiparallel spins.*

force between the two particles. Classically, we can visualize the electron and the proton as spinning charged particles; as they spin about their own axes, they produce current loops that act like tiny magnets. But the atom is not a classical object, and to describe it correctly, quantum mechanics is required. In particular, the electron does not follow some well-defined path: it can be found in effect everywhere, even 'inside' the proton itself. When the hydrogen is in its ground state, the probability of finding the electron at a given distance from the proton is the same in any direction; such a probability distribution is said to be spherically symmetric. The important point is that the magnetic force acting on the electron depends on the proton magnetic field, B, which varies in space as shown in Fig. V.4a and, in particular, changes sign when B reverses orientation. Thus, for a fixed electron and proton spin orientation, the magnetic energy vanishes when averaged over all positions of the electron outside the proton. The only nonvanishing contribution comes from the electron distribution inside the proton (where the magnetic field has the same general orientation throughout), its sign being determined by the relative spin orientation of the two particles, positive for parallel spins, negative for antiparallel spins. The ground state of hydrogen, predicted by the electrostatic interaction to be at $-13.6\,\text{eV}$, shifts up by a tiny amount $(3 \times 10^{-6}\ \text{eV})$ and the anti-aligned spin configuration shifts down by the same amount. It follows that if the electron in an aligned spin configuration flips its spin, it ends up in the lower-energy state, emitting a photon with energy

6×10^{-6} eV, or wavelength 21 cm (Fig. V.4b). The only trouble with this explanation is that this spin-flip transition is highly forbidden; for each atom, it occurs spontaneously only once every 10 million years, a long time even on the astronomical timescale.

It was astrophysicist Hendrik van de Hulst who suggested in 1944 that collisions between atoms could occur often enough in interstellar clouds to speed up the transition rate to a detectable level. A rough calculation shows that in each cm-cube of the cloud there is an interaction and, therefore, emission of a 21-cm photon once every few hundred years, still a relatively rare event but frequent enough to be detectable. Does it mean that by this time, after billions of years of existence, no hydrogen atoms are left in the aligned states and no radiation is possible? No. The reason is that the hydrogen atoms are continuously ionized by high energy radiation and cosmic rays. When the electrons and protons recombine into hydrogen, aligned configurations are formed three out of four times. The upper energy states will thus be continuously created, and there will always be 21-cm radiation.

2.4 The Spiral Structure of the Galaxy

Astronomers have known for some time that the Sun lies in an off-center position in the Galaxy, at about 30000 ly from the Galactic center, and that it rotates around the center at the fantastic speed of 250 km/s. From these data and Newton's gravitational law of motion (see Appendix A), one can obtain an estimate of the mass of the Galaxy interior to the Sun's orbit, which turns out to be $1.3 \times 10^{11} M_\odot$. The same law of motion also indicates that matter in the Galaxy does not rotate uniformly but rather *differentially*, the inner parts moving more rapidly than the outer parts. (To a first approximation, the squared velocity is given by $v^2 = GM/r$, where G is Newton's universal gravitational constant, r the distance from the galactic center, and M the mass of matter inside that radius). It follows that if we direct a radio telescope toward the outer rim of the Galaxy as it is rotating in the same direction but at a slower rate than our solar system, all the 21-cm radiation we receive will be shifted toward shorter wavelengths. The shift will be greatest and the light intensity weakest for the farthest emitters. If hydrogen is uniformly distributed throughout the Galaxy, we expect the radio signals to have a continuous, slightly asymmetric distribution, with a peak at 21 cm in wavelength. What is actually found as the telescope scans the galactic

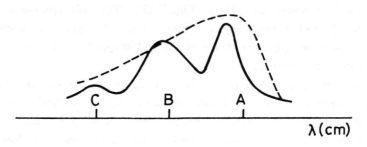

FIG. V.5 *Sketch of the intensity of 21-cm radiation as would be received by an Earth-bound detector looking toward the outer rim of the Galaxy in the direction where the rim is rotating away from us. The spectrum shows that the hydrogen distribution is divided into three separate regions.*

outer edge comes as a big surprise: both the Doppler effect and the distance effect are there, as predicted, but the spectrum is not continuous but rather divided into three distinct maxima, centered at slightly different wavelengths (Fig. V.5). The three separate distributions of radiation clearly belong to different HI regions, which would form three continuous arcs, or arms, that curve about the center of the Galaxy. If we now turn the radio telescope toward the Galactic center, fragments of three more arms can be seen. This confirms a fact long suspected: the Milky Way has a spiral structure similar to that observed in other disk galaxies.

What really is this curious and beautiful structure? Astronomers have known for some time that gas flows out of the galactic nucleus at a rate of about one solar mass per year. Nobody understands how or why. It was thought that this outflowing material would stretch and spread along some curving arc by the differential rotation of the Galaxy, thus forming a spiral arm. But this attempt to understand the spiral structure in terms of arms made up of a permanent material is simply untenable: material arms, rotating with a period of some 240 million years around the galactic center, would have been through 50 revolutions since the formation of the Galaxy, 12 billion years ago; they would be wound tighter and tighter by the differential rotation and would long have disintegrated. A possible way out of this difficulty was shown by Chia Chiao Lin, Bertil Lindblad, and Frank Shu. The key idea is that every spiral arm represents not a material formation but a *wave*. Because an interstellar cloud flowing out of the galactic center is moving at a higher velocity than a spiral arm found on its path, it will overtake it on the inside, just as the arm sweeps past matter on the outside. As it enters the arm, the cloud is slowed down abruptly by the denser gas already there, while the gas just behind, continuing at its normal speed, runs into the leading particles. The net effect is a pile-up of gas and dust into higher densities on the inner edge of the arm, leading to conditions that could trigger collapse of matter and formation of stars. The cloud travels within the arm for a relatively long time, then emerges from the arm to resume its more rapid motion until it overtakes another arm. Stars similarly move on through and out of the spiral arms. Thus, spiral arms are really *density waves*. Just as we do not see the same water and the same foamy tops on the crest of a breaking water wave at two different instants, we do not see the same clouds and the same stars in a density wave when we look at a later time. It is for this reason that the spiral pattern appears so stable despite the rapid differential galactic rotation.

About half of the interstellar gas, composed mostly of low-density hydrogen ($n_H = 0.1\,\mathrm{cm}^{-3}$), is distributed uniformly throughout the Galaxy in a relatively thin layer in the galactic plane. The rest is found in the regions of the spiral arms as diffuse clouds of atomic hydrogen ($n_H = 10\,\mathrm{cm}^{-3}$), emission nebulae of ionized hydrogen ($n_H = 100\,\mathrm{cm}^{-3}$), or dense clouds of molecular hydrogen ($n_H = 10^6\,\mathrm{cm}^{-3}$). Thick clouds of dust ($n_H = 1000\,\mathrm{cm}^{-3}$) also darken vast regions of space. Most of the stars that populate the Galaxy are either lone or binary stars. But exclusive to the spiral arms are groups of $10 - 50$ stars, called *associations*, all of them very hot, with surface temperatures exceeding 30000 K. Larger, rather ill-defined groups (open clusters) containing $100 - 1000$ stars, with surface temperatures $\sim 6000 - 30000\,\mathrm{K}$, can also be found in the spiral arms as well as in other parts of the Galaxy.

2.5 Interstellar Dust

More than 200 years ago, William Hershel described vast regions in the sky, such as the Coalsack nebula and the Horsehead nebula, where there is a marked absence of stars, as 'holes in the sky'. At first, astronomers were rather confused by the reddening of spectra from stars whose surfaces they knew to be at temperatures in the order of 30000 K from the presence of ionized helium. The reddened starlight was also found to be polarized. All rather puzzling. Astronomers now think that the most plausible explanation for all these disparate observational facts is the existence of intervening clouds of an obscuring material called *interstellar dust.*

To understand why starlight from some directions in the sky is *reddened*, one just needs to consider a local phenomenon, the reddening of the setting Sun. Essentially what happens here is that the blue component of the light beam directed toward us is scattered more out of the beam than its red component. We know sunlight is composed of various color components whose intensities peak near yellow. As the electric field of the incoming radiation tickles the electrons of the air molecules into rapid oscillations, the disturbed electrons emit radiation. This field represents the light scattered by atmospheric molecules. Its amplitude is proportional to the electron acceleration which is itself proportional to the square of the frequency of the incident light. The intensity of the scattered light, given by the square of the scattered electric field, must then vary as the *fourth power* of the frequency. Thus, the number of photons scattered depends strongly on their frequencies. For example, the blue component of an incident white light is scattered five times more than its red component. When we look at the sky, we look away from the Sun at the scattered radiation, we see the blue sky. When we look at the setting (or rising) Sun, we look directly at the Sun, when sunlight has to penetrate a greater column of air and has to undergo more scattering than light from the noon Sun, we see red.

Actually, air molecules are rather inefficient agents in scattering visible light; it would require an incredible mass of air to produce anything resembling an obscuration of sunlight. This is so because the typical size of air molecules (about 1 Å) is tiny compared to the wavelength of visible light (about 5500 Å); to efficiently block any wave, the blocking agent must be at least of a size comparable to the wavelength of the wave to be blocked. That is why the walls of ordinary buildings cannot obstruct radio signals. Another related question concerns the color of the clouds in our sky: why do they look white? Clouds are formed when molecules condense into water droplets very much larger than the molecules themselves. The basic reason again lies in the relative sizes of the scatterers and of the waves being scattered. Calculations show that the intensity of the light scattered from clouds is then *independent* of the light frequency, all colors are equally scattered: the clouds do not look blue, they look white.

Now, careful measurements show that starlight undergoes a much less intense reddening through scattering by interstellar medium than sunlight through the Earth's atmosphere. The intensity of the scattered radiation is found to vary *linearly* with light frequency; in fact, the situation is closer to the scattering from clouds than from the atmosphere. So it is reasonable to suspect that the scatterers in the interstellar medium are not atoms or molecules but grains of material, and these grains are smaller than water droplets in clouds. A detailed calculation shows the observed dependence of intensity on frequency can be reproduced when the size of the grains is in the order of the wavelength of light, that is, about 5000 Å. Such a grain could contain well over 10^{11} molecules of typical size of 1 Å. Astronomers refer to this kind of material as interstellar dust. Dust is everywhere in space, but it is particularly concentrated in some regions, called *dark clouds*. Since the size of a dust grain is comparable to the light wavelength, dust is most effective in scattering visible and ultraviolet light. It acts as the primary agent for the visual extinction of starlight we see in 'holes' of the Galaxy. These dark clouds are typically about 10 ly across. A complete extinction over 10 ly of dust implies a density of 10^{-10} grains cm^{-3}. It is estimated that such clouds could pack hydrogen to a density over 10^3 atoms cm^{-3} (*i.e.*, 10^{13} times the grain density). There also exist smaller, more regularly-shaped concentrations of dust, called *globules*, which are suspected to be undergoing gravitational collapse. A typical globule has a size of about 2 ly in diameter, a mass of 60 $M_\odot$, a temperature of the order of 10 K, and a hydrogen density of $\sim 10^3$ cm^{-3}. About 25000 such globules populate the Galaxy.

As all other forms of matter in space, dust grains are exposed to all sorts of radiation which heat them up. Thermal equilibrium is reached when the resulting energy emission is balanced by the radiation absorption. The equilibrium temperature may reach hundreds or thousands of degrees in the vicinity of stars, but may plunge to 100 K in interstellar clouds, and even 10 K in large dark clouds. Why are dust clouds ten times colder than diffuse clouds? In a gas cloud, atoms are constantly ionized, mainly by cosmic rays, releasing electrons of fairly high kinetic energy. The free electrons cool off by losing energy through repeated collisions with atoms until they have not enough energy left to interact. At this point a balance between heating and cooling is attained and a stable temperature established (all radiation emitted escapes the diffuse cloud). In this cooling mechanism, the most effective coolant is an atom with the lowest possible first excited state; its presence is the major factor in determining the ambient temperature. In diffuse clouds, ionized carbon plays that role: its first excited state lies at only 0.01 eV above its ground state, which means a possible average kinetic temperature of 50 K for electrons. On the other hand, dust absorbs soft X-rays and ultraviolet radiation, thus reducing both the ionization rate of the atoms and the heating of the medium, and keeping the carbon (as C or CO) neutral. Since the excited levels lie closer to the ground state in neutral carbon atoms than in ionized carbon atoms, thermal equilibrium sets in at a lower temperature in a dust cloud than in a gas cloud.

Now, the starlight that is reddened as it passes through the interstellar medium is also found to be *polarized* in the plane of the Galaxy. Therefore, the interstellar dust that causes the reddening of light must also act like a polarizing agent partially cancelling the radiation electric field component perpendicular to the galactic plane. In other words, it must be composed of elongated conducting grains lined up at right angles to the galactic disk, as if forming a cosmic-wide conducting grid. Of course, normally the grains move to and fro, but one could visualize them as tiny permanent magnets locked in place by a weak magnetic field that is known to exist throughout the Galaxy. To produce the polarization observed, the field must have a strength of about $10^{-6} - 10^{-5}$ gauss, an estimate confirmed by more refined methods. Interstellar dust apparently must have *paramagnetic* properties to explain both the direction of polarization of the filtered starlight and the relative parallel orientation of the galactic plane and magnetic field observed in our region of space.

The chemical composition of interstellar dust grains remains uncertain but is generally thought to be silicates (like sand) or carbon compounds (like graphite or silicon carbide), both conducting and both paramagnetic. Where does this dust come from? Here again, we do not know for sure. But it is a well-known fact that stars in the late stages of their evolutionary lives shed large amounts of material via stellar winds in space. The dust grains might have condensed from heavy elements in the gas when it is cooled in the stellar atmosphere or upon expansion in the wind. In oxygen-rich stars, where oxygen is fixed by carbon in the very stable gaseous compound carbon monoxide, silicates would be the most common form of dust, whereas in carbon stars silicon carbide and graphite would be more abundantly produced. Another possible source of dust, especially as grains with unusual isotopic abundance, is matter ejected during nova and supernova explosions.

Interstellar dust appears to be the link between the old and the new in the Galaxy: it serves not only as a repository of remnants of dead stars but also as a site for the formation of molecules, raw material for stars to be born. The presence of dense gas of hydrogen and the prevalence of low temperatures in dust clouds are both conducive to the formation of molecular hydrogen via the chemical reaction $H + H \rightarrow H_2$. The hydrogen atoms are, of course, neutral but in a dense medium their charge distributions are deformed so that the atoms look electrically charged from some directions. We say that they are *polarized*. Polarized atoms interact through an attractive force of electric origin, the van der Waal's force. The dust grains act as catalysts in such reactions, serving first as sites for gathering atoms and then as absorbers of the excess momentum and energy released by the reactions. Further, if enough of this energy is transferred as kinetic energy to the newly-formed molecule, the molecule may lift off the grain surface. Or else the molecule may receive enough energy through absorption of photons and gets released in space. This process of thermal ejection is particularly efficient when the dust cloud is juxtaposed to an *emission nebula*, a region of ionized hydrogen (HII) surrounding a hot star and a source of copious radiation. Once free in space, molecules are submitted to bombardments by cosmic rays and high-energy radiation. Fortunately, they are partially protected from destruction by the dust which acts as an effective shield against the ultraviolet radiation. That is why molecules vary in abundance and kinds at different levels within a cloud, with only the most stable compounds surviving in the outer regions and with the greatest number and the greatest variety of molecules deep inside.

2.6 Giant Molecular Clouds

The process of molecular formation described above plausibly explains the juxtaposition of dust clouds and dense concentrations of molecules often observed in the sky. Such large, massive aggregates of cold interstellar matter, are sometimes called *cloud complexes* or *Giant Molecular Clouds* (GMCs). One of the closest to us is located in the constellation Orion; it contains besides an emission cloud, the famous Orion Nebula, a huge cloud of carbon monoxide and various other molecular clouds lying in regions not yet ionized by the stars, as well as several other interesting objects. Farther away, toward the center of the Galaxy, Sagittarius B2 is another remarkable dense and dusty complex with an unusual concentration of many rare molecular species.

Mean particle densities in GMCs range from $100 \, \text{cm}^{-3}$ to $1000 \, \text{cm}^{-3}$, and masses from a thousand to a million solar masses. The lower limit, $10^3 \, M_\odot$, is physically meaningful because massive (OB) stars can form in association only in clouds exceeding this mass. Over most of their volumes GMCs have the same physical properties as the cold dark clouds. But there are two important differences. The first is that GMCs almost always have clusters of OB stars adjacent to them or embedded in them. The second is that they contain hot, massive molecular cores, with temperatures as high as 1000 K, densities as high as $10^8 \, \text{cm}^{-3}$, and masses ranging from 50 to 1000 $M_\odot$. Are these cores, or 'clumps', the sites of eventual formation of single OB stars and star clusters?

In other galaxies, OB stars trace spiral arms. Since in our Galaxy OB stars are associated with GMCs, it is plausible that GMCs are associated with spiral arms too. Moreover, the observed temperatures and sizes of molecular clouds give direct evidence for GMCs in spiral arms. GMCs without OB stars may also occur between the arms; in any case, colder, smaller molecular clouds (SMCs) are certainly observed in regions near the solar system.

2.7 Interstellar Magnetic Field

We showed earlier that the polarization of starlight could be explained in a natural way by the polarizing action of dust grains lined up in a weak large-scale magnetic field in interstellar gas. Evidence of such a field comes from the observation of (synchrotron) radio emission by relativistic cosmic-ray electrons as the electrons spiral around the magnetic field lines. The strength of the interstellar magnetic field can be inferred from measurements of the fine-structure splitting of radio lines, indicative of the presence of a magnetic field (an effect observed in the laboratory in most atoms and that goes by the name of Zeeman). The average strength deduced is about ~ 10 microgauss in regions of densities smaller than 100 atoms cm^{-3}; it increases with density, with large variations from cloud to cloud. The galactic field is

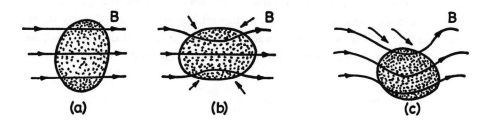

FIG. V.6 *(a) Magnetic field lines are frozen in a ball of charged particles in an electrically neutral admixture (ionized cloud or plasma). (b) When the ball is compressed, the field lines are similarly squeezed. (c) When the ball is moved to one side, it drags the lines along.*

weak (for comparison, the Sun's magnetic field is about 1 gauss), but covers immense regions of space.

If the origins of the galactic magnetic field are still shrouded in mystery, the importance of its role in the dynamics of interstellar medium is more evident. Since the gas clouds always contain free electric charges in varying degrees (resulting from ionization of atoms by cosmic rays and high-energy radiation), and therefore are electrically conducting, the gas and magnetic field are frozen together. As we have mentioned, the field lines generally run parallel to the plane of the galactic equator. If a cloud moves to one way, the magnetic lines of force elastically stretch out without breaking and move with the cloud. But the deformed field lines exert a counterforce that tends to pull the cloud back to its initial position. Similarly, if the cloud contracts, the field lines are compressed as well, but at the same time they exert a force that prevents further contraction (Fig. V.6). This is a manifestation of a well-known phenomenon, first studied by Michael Faraday in 1831.

But the co-existence of gas and magnetic field is unstable. The gas cannot cross magnetic lines but it can slide along the lines, draining into weaker spots. If through shock waves or perturbations of some kind a hollow develops in the magnetic field, gas will slide down into it, gravitationally attracted by matter already present there, sagging the depression even further. More gas will flow into the well until the field lines become so compressed that their elasticity can balance against the mass of the gas in presence. Many astronomers think that *giant complexes of clouds* are formed from diffuse HI clouds or SMCs as these clouds pass through the spiral wave density shocks and are further coalesced by the magnetic instabilities. Eventually, the gases and dust become dense and cold enough to self-gravitate. From this stuff, stars are born.

Summary

Space between stars is filled with rarefied and cold matter whose chemical composition reflects that of the Universe as a whole. The most striking feature of our Galaxy is its spiral pattern. This pattern is probably formed by matter density waves that the Galaxy builds up as it rotates rapidly and differentially about its own axis. Half of the interstellar matter is distributed very thinly throughout the Galaxy in the equatorial plane. The rest is found in the spiral arms as cloud formations of wide-ranging sizes and densities. Of greatest interest for their roles in the theory of the evolution of stars are the Small Molecular Clouds (SMCs) and the Giant Molecular Clouds (GMCs) which can become large, dense, and cold enough to self-gravitate: nurseries for future stars.

---

3. THE BIRTH OF STARS

As we have seen, the space between the stars is filled with a tenuous matter composed mainly of hydrogen and helium and, to a lesser extent, of other elements in atomic or molecular forms, in various concentrations, ranging from ten atoms per cm^3 in diffuse clouds to over a million of atoms per cm^3 in molecular clouds. A substantial part of this matter is contained in complexes of dust and molecular gas found in the spiral arms of the Galaxy. Embedded in these giant formations of clouds are hot, massive, and presumably young stars, grouped into extensive aggregates, called *associations*. The existence of matter of increasing densities and the presence of cold, dense clouds and large globules of gas along with hot, young stars in the same regions of space are two empirical facts which strongly suggest that stars condense in a gradual evolution from primal nebular matter. A close examination of many presumably star-forming regions tends to support this scenario. A good example is the Orion complex already mentioned before. It contains a giant emission cloud (the Orion Nebula) and a dense, cool cloud of gas, along with well-defined HII (ionized hydrogen) regions and stars in various phases of evolution. In the Nebula's luminous core, one can clearly see through a telescope clusters of faint, probably newly condensed stars amid expanding clouds of glowing gas and several dark clumps of dust apparently on the verge of collapse. The Nebula seems to be split into two by a giant molecular cloud (GMC) with a mass estimated at $2000\,M_\odot$. In the vicinity of this dense cloud are found two associations of hot, massive high-luminosity stars, the Orion OB association and the Orion Trapezium. Bright, compact HII regions occur throughout the molecular cloud along with many remarkable strong sources of infrared (IR) radiation.

In this section, we first discuss why we believe stars are continually born in the Galaxy, then give physical arguments in support of the above star-forming scenario. But to have a proper perspective of this phase in the general context of stellar evolution, let us first introduce an important tool in astronomy, the *Hertzsprung-Russell diagram*, which is a systematic representation of the basic properties of stars.

3.1 The Hertzsprung-Russell Diagram

During most of its lifetime, a star can be described roughly as a sphere of gas in thermal equilibrium, with luminosity L, surface (or effective) temperature T, mass M, and radius R.

A star radiates in all directions, so the fraction of its power reaching a detector with receiving surface area A a distance d away is $A/4\pi d^2$. Therefore, the *intrinsic luminosity* of the star, or the total energy it emits per unit time, is given by $L = \mathcal{L}4\pi d^2$, where $\mathcal{L}$ is the star *apparent brightness*, or the energy per unit time per unit area received by the Earth from the star at distance d away. To ascertain the luminosity, we need the apparent brightness and the distance to the star. The apparent brightness is measured by photometry and is usually available only within a range of wavelengths (for example, blue or yellow); corrections are then applied to get the apparent brightness for all wavelengths. The distances to nearby stars can be determined directly by the parallax method, which is reliable within 20 pc and helpful for estimating distances to 100 pc. For more remote stars, indirect methods have been devised, but they all depend ultimately on the accuracy of the parallax measures of nearby stars (see Chapter VII).

As we have explained, the surface temperature of a star can be obtained from its radiation spectrum: the temperature is deduced simply from the observed peak of the spectral distribution via Wien's displacement law (Sec. 2.1). Even more information can be extracted from an analysis of the absorption lines, which arise from processes occuring in the star outermost layers, or photosphere. The pattern of line spectrum depends on the temperature, the chemical composition, and the pressure prevailing in the photosphere. In this way, stars can be divided into various *spectral types*. In a sequence of decreasing effective temperatures, the spectral types read OBAFGKM.

Since a star can be approximately considered as a black body, its luminosity is also given by $L = 4\pi R^2 f = 4\pi R^2 \sigma T^4$(erg/s) where R is the radius of the star in cm and σ the Stefan-Boltzmann constant (Sec. 2.1). In cases where it is not possible to measure directly the radius of stars, this relation is used to find R after L and T have been determined by other means.

Finally, as for the star mass, again it is a difficult technical problem unless we have a double-star system. According to Kepler's and Newton's laws of motion (Appendix A), when two stars are in orbit around each other, the period of revolution increases as the distance between the two stars increases and as the total star mass decreases. So once we have measured the period of revolution and the distance between the two stars, we can calculate the sum of the masses of the two bodies. Furthermore, in a system of orbiting objects, each object revolves around the center of mass of the system. By measuring the distance of each star from the center of mass, we get the ratio of the masses. Now that we know both the sum and the ratio of the masses, we can obtain the individual masses. An important conclusion can be drawn from study of stars amenable to this technique: stars of the same spectral types have nearly identical masses. Thus, if we extend this empirical result to all stars, we get in a first approximation an estimate of the mass of a star from its spectral properties.

We see then, of the *four* basic properties of stars, only *two* are independent. These two are generally chosen to be L and T. If every star is

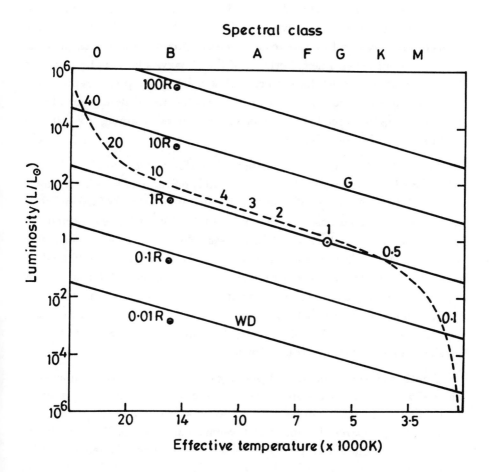

FIG. V.7 *Hertzsprung-Russell diagram. The dashed line represents the main sequence; along this line, numbers give stellar masses in solar units. G indicates the general region of red giants and WD, that of white dwarfs.*

represented by a point on a two-dimensional logarithmic plot, luminosity versus surface temperature, with the temperature scale running the 'wrong' way, we have the *Hertzsprung-Russell* (HR) *diagram* (Fig. V.7), named after its creators. Stars within 20 ly of the Sun are close enough for us to have accurate distances and may be considered as a representative sample of all the stars in our general neighborhood of the Galaxy. The striking feature of the HR grouping of such local stars is that most of the points representing these stars are not scattered all over the graph but are strung out along a slightly curved band (called the *main sequence* band) running from the hot, bright stars to the cool, dim ones.

Since $L \sim R^2 T^4$, curves of constant R for different L and T can be plotted. Given L and T, we can then simply read R off the graph without making any calculation. If now we also add in the measured masses of normal stars, we realize that a relationship between L and M also exists, as expected. We find L varies as M^n, where $n \approx 3$ for masses below $M_\odot$ and $n \approx 4$ for masses above $M_\odot$. Arthur Eddington suggested that these relations should hold not only for binary stars, whose masses can be directly measured, but also for all stars on the main sequence, including stars of unknown masses. Therefore, a measurement of L also yields M, thus expanding our mass knowledge to a vast number of stars. We will refer to stars on the main sequence as main sequence stars or, simply, ordinary or normal stars.

Another extension of the HR plotting can be made for stars in *clusters*, many of which at great distances from us. A cluster's angular size is always small, so all of its stars can be assumed to lie at the same distance from us. A plot of their *apparent* luminosities $\mathcal{L}$ versus temperatures T can be drawn; it has the same shape as a plot of L versus T, except for the labeling of the vertical scale. Sometimes the distances to the clusters can be determined from the luminosity of the variables they contain. An instructive example is the globular cluster M3 (Fig. V.8). Here, the distance to M3 is derived from the luminosity of a class of pulsating stars, RR Lyrae variables, which are closely analogous to Cepheids (see Chapter VII). The HR diagram shows few M3 stars on the main sequence. Most 'turn off' to the right toward larger radii and lower surface temperatures (the *red giants* region), then turn back toward higher temperatures at a constant, high luminosity to form a *horizontal branch*. Detailed calculations and observations yield typical estimates

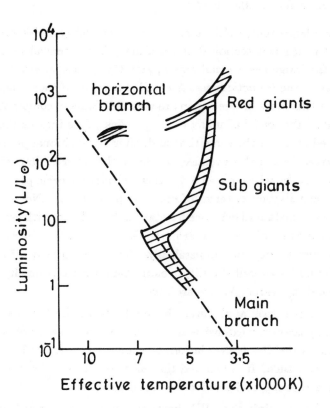

FIG. V.8 *Distribution of stars of a globular cluster on the Hertzsprung-Russell diagram.*

of about 12 billion years for the age of the globular clusters. Since the globular cluster stars are the oldest known objects in our galaxy, this number is generally considered as giving also (the lower limit of) the age of the Galaxy.

In the lower left-hand corner of the HR diagram, another group of stars with very low luminosities and very high surface temperatures can be clearly delineated: they are the *white dwarf* stars.

Why are the stars distributed in the HR diagram the way they are? What track in the diagram will a newly formed star follow as it evolves with time? Will it ever become a red giant or a white dwarf?

3.2 Evidence for Stellar Birth

There is evidence that, unlike stars, galaxies are fairly stable objects. We mentioned in the last section that the Galaxy has a central bulge. At the center of the bulge lies a small region, just 150 ly across, which contains a thousandth of the Galactic mass. A detailed analysis of theory and observations shows that stars in this region are very old. Stars of comparable age are also found in the nuclei of many other galaxies. This implies that galaxies were formed at about the same time and, after a transitional period, did not evolve further. From this premise, it follows that over billions of years the Galaxy has on the whole preserved without change its basic properties, such as its star distribution in terms of mass or spectral type. Now, whereas we have only circumstantial evidence for star birth, we know with certainty that stars die at the rate of one per year in the Galaxy. We know, because stars do not die in secret, they die unmistakably, often spectacularly. If the Galaxy is to maintain its overall stability, young stars must continually appear to replace the disappeared. Where are they?

Let us again look at the star clusters found in the Orion star-forming region. They are either extensive associations of a few dozens of hot, massive stars, such as the Orion association, or compact clusters of low-mass stars in the Orion nebula. It is believed that all these stars and others in similar situations are *young*, and this for several reasons.

First, by examining the HR diagram, one can get a rough estimate of the star age. The stars in question have perceptibly higher luminosities than the majority of the observed stars having the same surface temperatures; they lie slightly to the right of the main sequence band. It is estimated that the low-mass stars average only about a million years in age, while the massive supergiants are a few million years old. Compared to the Galaxy, they are indeed young. Secondly, stars in an association spread over a region a few hundred light years across and thus interact weakly with each other. They evolve in their respective orbits about the galactic center at different speeds; stars closer to the center move faster, and will quickly move ahead of stars farther out (see Sec. 2.4). Soon, well before a revolution (about 250 million years) is completed, the stars will be farther apart and will not form a recognizable association. Thus, stars we see in an association must be necessarily much younger than the Galaxy. And, finally, there is an argument based on dynamics which we will explain more in detail later on (see Sec. 4.1). For now, let us simply say that the main source of stellar energy is the thermonuclear process taking place in the interior of stars. Most visible stars radiate because four protons combine, through a series of steps, into an α-particle (a helium nucleus). Since a proton has a mass of 1.0080 (atomic mass units, with 1 AMU= 931.1 MeV) and an α-particle a mass of 4.0027, the excess mass of 0.0073 per proton should be available as energy for the star to expend as radiation. In the most favorable case of a star made of pure hydrogen, the initial energy supply is given simply by the product of that excess mass by the energy equivalence of the star mass (Mc^2) or, alternatively, $10^{52}(M/M_\odot)$ erg. A star with a mass of $20M_\odot$, a typical example of an OB star in associations, could radiate at the rate of 10^{38} erg/s, so that its nuclear energy reserve would run out in less than 100 million years. Actually, this is an overestimate because the star structure should change drastically well before its stock of fuel is exhausted. A real star with such a mass can expect a lifetime of only 10 million years. Because they evolve so fast, supergiant stars must have burned out by the time they are dispersed from their birth places in a spiral arm; that is why they are rare among representative field stars. Clearly, massive stars with high luminosity could not have been in the Galaxy from the beginning: we still see them, they must be continually formed.

All the stars thought to be young for the above reasons are surrounded by remnant dust and molecular clouds, some even still in their dusty cocoons. On the other hand, stars known to be old — the Sun and the globular-cluster stars — have no gas or dust around them. Therefore, we are justified in believing that star formation is a permanent process and that massive stars evolve out of interstellar medium in the regions of dense matter in the spiral arms of the Galaxy while low-mass stars may be formed throughout the galactic disk.

3.3 Formation of Protostars

In any extensive cloud formation, the galactic differential rotation creates large scale turbulence, which in turn acts to stabilize the cloud against gravity. Thus, two opposing forces are present in a cloud: *internal pressure* that tends to disperse matter and *gravity* that pulls it together. The relative sizes of their contributions to the internal energy of a cloud — the thermal and turbulent energy of the system and its potential gravitational energy — will determine whether the clump will eventually collapse or dissipate. To make rough estimates, let us ignore the turbulent motions. Since the total thermal energy arises simply from the contributions of the individual particles, it is given by $KE = \frac{1}{2}Mv^2$, where M is the mass of the cloud and v the particle mean velocity related to the *mean temperature* by $v^2 \sim T$. On the other hand, the gravitational energy is the energy of interaction between all pairs of particles, which is given by $PE \sim -\frac{3}{5}GM^2/R$, G being the gravitational constant and R the radius of the clump, assumed spherical and of uniform density throughout. The negative sign in PE reflects the attractive nature of the gravitational force, and the factor $\frac{3}{5}$ is a detail arising from calculus. Collapse occurs when $KE + PE = 0$. For given mass and temperature, the critical radius at which the cloud collapses is given by $R_c \approx 0.2M/TM_\odot$ pc .

Whether as an SMC or a GMC, a molecular cloud is submitted to a global turbulence generated by the galactic rotation. Matter in the cloud moves about at high (supersonic) speeds and undergoes compressions by shock waves. Sheets, filaments, clumps will condense. If these structures remain large enough, their internal motions will remain supersonic. Successive substructures will form until they become small enough for their internal motions to be subsonic. Subsonic structures suffer no internal shocks and so cannot possess or develop further structures. Such a small, subsonic clump tends to be very dense in the center and very diffuse near the surface. As these density differentials, or gradients, develop, the flow of energy into the clump from outside stops. Once the external turbulence is *decoupled*, the clump is free to dissipate its remaining internal turbulence and the internal motions become purely thermal. It turns out that a typical cloud when it becomes subsonic has a mass sufficiently large for gravity to overcome internal pressure. With no stabilizing force, the structure collapses spontaneously, and a single *low-mass star* will form. Since small, dense, subsonic clumps can be produced by supersonic motions in any type of molecular cloud, low-mass stars should form widely, as observed.

The process of formation of massive cores is quite different, as is suggested by observations. *Massive stars* are seen near other massive stars. In associations as well as in older clusters, the most massive stars occur in the more condensed regions. Within a large cloud system, massive stars are younger than smaller stars. Thus, the general picture is that the gas within a star-forming region becomes more centrally concentrated with time, either by a central condensation through supersonic collisions between clumps or by successive generations of low-mass stars in the same region of space. In these conditions, supersonic clumps will pick up matter from the surrounding medium at *enhanced rates* and become massively accreted clumps of heated gas within a few hundred millions of years. When a *density wave shock* from a nearby expanding HII region or a supernova explosion hits such a clump, the gas in the cloud is compressed and its volume decreases. Past a certain critical radius, the massive clump is decoupled from the surrounding turbulence by its large density gradients and becomes dynamically unstable. Collapse begins. But if no external shocks occur, the supersonic core will eventually subdivide into small subsonic fragments, and only low-mass stars can form.

To summarize, SMCs can promote only subsonic clumps, out of which small stars form. GMCs can develop successively many massive supersonic cores, which must accrete matter at enhanced rates and must be provided with strong external pressure, for example from external shocks, to become self-gravitating. An external shock, when it comes, triggers a process that leads to a *serial formation* of massive stars.

Once initiated, the dynamical collapse proceeds quickly and very non-homologously. Within thousands of years, the collapsing globule forms a dense central region which tends, if there is enough mass, to fall away from the outer regions to make a progressively opaque core. Throughout the free fall stage the core accretes infalling material from the outer parts to build up an object having a typical stellar mass.

When the globule is finally decoupled from external forces it can be considered as a statistically steady object. There is then a useful statement about its energy, based on the *virial theorem* (see Appendix A), which says that its internal kinetic energy is equal in magnitude and opposite in sign to half its potential gravitational energy, $KE = -\frac{1}{2}PE$. Initially, when the cloud is dispersed to infinity, both energy components vanish, and the total energy, which we define as the sum of the internal and potential energies, is therefore equal to zero, $E = 0$. When the cloud has condensed to a finite radius and reached equilibrium, it has potential energy PE and internal energy KE, and its total energy has decreased to $E = \frac{1}{2}PE$. Since energy is conserved overall, an equal amount of energy must appear elsewhere: it appears in the form of radiation emitted by the cloud as it contracts from infinite dispersion to its present size.

Under compression, the gas particles in the condensing core are crowded together and come sufficiently close to one another to experience mutual *electromagnetic* interactions. They speed up or change states and, inevitably, emit photons. Half of this energy leaves the cloud in a flare of *infrared* radiation (because $T \sim 50\,K$), invisible to the optical astronomer. The other half remains. Although some of this remaining radiation is absorbed by particles, and the kinetic energy thus gained goes into slowly heating up the gas, most of the energy available within the core is expended in internal work, first in breaking up molecular hydrogen into atoms ($H_2 \rightarrow H + H$), then in ionizing hydrogen atoms ($H \rightarrow H^+ + e^-$). This substantial diversion of energy into work prevents the internal pressure from maintaining itself at the level required to support the weight of the outer parts of the cloud, and the whole structure keeps on falling inward with nearly the speed of free fall. In the short lapse of a few thousand years, the cloud shrinks to a thousandth of its original size. Its luminosity drops ever lower because it is contracting and the surface temperature remains the same. The surface temperature is unchanged, at 50 K or so, because free fall only increases the translational velocities of the particles but leaves their random motions unaffected. The cloud's track plunges straight down on the HR diagram (Fig. V.9).

When hydrogen disassociation is completed, the conversion of gravitational energy into thermal energy can proceed more efficiently. Soon the system reaches a state which may be described as 'almost static' or, more precisely, as one in which the thermal pressure at every level in the cloud exactly balances the weight of a column of unit cross-sectional area above that level. Matter within the body may still be moving around, but its local acceleration is so small that no motion is perceptible in our timeframe. The cloud has reached *quasihydrostatic equilibrium* (QHE) and attained the status of a *protostar*.

At these earliest stages of their evolution, the protostars are probably surrounded by envelopes of infalling dust, by the remains of the original protostellar clouds, and by the lower-density gas in the molecular complex. The basic tool in the theoretical study of stars and protostars is the *stellar*

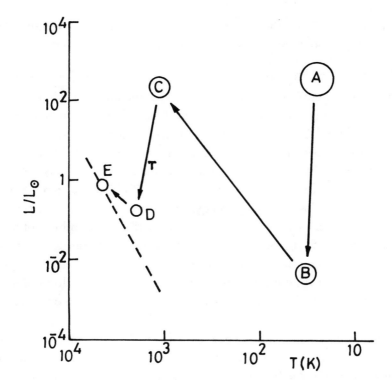

FIG. V.9 *Evolutionary track of a low-mass protostar on the Hertzsprung-Russell diagram: (A) Contraction of cloud and formation of a stellar nucleus; (B) Loss of internal energy through atomic and molecular disassociations, and collapse of the cloud core; (C) End of disassociation, convective transfer of energy, intense infrared radiation; region of the T Tauri stars; (D) Radiative transfer of energy and initial phase of nuclear fusion; (E) Arrival on the main sequence.*

models. A stellar model is an idealized mathematical representation of the correlations between the basic properties of the star based on physical laws and constrained by observations. In current models, the protostars should have energy distributions peaked at wavelength 10^{-4} cm and should show evidence of infall of the surrounding gas. A true protostar would be observed as an infrared (IR) object, with no radio continuum emission or IR recombination lines, and with incontrovertible evidence that contraction is its source of heat. The new generation of space telescopes planned for the

1990s (*e.g.*, SIRTF and the Large Deployable Reflector) will certainly extend our knowledge about this phase of star formation.

3.4 Evolution of Protostars

To see how the surface of the protostar can remain cool while its center is burning, we must understand how heat is transferred in matter. The photons produced in the central parts of the protostar are absorbed by the gas particles. These particles, in turn, reemit photons in all directions within the system, and the cycle of absorption and emission is repeated. A photon — starting from the center, making billions of billions of steps, back and forth, in and out, losing a bit of energy here, regaining some of it there, and raising the ambient temperature as it goes — will take hundreds of millions of years to work its zigzag way out to the surface. The transfer of energy in matter through radiation is a very, very slow process. Does it mean that the protostar will remain dim, invisible to observations, for hundreds of millions of years to come? The answer is no: heat can also be moved about by convection.

Convection is the transference of heat by the bulk motion of matter. It arises when there exists a large difference in density or temperature in different parts of a gas (or fluid); the gas, propelled by a force produced by this density or temperature gradient, carries the stored heat from one place to another, cooler place. Convection is a fast and extremely efficient means of transferring heat; the mass of the moving matter need not even be very large. It may take some time to bring a large kettle of water to the boil, but once bubbles are formed, water becomes quickly and uniformly hot.

In a protostar the convective transfer of energy begins when molecular disassociation just starts. As in boiling water, the disassociated molecules expand the gas, forming bubbles which rise through the denser layers to the surface. Once the collapse is over and the interior has become hotter, the heat transfer proceeds at a brisk pace. The same large temperature gradient that successfully counteracts gravity and brings the collapse to a stop also sets in motion the convective currents that carry hot gases up to the surface. In just a few months, the exterior temperature shoots up sharply.

The luminosity of a young stellar object having, say, a solar mass, increases dramatically to the equivalent of one hundred Suns in another flare of IR and near-IR radiation; meanwhile, its size remains essentially unaltered,

TABLE 2

Some signposts on protostar evolutionary tracks

Embedded infrared sources*: Luminous protostars surrounded by envelopes of infalling circumstellar dust and by the dark cloud material in the cloud complex. Age: 10^5 years.*
OB Star clusters and associations*: High mass stars approaching the main sequence along radiative tracks. Age: 10^6 to 10^7 years.*
Herbig emission stars*: Irregular variables of early type, closely associated with the dark-cloud material out of which they were formed. Will evolve into massive stars. Age: 10^5 to 10^6 years.*
T Tauri stars*: Young objects characterized by irregular optical variability and large mass outflow. Represent low mass protostars approaching the main sequence. Age: $10^6 - 10^7$ years when observed near OB complexes, $10^5 - 10^6$ years when near Herbig emission stars.*

---

about forty times that of the Sun. Once the protostar has become fully convective and the surface temperature has reached 4000 K, or so, the internal temperature gradient drops, the convective currents gradually weaken, and the contraction starts anew but at a much more leisurely pace. The system is in a state of quasi-equilibrium; the gravitational and pressure energies change in locked steps. As the radius decreases, the average temperature and the core temperature, although not yet the surface temperature, increase accordingly. When gravity has compressed the protostar to a ball twice the size of the Sun, the first photons have just made their way out to the exterior and radiation becomes the dominant mechanism for transferring energy throughout the system. As the protostar becomes fully *radiative*, its outer layers respond directly to the outgoing photons. The surface temperature and luminosity start to rise, and keep on rising while the system goes on contracting until it reaches one solar radius or whatever value consistent with the size of main sequence stars of the same type.

In this phase of their evolution, the protostars should radiate in the far and near infrared regions of the spectrum, possibly with sudden increase in luminosity. Many IR objects discovered in the past few decades appear to be protostars at their flare stage. One such object is FU Orionis, which flared up suddenly in 1936, increasing its brightness by tens of times, but the brightness has inexplicably remained almost unchanged ever since.

Another signpost of this phase of the protostellar evolution is a class of very interesting stars, called the *T Tauri stars*, so named after the variable star T in the constellation Taurus. Generally they are cool objects ($T \sim 4000\,\mathrm{K}$) whose brightness varies rapidly and irregularly. These variations stem from instability both in the stars upper layers and the circumstellar envelope. Observed luminosity and temperature place these stars to the right of the main sequence in the HR diagram, where the young stars are supposed to lie. The T Tauri are always observed in closely-packed groups, called T associations, embedded in dense clouds of dust and gas. They occur often, but not always, near OB associations, which are groups of younger, more massive hot stars. Many of the T Tauri, viewed at resolution of one second of arc or less, reveal themselves as double or multiple systems, usually including low-mass companions and extending over regions some tens of times the size of the solar system. The most dramatic example is the case of the prototype T Tauri star itself: it has a faint infrared companion of very small mass orbiting some 100 Au away (1 Au is equal to the distance Earth to Sun); some astronomers suspect that this companion may be a newly forming giant planet.

The spectra of the T Tauri have three very significant features. First, they exhibit lithium absorption lines, which indicates that lithium is hundreds of times as common in the T Tauri stars as in the solar atmosphere. Since lithium is among the first elements to burn in nuclear fusion reactions when stars reach the main sequence, the abundant presence of lithium in these stars implies that nuclear burning has not yet begun. Secondly, the spectra also exhibit emission lines of hydrogen and ionized calcium, among others. A comparison of the details of the observed lines and calculations shows that temperature increases with altitude in the atmosphere of the stars, a situation reminiscent of that observed in the upper layers of the solar atmosphere. Everything suggests the presence of rapid convective motions. Finally, many absorption line components are shifted toward the blue, which is reasonably interpreted as an effect due to a continuous ejection of matter from the stellar surface at the estimated rate of $10^{-7}\,M_{\odot}$ per year. This mass outflow may represent the mechanism by which stars of near-solar mass slow their rotation to values consistent with the low rotational velocities observed in stars of this class.

Observational evidence strongly suggests that T Tauri stars are low-mass protostars which have gone past their convective contraction stage and are beginning their radiative phase on their way to the main sequence.

More massive protostars evolve somewhat differently. In such a protostar, the gravitational contraction proceeds at a faster pace and the temperature in the interior rises more rapidly, which would cause the stellar opacity to drop sharply. Well before the protostar has reached the main sequence, the convective transfer of energy is replaced completely by the radiative transfer. When the mode of energy transfer changes, the evolutionary track of the protostar turns leftward in the HR diagram. As the protostar continues to contract, its luminosity remains constant but its temperature keeps on rising until it reaches the main sequence. In our effort to understand massive protostars, the *Herbig emission stars* play an essential role. Their spectra show both emission and absorption lines. The emission lines — at times simple with only H_α lines, at times much more complex indicating the presence of heavier elements — are produced in the stars themselves. Estimates of surface gravity, together with the stars' juxtaposition with dark clouds, would place the Herbig emission stars in the region of the HR diagram occupied by high-mass stars approaching the main sequence along equilibrium radiative tracks. The absorption lines result from attenuation of the stellar radiation by gases surrounding the stars. They are violet-shifted because the intervening gases are streaming rapidly away from the stars. These features indicate that the Herbig emission stars are surrounded by a hot ($T \sim 2 \times 10^4\,K$) expanding ionized gas. The observed optical variability may stem from variations either in the intrinsic brightness or in the physical size of the circumstellar envelope.

From the end of the free-fall period onward, the protostar is slowly contracting; its gravitational energy slowly decreases (*i.e.*, increases in magnitude). Again by the virial theorem, half of the energy released during this contraction process will be transformed into heat and the other half will leave the star as radiation. If gravitational energy is the only source of the star's radiant energy, the total amount of energy the star radiates over its life span must be equal to its thermal energy supply. For a star like the Sun, with an average temperature of $10^7\,K$, the mean energy per particle is

$\frac{3}{2}kT = 2 \times 10^{-9}$ erg. Assuming the star to consist only of protons and electrons, we can estimate from its mass the total number of particles it contains to be 2×10^{57}. So its thermal energy amounts to $\approx 5 \times 10^{48}$ erg; an equal amount of energy is available for radiation. At the Sun's observed radiant power of $L_\odot = 4 \times 10^{33}$ erg s^{-1}, this energy supply would last some ten million years. Now, the Sun must have gone safely through its own protostar stage: it has existed as a true star for about 4.5 billion years, as geological evidence on Earth indicates. How has it managed to escape the seemingly inexorable grip of gravity? What is the alternative, so much more prodigious, source of energy that would prolong its life a thousandfold?

Summary

One of the most useful tools in astronomy is the Hertzsprung-Russell diagram. It is more than just a plot of luminosity versus temperature of stars: it is a systematic representation of the correlations between the basic properties of normal stars. From the use of this tool as well as from other observational evidence and calculations based on stellar models, we are led to conclude that many stars in the sky are young.

Star-forming regions are generally found embedded in dense cloud complexes, Small Molecular Clouds or Giant Molecular Clouds. A clump of cloud becomes self-gravitating when it has enough gravitational energy to overcome the dispersive internal turbulent motions. It can gain gravitational energy either by accreting mass from the surrounding medium or by undergoing a sudden compression under the action of some powerful external force. Within a short time, the collapsing globule of gas will build up a dense central nucleus, which tends to fall away from the outer parts but which may continue to pick up infalling material from the envelopes to form a progressively opaque core having a typical stellar mass. The evolution of this object, or protostar, will take it through several successive phases (free-fall, convection, and radiation), characterized by increasingly higher internal temperatures, before it reaches its position on the main sequence line of the Hertzsprung-Russell diagram.

These early evolutionary phases of stars are marked by several events: flares of infra-red radiation, formation of the T Tauri stars and the Herbig emission stars.

4. BRIGHT, SHINING STARS

At the end of the previous section, we came to the conclusion that both gravitational and thermal energies were inadequate to sustain prolonged radiation for the Sun and for most of other stars over their lifetimes. There must necessarily exist other sources of energy inside ordinary stars.

Up until this point in the stellar evolution, two kinds of forces are involved: the *gravitational force* and the *electromagnetic force*. As we shall see in greater detail in Chapter VI, there exist in Nature two other fundamental forces, the *strong force* and the *weak force*, so called because the former is the strongest of all known forces and the latter, much weaker than the electromagnetic force. In contrast to the other two more familiar forces, they both have a short range, extending over about 10^{-13} and 10^{-15} cm, respectively. Because their action is mainly confined within this range, they do not manifest themselves in everyday life but are nevertheless essential to the structure and the dynamics of matter at a deeper level, the level of atomic nuclei and elementary particles. These are precisely the two forces thought capable of producing, via a process called the *thermonuclear fusion*, the energy necessary to sustain the long-lasting luminosity observed in most stars. Although there is yet no direct empirical proof, the evidence for such forces being the generator of energy inside ordinary stars is compelling. In addition to the stellar luminosity, these forces can explain the chemical composition found in the celestial bodies. An interplay between the strong nuclear force and the gravitational force provides the most plausible explanation for a vast body of data, from the remarkable stability of stars on the main sequence to various phenomena marking the later stages of the stellar evolution. To verify the theory more directly, large scale attempts have been or are being made to detect products of the nuclear reactions occuring inside stars.

4.1 Nuclear Sources of Stellar Energy

Except for the nucleus of the ordinary hydrogen atom, which consists of a single proton, nuclei of atoms, including other isotopes of hydrogen, contain both protons and neutrons, collectively called the *nucleons*.† Nucleons interact with one another both through a mainly attractive nuclear force and a

---

† The nucleus of ordinary hydrogen (H) contains one proton; the nuclei of its isotopes, the deuterium (^{2}H) and the tritium (^{3}H), contain besides the

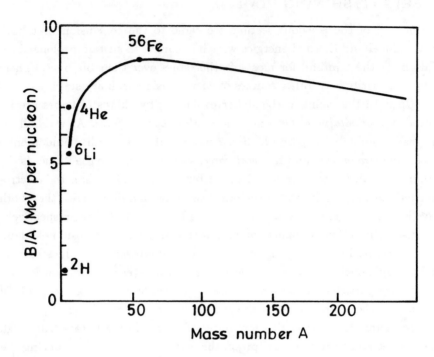

FIG. V.10 *The nuclear binding energy per nucleon. Note that ^4He is more strongly bound than the neighboring-mass nuclei and that ^{56}Fe has the largest average binding energy among all of the nuclei.*

repulsive electric force. Since the former is a hundred times stronger than the latter, it contributes the largest part to the nuclear potential energy.

The nucleus is of course a bound system of nucleons, and so its total mass is always less than the sum of the masses of the constituent nucleons. The mass deficit represents the *binding energy B* of the nucleus, *i.e.*, the energy that must be provided to split the nucleus apart so that its nucleons are at

---

requisite proton one neutron and two neutrons, respectively. Ordinary helium (^{4}He) has two protons and two neutrons; its isotope ^{3}He has two protons and one neutron. Generally a nucleus is designated by the symbol ^{A}Z, where Z stands for the name of the element and A the mass number, or the number of nucleons it contains.

rest and out of reach of each other's forces. Defined in this way, the binding energy is a positive quantity. The average binding energy of a nucleus, B/A, is its total binding energy, B, divided by its nucleon number, A. Figure V.10 shows the variation of B/A with nucleon number. A remarkable feature of the curve is that it remains relatively flat except for the very light nuclei: the average binding energy for most nuclei is about 8 MeV per nucleon.

The binding energy curve can be understood as follows. If every nucleon in the nucleus attracted all of the others, the binding energy B would vary with the number of pairs, or roughly with the square of the particle number, A^2. The fact that B is observed to vary not as A^2 but linearly with A must mean that each nucleon attracts not all of the others but only its nearest neighbors. Experiments show that the nuclear density in most nuclei is nearly constant, and thus each nucleon has about the same number of neighbors. And so each nucleon contributes about equally to the binding energy, as observed. But, evidently, nucleons on the nuclear surface have fewer neighbors and so are less tightly bound than those near the nuclear center. A correction taking into account surface tension must be applied to the linear approximation. It is proportional to the surface area, and is largest in the very light nuclei (where practically all nucleons are on the surface). Thus, B/A should be small for light nuclei and should increase rapidly to reach a plateau in heavy nuclei. The binding energy is further reduced by the Coulomb repulsion of the protons. This effect, while small in light nuclei, becomes substantial as the proton number increases, which makes the plateau tilt downward for A greater than 60. At the peak of the curve, when $A \sim 60$, nuclei are most tightly bound.

The behavior of the binding energy curve suggests that nuclear energy can be extracted in two ways. First, in *nuclear fusion* two light nuclei combine into a single nucleus with A still below 60; the final average binding energy will be higher than the initial average binding energy and energy is released. For example, in controlled fusion reactors and in fusion weapons, a deuteron and a triton (nuclei of deuterium and tritium) are made to fuse into an α-particle (nucleus of helium) and a neutron, resulting in a release of 3.5 MeV per nucleon. In the second method, known as *nuclear fission*, a nucleus much heavier than iron splits up into smaller nuclear fragments and, again, the energy unlocked from binding is converted into kinetic energy for the final products. In controlled fission reactors and in fission-based weapons, an isotope of uranium (^{235}U) is often used as fuel; it breaks up into a variety of lighter nuclei, yielding about 0.9 MeV per nucleon. Needless to say, there is no real overall gain of energy, only a transfer of one form of energy and mass into another.

Stars approaching the main sequence contain few if any heavy elements but plenty of ionized hydrogen atoms (*i.e.*, protons). Fission is clearly improbable but fusion is quite possible. The process of fusion occuring in the stellar interior is *thermonuclear*, that is, it involves highly energetic but still thermal particles, with a Maxwellian velocity distribution characteristic of the star temperature. At temperatures of the order of ten million degrees (10^7 K) prevailing in the center of stars at this stage, the particles in these regions have a mean kinetic energy less than a thousand electron volts (1 keV). This energy is too small for two protons or any two nuclei to overcome their mutual electric repulsion, or *the Coulomb barrier* (which generally exceeds 1 MeV), and approach to within the range (10^{-13} cm) of their mutual nuclear interactions. But for a gas in thermal motion, there will always be some particles with energies much higher than the mean value and capable of initiating nuclear reactions. Of course, as the gas temperature goes up, the number of energetic particles increases accordingly, and so does the rate of reaction. Moreover, even particles with energies much smaller than 1 MeV might be able, nevertheless, to penetrate their mutual Coulomb barrier and interact. This phenomenon unthinkable in terms of classical physics can occur by virtue of the wave properties of particles; it is referred to as the *quantum tunnel effect* (see Appendix B). Calculations show that in the central regions of stars approaching the main sequence, these two factors allow protons with an energy as low as 20 keV to ignite proton-proton fusion. Although the number of protons having such an energy is still a tiny fraction of all particles inside stars at this stage, it is sufficiently large in absolute terms for nuclear reactions to proceed and generate energy. As is clear from our discussion, the rate of nuclear reactions and hence the rate of nuclear energy production depend strongly on both the Maxwell-Boltzmann distribution and the quantum tunnel effect — both nonlinear functions of temperature — and so are *strongly influenced by the temperature*: fusion in stars will turn on and off quickly with small changes in temperature.

4.2 On the Main Sequence

As a protostar is approaching the main sequence, it goes on contracting gravitationally, raising the temperature in its core to above ten million degrees.
At this point, with the onset of nuclear fusion, the protostar becomes a *star*,
a much more stable object, driven by the play of nuclear and gravitational
forces. As the rate of nuclear reactions is strongly affected by the temperature, the production of nuclear energy might for an extremely short time
exceed that of gravitational energy in the star central regions. The heated
stellar material will expand *locally* to some *infinitesimally small* degree. Since
the whole system is in quasi-equilibrium, the expansion causes both pressure
and temperature to fall ever so slightly, which in turn immediately dampens
the nuclear fusion rate. The gravitational pull will quickly take over as the
dominant source of energy, and the resulting compression instantly raises the
temperature, restoring the system to its original equilibrium state which has
been temporarily broken by local heating. This local variation in the state
of matter should not be construed as pulsations of any kind for the star as a
whole. On the contrary, the harmonious interplay between contraction and
fusion, together with the abundant supply of hydrogen as fuel, contributes
to maintain the star in an overall *stable structural equilibrium*, at constant
radius and temperature, that can last for a considerable length of time. Most
luminous stars we see in the sky belong to this period of their evolution, by
far the longest in their active life.

At this stage, nuclear energy is generated in the stellar interior mainly
by the conversion of hydrogen into helium, which can proceed through two
different sequences of nuclear reactions, called respectively the *proton-proton
cycle* and the *carbon-nitrogen cycle*.

• *The proton-proton cycle*

The first step in the proton-proton cycle is the combination of two protons
into a deuteron (nucleus of ^{2}H), producing a positron (e^+) and a neutrino (ν):
$p+p \rightarrow {}^2\text{H}+e^+ +\nu$. This reaction has a very low probability of occuring for
two reasons. First, it involves the weak interaction in the inverse β-decay, or
transformation of a high-energy proton (p) into a neutron (n): $p \rightarrow n+e^+ +\nu$;
the weakness of the interaction makes the decay slow to proceed. Second,
the number of protons having the relatively high energy needed to overcome
the Coulomb barrier is very small, a tiny fraction ($\sim 10^{-8}$) of all protons in

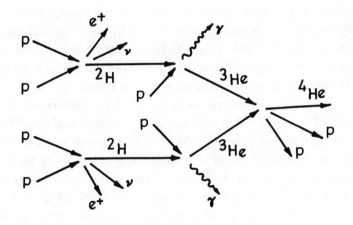

FIG. V.11 *Sequence of reactions in the main branch of the proton-proton chain of fusion.*

a star. The reaction rate is thus very small, about 5×10^{-18} per second per proton. Nevertheless, there are so many protons around (of the order of 10^{57} in the Sun) that the reaction can indeed take place at a sufficient rate to get the chain of reactions going.

Once a deuteron is formed, it is very unlikely to react with another deuteron because there are relatively few of its kind around (about one deuteron for every 10^{18} protons); but it is very likely to combine quickly with another proton to produce helium isotope ^{3}He. The next step is the fusion of two ^{3}He to produce ordinary helium (^{4}He or α-particle) together with two protons. Since ^{3}He is scarce, it will take a few million years for two of them to combine. This sequence of reactions, known as the *proton-proton cycle*, is shown in Fig. V.11. The net reaction is the conversion of four protons to helium:

$$4\,p \rightarrow {}^4\text{He} + 2\,e^+ + 2\,\nu,$$

releasing 26.7 MeV. The energy effectively converted to heating and radiation is somewhat less because the neutrinos, which interact very weakly with matter, emerge directly from the star into space carrying away 0.26 MeV of energy each. As we already mentioned, the rate of nuclear fusion depends

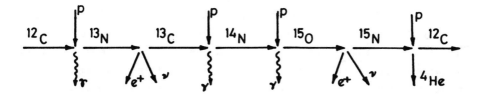

FIG. V.12 *Sequence of reactions in the carbon-nitrogen cycle of fusion.*

strongly on temperature; for the proton-proton chain, it varies as the fourth power of the temperature, $\sim T^4$.

There are other ways for ^{3}He to react: it could combine with protons, with deuterons, or with α-particles. But reactions with protons are not possible because they would produce unstable systems which immediately break up on formation, while fusion with deuterons is very improbable because deuterons are scarce, being converted into ^{3}He as soon as formed. The third alternative is the only one possible when a ^{3}He nucleus combines with a ^{4}He nucleus to form the nucleus of the beryllium isotope ^{7}Be. Next, two paths can be considered. Either the ^{7}Be captures a proton to produce ^{8}B which β-decays according to ^{8}B $\rightarrow$ ^{8}Be + e$^+$ + ν, and the ^{8}Be splits into two α-particles. Or the ^{7}Be captures an electron and turns into a lithium nucleus (plus a neutrino), which then grabs a proton to become an unstable ^{8}Be which quickly decays into two α-particles as in the previous case. Of the three possible branches of the proton-proton chain, which in fact yield the same net reaction and release the same energy, the first, as described in the previous paragraph, is the dominant one. But the second branch is interesting in that it produces fairly energetic neutrinos having a continuous energy distribution with an endpoint at 14 MeV, quite different from the energy profile of neutrinos produced in the first branch. (Neutrinos are produced in the third branch too, but they are monoenergetic).

• *The carbon-nitrogen cycle*

If in addition to hydrogen and helium there are heavier elements inside a star, nuclear fusion can follow other paths. One such path, called the *carbon-nitrogen (CN) cycle*, is very important in the evolution of massive stars. It consists of a series of six reactions — proton captures and beta decays — as shown schematically in Fig. V.12. The net process is $4\,p \rightarrow$ ^{4}He $+\,2\,e^+ +2\,\nu$, exactly as in the pp chain, and the energy released is also about the same. The ^{12}C is needed to start the chain and is recovered at the end; it plays a catalytic role in the reaction cycle. Once initiated, this cycle can proceed more rapidly than the pp cycle because it does not involve a slow process analogous to the deuteron formation. However, as the Coulomb barrier is six times higher for proton-carbon reactions than for proton-proton reactions, the CN chain will not be competitive with the pp chain until the central temperature of the star is near 20 million K. The center of the Sun is not that hot, and so the pp chain dominates. But stars with mass $M > 3M_\odot$ can reach this level of temperature and the CN chain will be their main source of nuclear energy. This chain depends very strongly on the temperature, $\sim T^{16}$.

Finally, protons can also combine with the nuclei of light elements, such as lithium, beryllium, and boron, if they happen to exist in the star. These reactions can proceed at relatively high rates, even at temperatures as low as 10^6 K, because the Coulomb repulsion between the proton and the target nucleus is not as strong here as in collisions of protons with carbon or nitrogen. The particularity of these reactions is that the reacting elements are not restored to the system but are irretrievably lost. For this reason few light elements are found in the Sun and other normal stars. On the other hand, their presence in a star is an excellent indicator of its very young age. T Tauri stars are a case in point (see Sec. 3.4).

• *The solar neutrinos*

Nuclear reactions in the center of the Sun produce both photons and neutrinos. The photons interact with the solar material and will take millions of years to reach the photosphere. The light that reaches us from the Sun comes from the energy released in reactions occuring in the photosphere, and is characteristic of the Sun's surface, not of its interior. Neutrinos, in contrast, are weakly interacting particles and rarely scatter off anything. A neutrino produced in the Sun would go directly to the solar surface. Thus, if we could detect solar neutrinos, we would have direct evidence of the nuclear processes inside the Sun and direct information about the conditions there now. Unfortunately, neutrinos are elusive particles, hard to detect for precisely the same reason they interact weakly with matter.

Raymond Davis and collaborators have devised a sensitive experiment to detect solar neutrinos by collecting argon-37 atoms produced in the reaction

$$\nu_e + {}^{37}\text{Cl} \rightarrow {}^{37}\text{Ar} + e^-.$$

The threshold energy for this reaction, or the minimum neutrino energy required for the chlorine-argon transmutation, is 0.814 MeV, well above the average energy of the bulk of neutrinos produced by the Sun. Thus, Davis's experiment was primarily designed to detect the relatively few high-energy neutrinos emitted in the secondary branch (${}^8\text{B} \rightarrow {}^8\text{Be} + e^+ + \nu_e$). The experiment's long term average capture rate from 1970 through 1985 was 0.472 ± 0.037 solar neutrinos captured per day, compared to the theoretical capture rate of 1.2 captures per day predicted by the standard solar model. Experiments performed in the Kamioka lead-zinc mine in Japan have confirmed this deficit. The persistent disagreement between calculated and observed fluxes is troubling because it could mean that we understand solar physics or neutrino physics less well than we think. But, on the other hand, we could argue that since the rate of ${}^8\text{B}$ production is strongly affected by conditions in the solar core which we know with less than complete confidence, the apparent disagreement between theory and experiment might be due a lack of understanding of the production process of neutrinos or their subsequent interactions with the solar material.

Therefore, it is important to detect the more abundant low-energy neutrinos produced in the deuteron-formation reaction of the main branch of the *pp* chain, which is also physically better understood. One possible way would be to let the incoming neutrinos interact with liquid gallium, transmuting it into germanium which could be extracted and counted. The relevant reaction, $\nu_e + {}^{71}\text{Ga} \rightarrow e^- + {}^{71}\text{Ge}$, has the energy threshold of 0.23 MeV. A gallium detector could monitor many of the *pp* solar neutrinos which, we recall, have an average energy of 0.26 MeV. Two experiments of this kind, the Soviet-American Gallium Experiment and the European Gallex collaboration, have begun gathering data. After the first few months of observation, the neutrino deficit looks even worse at low energy.

Suppose further data confirm these preliminary results, what could be the explanation for the shortage of solar neutrinos without calling into question the essentials of the standard solar model? First, note that the chlorine and gallium detectors can see only *electron neutrinos, ν_e*. But according to the theory of particles (Chapter VI), neutrinos come in three kinds, or 'flavors', associated respectively with the electron, the muon, and the tau-lepton. Neutrinos of different flavors can be converted into one another if there is a mass difference and a mixing between the flavor eigenstates, and there is no reason to believe that there is not. So it could be that the electron neutrinos, once produced by perfectly well understood mechanisms in the Sun, are converted, during their outward trip, into muon neutrinos by some sort of resonant interaction with electrons in solar plasma at a sufficient rate to reduce the flux of ν_e reaching the Earth. Whether this mechanism by itself could resolve the twenty-year old mystery still remains to be seen. The last episode in the long saga of these fascinating little ones has not been told. The Sudbury Neutrino Observatory, soon to be built deep in a nickel mine in Canada with its large heavy water detector that can detect the passage of neutrinos of all flavors (as they break deuterons apart), will certainly have something interesting to tell us.

• *Properties of stars on the main sequence*

As we have seen, most observed stars for which the luminosity and temperature can be deduced observationally lie on a diagonal band of the HR diagram called the main sequence. The theory of stellar evolution indicates that stars on the main sequence are chemically homogeneous, with constant fractional abundances of elements throughout most of their volume, and are burning hydrogen to helium in a relatively small core. The position of a star on the band varies little with the amount of hydrogen burned up as long as the stock of hydrogen in the core has not been completely exhausted. The most important factor in determining a star's position on the main sequence is its mass.

From observations of the orbital motions of binary stars, we know that the masses of ordinary stars range from about 0.1 $M_\odot$ for the lowest to somewhat more than 50 $M_\odot$ for the highest. This empirical fact can be understood as follows. A protostar with a mass less than 0.08 $M_\odot$ (called a *brown dwarf*) does not have enough self-gravity to compress its center to the high temperatures needed to achieve thermal equilibrium via hydrogen burning; after a deuterium burning phase, it will cool to its fully degenerate configuration. In contrast, an object with a mass greater than 60 $M_\odot$ must have its interior compressed to such high temperatures that nuclear burning can proceed at greatly enhanced rates and produce a significant excess of thermal energy and pressure to the extent that radiation pressure ($P \sim T^4$) begins to dominate over matter pressure ($P \sim T$). Eventually, the star either explodes, or ejects enough matter to fall below the critical limit of $M \sim 60\,M_\odot$. However, this does not mean that massive or even *supermassive* stars might not form and live for a short time before disrupting themselves. Stars with masses in the thousands or even millions of solar units may have an exceedingly short life but, with their huge masses and their huge energy release, they would not pass without disturbing the universe.

Basically a star shines because heat can leak out from the interior. The rate of the radiative leakage determines the star luminosity L, which turns out to have a simple dependence on the other basic characteristics of the star. As a function of mass, it varies roughly as $L =$ constant $\times M^4$ (as an average over all stars on the main sequence). This relation means that a main sequence star of ten solar masses radiates ten thousand times more energy per second than the Sun. Yet it has only ten times more fuel to burn than the Sun, because both stars can use only ten percent of their total hydrogen for core-burning on the main sequence. So the more massive star has a lifespan on the main sequence of about ten million years, or one thousand times shorter than the Sun's lifespan.

The temperature within a star can be estimated from the virial equation, $KE = \frac{1}{2}PE$, which yields T(interior)$\sim MR^{-1}$. Since nuclear fusion is ignited at a certain threshold temperature, regardless of other stellar properties, all stars reaching the main sequence should have roughly the same core temperature and hence the same mass-to-radius ratio. Putting this result together with the two relations on luminosity ($L \sim M^4$ and $L \sim R^2T^4$), we see that the surface temperature of stable stars increases weakly with mass (something like $T =$ constant $\times \sqrt{M}$). For example, given that the Sun has a surface temperature of 5800 K, a ten-solar mass stable star would have an effective temperature of about 20000 K. Massive main sequence stars not only have higher luminosities but also hotter surfaces than low-mass main sequence stars. That is the reason for the distribution of stable stars along the diagonal band of the HR diagram.

Not all stars on the main sequence have the same internal structure,

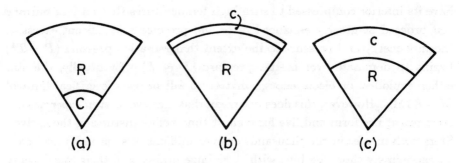

FIG. V.13 *Internal structure of stars in terms of prevailing modes of energy transfer, by convection (C) or by radiation (R): (a) High-mass star; (b) Solar-mass star; (c) Subsolar mass star.*

as theoretical astrophysicists have learned through their extensive numerical calculations of star models. For one thing, small stars are denser than big stars. This is easy to understand if we just remember that on the main sequence MR^{-1} is roughly constant and the density is given by $\rho \sim MR^{-3}$, and so $\rho \sim M^{-2}$. But more importantly, stars may also differ in their internal make-up. Hot, massive stars have a convective inner zone surrounded by a radiative region where energy is transported by radiation. For example, for a star ten times the mass of the Sun and four to five times as hot, the convective core may occupy one-fourth of the star's volume and contains the same fraction of its mass (Fig. V.13a). The temperature there is likely about 25×10^6 K. At this temperature, the carbon-nitrogen cycle is the star's main source of energy. As energy is generated at a prodigious rate in this process, radiative transfer alone is not adequate to remove the heat produced, and convection becomes necessary.

But for *red dwarfs*, those stars on the lower main sequence, the reverse seems to hold. A red dwarf whose mass is 0.6 solar units and luminosity 0.56, has such a low self-gravity that its center can be heated only to about 9×10^6 K, and so only the proton-proton chain can be initiated. Since a lower temperature makes the stellar material more opaque, it is difficult for radiation alone to move energy produced at the center all the way to the surface; convection must do the job. A typical red dwarf has a radiative intermediate region enveloping a small hydrogen-burning core, and is in turn surrounded by a convective zone which contains about one-tenth of all the star's mass (Fig. V.13c).

How about the Sun? This mid-main-sequence star has a central temperature of about 14×10^6 K. It draws energy from reactions in the proton-proton cycle and, for some small fraction, in the carbon-nitrogen cycle. It has a structure similar to that of a red dwarf, but its outer convective layer is considerably thinner, containing only about 2 percent of the Sun's mass (Fig. V.13b). Interestingly, its central density is higher than that found in stars in other parts of the main sequence, about $135\,\mathrm{g\,cm^{-3}}$, or 100 times the mean solar density. This high concentration of mass at the center arises from a considerable amount of nuclear 'ashes' accumulated ever since the Sun reached the main sequence, some 4.5 billion years ago.

4.3 Away from the Main Sequence

Although a star remains on the main sequence for a very long time with hardly any changes in its basic properties, this is only a temporary pause, a sort of truce imposed by a balance of the forces in cause. When the stores of nuclear energy are used up, the star faces an inevitable end to its career. The slide toward this end starts when nearly all the hydrogen in the stellar *core* — defined roughly as that central part containing one tenth of its mass — has burned to helium. For each α-particle formed, four protons disappear. Thus, the central pressure (which is proportional to the number of particles in the core) gradually decreases, and eventually becomes inadequate to support the outer inert regions. The core begins to contract. As the core *contracts*, gravitational energy is released, adding to the nuclear energy already present. The temperature outside the core goes up together with the temperature gradient throughout the rest of the star, causing the regions outside the core to *expand*. As no energy reaches the surface of the star and as the potential energy of the particles at the surface increases, the particle kinetic energy and hence the *surface* temperature must decrease to keep the total energy at the surface constant. Thus, the general trend is for the star to move off the main sequence *rightward* toward lower (surface) temperatures on the HR diagram (Fig. V.14).

• *Up on the giant branch*

For a low-mass star, $M < 3M_\odot$, the nuclear energy comes from the *pp* cycle. Recall that the reaction rate depends on both the number of protons present

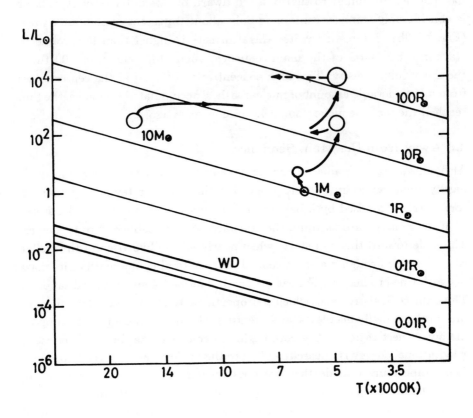

FIG. V.14 *Post main sequence evolutions of a* $1M_\odot$ *star and a* $10M_\odot$ *star.*

and the temperature ($\sim T^4$). So for the star to maintain its structural equilibrium, the temperature in the core must increase to compensate for the depletion of protons, which entails a general upward trend for the star's average temperature, and hence a brighter star. On the HR diagram, the star's track goes *upward* and rightward. For a higher-mass star, $M > 3M_\odot$, the CN cycle is the source of nuclear energy. Since the rate of fusion in this cycle goes steeply up with temperature, as T^{16}, only a slightest increase in temperature in the core will be needed to make up for the drop in the particle number. There is then no significant temperature change in the star as a whole, and hence the surface luminosity remains almost unchanged. On the HR diagram, the star follows a nearly *horizontal* rightward track.

Once all the hydrogen in the core has been converted to helium, the core becomes an inert solid helium sphere, not yet hot enough to ignite helium fusion. With no new energy inflow, the core becomes *isothermal*, with a uniformly constant temperature throughout, and no longer can support the outer layers of the star. At this point, whether the whole star contracts or not depends mainly on its mass. For a low-mass star, the temperature in the core has risen enough so that when the core hydrogen completely runs out, the temperature just outside the core becomes sufficiently high to ignite a thin shell of burning hydrogen. Thus, there is a smooth transition between core exhaustion and shell burning: no need for the whole star to contract. The same cannot be said of a high-mass star. Since its core temperature has remained more or less constant, the star must contract a great deal before the layers outside the core become hot enough to start hydrogen fusion. Its evolutionary track must turn abruptly leftward just before shell burning.

Meanwhile, the helium core still produces no new nuclear energy and continues to lose heat to the cooler surrounding regions. It must go on shrinking. The contraction of the core, accompanied by the addition of more mass falling down as helium ash from the burning shell, raises the density at the center of the star. Part of the high energy generated by the shell finds its way to the surface, but the luminosity carried outward by a radiative envelope is limited by the photon diffusion law and is nearly fixed by the star's mass. The major part of the energy produced by the shell goes into heating up the envelope, causing it to expand to a greater and greater radius. The surface temperature decreases for the same reason (conservation of energy) we described before. Now with R going up and T going down, the surface luminosity $L \sim R^2 T^4$ remains more or less constant, as we expected.

As the star expands, the surface temperature keeps falling, but it cannot fall indefinitely. In regions below the stellar surface, where the temperature remains low, hydrogen exists as neutral atoms. But as we go deeper, we find a zone where the temperature is high enough to ionize hydrogen. Bubbles will form, rising rapidly through the cooler material to the surface, and when the temperature at the surface drops below a certain limit, the whole envelope becomes convective. Hot gases flow upward, and the luminosity increases a thousandfold in about half a million years. As new energy now arrives at the surface, the temperature which has been falling there lately stabilizes around 4000 K. This temperature barrier forces the evolutionary tracks to turn almost vertically upward for all but the most massive stars. We now have a red and very large star, a *red giant*. To have a picture, imagine a star as massive as the Sun but 20 times larger and 200 times brighter. At the center of this giant star is a tiny core two thousand times smaller than the star itself but containing more than 10% of the star's mass, packed to a density around $1.4 \times 10^5 \, \mathrm{g \, cm^{-3}}$ or 2×10^{28} nuclei $\mathrm{cm^{-3}}$, in the form of helium. The temperature in the core is uniformly high, around 20 million degrees, yet not high enough to burn helium. Convection controls the energy transfer in all the surrounding envelope, except in the thin shell just outside the core and the thin transparent outermost region where radiation is the rule.

About 10^8 years have elapsed since we left the main sequence. Since that time stars of all masses (greater than $0.8 M_\odot$) have taken rather similar routes on the HR diagram, all ending in the formation of red giants. From now on, however, they will follow very dissimilar evolutionary paths, each driven by their own mass or, more exactly, the mass of their core.

• *From the helium flash to a quieter end*

Let us first consider low-mass stars. As the core continues to contract, it becomes hotter and denser. The electrons that have long been pulled out of their atomic orbits are now so tightly packed that they evolve into a new state of matter, called the *degenerate state*. Many aspects of degenerate matter are discussed in Chapter II and Appendix C, but it might be useful to recall here a few basic facts relevant to stars.

The degenerate state of a system of N identical fermions (such as electrons or neutrons) at temperature $T = 0$ can be defined as that configuration in which all the quantum-mechanical particle states, with energy extending from zero to a maximum E_F, are occupied, one particle per state. This is the familiar Fermi distribution at $T = 0$. The maximum energy, called the *Fermi energy*, represents the largest kinetic energy a particle in the system can have; it exists because the Pauli principle requires that no two fermions can have exactly the same quantum state and so the particles must fill up successively higher energy levels. The corresponding momentum, called the *Fermi momentum*, p_F, can be estimated, by dimensional analysis or by using the uncertainty principle, to be related* to the particle density n by $p_F \sim n^{1/3}$. The Fermi energy is accordingly given† either by

$$E_F = p_F^2/2m \sim n^{2/3}/mc^2 \text{ in nonrelativistic limit,}$$

where m is the particle mass and c the velocity of light, or by

$$E_F = p_F c \sim n^{1/3} \text{ in relativistic limit.}$$

Three points are worth noting. First, p_F does not depend on the type of fermion nor on whether it is relativistic or not; it depends only on the particle density n. This dependence can be easily understood. As the fermions are pushed in closer together, their relative distance decreases in proportion to the increase in their relative momentum in accordance with the uncertainty principle. Thus, an increasingly compressed state of matter gives rise to a larger degeneracy pressure. Contrast this situation with an ordinary gas, where the motions of the particles become much more sluggish as the temperature is lowered, and the pressure drops accordingly. Second, E_F depends on the particle density in two ways: $E_F \sim n^{2/3}$ in the nonrelativistic regime and $E_F \sim n^{1/3}$ in the relativistic regime. This different n dependence will

---

* More exactly, $p_F = h(3n/8\pi)^{1/3}$ or $p_F(MeV) \approx 6 \times 10^{-11} n^{1/3}$ where n is in cm^{-3} and h the Planck constant.

† The corresponding exact formulas are $E_F(MeV) = 18 \times 10^{-22} n^{2/3}/mc^2$ in nonrelativistic regime and $E_F(MeV) = 6 \times 10^{-11} n^{1/3}$ in relativistic regime. The energy equivalence of the particle mass, mc^2, is in MeV; the particle density, n, in cm^{-3}.

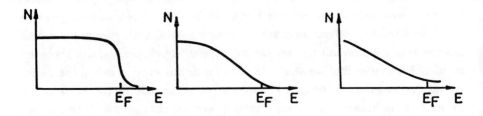

FIG. V.15 *Thermal distribution of a fermion gas at finite temperature in three situations: (a) Almost complete degeneracy; (b) Partial degeneracy; (c) Nondegeneracy.*

reflect on the average kinetic energy, calculated over the fermion distribution. In the nonrelativistic case, this average energy is $\frac{3}{5}E_F$ (nonrelativistic), and in the relativistic case, $\frac{3}{4}E_F$ (relativistic). The third and final point to remember is that transitions to lower (occupied) kinetic energy states in a degenerate fermion system at $T = 0$ are absolutely forbidden by the Pauli principle: a degenerate fermion gas at $T = 0$ cannot lose energy by radiation.

When $T \neq 0$, particles just below the Fermi level will be excited to states above, and the characteristic sharp drop at the endpoint of the Fermi distribution is smeared into an exponential tail similar to a classical distribution. The system then acquires a mixed behavior. Particles below the Fermi level are degenerate and obey the laws of quantum mechanics, those above are nondegenerate and may be treated as classical. Thus, a practical criterion for a system at $T \neq 0$ to be fully degenerate is E_F much greater than kT, and to be nondegenerate, just the reverse, E_F much smaller than kT (see Fig. V.15).

Let us return to our solar-mass red giant given as illustration above. Its core has a temperature of $T = 2 \times 10^7$ K, a helium density $n_{He} = 2 \times 10^{28}$ cm^{-3}, and an electron density $n_e = 2n_{He}$. We find the energy equivalence of the electron momentum to be $p_F c = 2 \times 10^5$ eV, somewhat less than the energy equivalence of the electron mass ($mc^2 = 5 \times 10^5$ eV). Applying the nonrelativistic formula we get the electron's Fermi energy $E_F = 4 \times 10^4$ eV, then its average energy (2.4×10^4 eV), which is considerably greater than both the binding energy of the helium atom and the thermal energy, $kT = 2 \times 10^3$ eV. The electrons in the core are indeed unbound and partially degenerate. The more massive helium nuclei evidently remain nondegenerate.

The degenerate electrons will exert an outward pressure on the base of the star's envelope. This pressure can be evaluated by using an elementary result of thermal physics. The pressure of any gas is defined as the force per unit area that it exerts on a real or imaginary wall. For each degree of freedom, it is given by $P = n < vp >$, that is, the product of the particle number density n by the average of the product of the particle velocity v and momentum p. Therefore, for an isotropic gas, we have $P = \frac{2}{3}n < E >$, which gives $P = \frac{2}{5}nE_F$. Whether this pressure can support the envelope or not depends on the inward gravitational force, which increases with the mass of the envelope. It turns out that when the star mass is greater than $0.8M_\odot$, the total pressure in the core is not sufficient to support the outer layers, and the core continues to contract and heat up. When the temperature reaches 10^8 K, the *triple α fusion* begins:

$$^4\text{He} + {}^4\text{He} \rightarrow {}^8\text{Be} + \gamma$$
$$^8\text{Be} + {}^4\text{He} \rightarrow {}^{12}\text{C}^* + \gamma.$$

Carbon-12 is formed in an excited nuclear state (an asterisk is used to indicate this fact). It readily combines with helium to form oxygen. Or it can react with a proton to give ^{13}N. This isotope of nitrogen is unstable and decays to ^{13}C, which undergoes the reaction: $^{13}\text{C} + \alpha \rightarrow {}^{16}\text{O} + \text{n}$. This reaction is an abundant source of neutrons and will be called on, at least in high-mass stars, to play a key role in nucleosynthesis.

The triple α chain yields a total energy of 7.5 MeV at a rate that depends strongly on temperature ($\sim T^{40}$). Whereas in a normal star, any slightest local increase in the core temperature and pressure is responded by a local expansion and a reduction of temperature, here the degenerate pressure is completely *decoupled* from thermal motions. So any increase in the core temperature leads to an accelerated energy production without the compensating effects of a pressure increase and a stellar expansion. A runaway production of nuclear energy ensues: after ignition, helium burns with a 'flash'. This sudden massive release of energy generates enough internal luminosity to expand the core, lowering both density and temperature. The explosion is thereby contained, with no catastrophic effects for the star, and the electrons return to a nondegenerate state. With a lowered temperature, the triple alpha process shuts off, and even the hydrogen-burning in the outer shell slows down. Both temperature and pressure gradients drop throughout the star, and the surface will contract as the core expands.

Eventually, the core expansion damps out and, abetted by the infall of the outer envelope, the core starts contracting again, raising its temperature back to 10^8 K. Helium burns anew but now under the control of the physics of a nondegenerate gas. When the helium in the core is finally used up and the nuclear energy production there comes to an end, a familiar sequence of events takes place. The outer layers collapse, releasing enough gravitational energy to ignite helium fusion in a shell outside the now dormant carbon core. And in due course, convection will set in, sending the star up to the *supergiant* status, a red globe with a radius of $200R_\odot$. The surface luminosity, now fed by the double-shell sources of hydrogen and helium nuclear energy, rises to thousands of times the solar luminosity. Meanwhile, the core is compressed, and the electrons again become degenerate. The degenerate electron pressure should be high enough to support the envelopes of low-mass stars, so that the core, now a solid carbon-oxygen sphere, remains inert at a temperature below the next fusion point.

One of the most beautiful sights in the heavens is that of the *planetary nebulae*, those diaphanous rings of gases surrounding very hot but fairly faint stars. Among those objects, the ones thought to be young comprise an extensive, cool envelope enclosing an HII region observed as the optical ring, but the older ones seem to have lost their outer envelopes. Most are observed to be slowly expanding. Apparently unrestrained by any strong attractive force, they will expand without limit and, after some tens of thousands of years, will disperse in space. Some of them have at their center a faint nucleus, roughly the size of the Earth. If this object has a mass of one solar unit, we get an average density of about a million grams per cm^3. This is about the density of a white dwarf. Is it mere coincidence? Planetary nebulae are concentrated toward the galactic plane, just in the same region preferred by many red giants of high luminosity. Finally, statistical data show that a planetary nebula is formed somewhere in our Galaxy about once a year, at the same rate as the formation of white dwarfs. On the other hand, red giants number about a million in our Galaxy, which tells us that if red giants are to be the parents of planetary nebulae, this phase of evolution should last

TABLE 3

Post-main sequence evolution for stars of different masses.

$M = 0.1\,M_{\odot}$	*Main sequence (type M) $\rightarrow$ White Dwarf*
$M = 1\,M_{\odot}$	*Main sequence (type G) $\rightarrow$ Red Giant*
	$\rightarrow$ *White dwarf*
$M = 4\,M_{\odot}$	*Main sequence (type A) $\rightarrow$ Red Supergiant*
	$\rightarrow$ *Supernova $\rightarrow$ Neutron Star*
$M = 10 M_{\odot}$	*Main sequence (type O) $\rightarrow$ Red Supergiant*
	$\rightarrow$ *Supernova $\rightarrow$ Black Hole*

a million years, in agreement with other estimations.

Those are the observational underpinnings for the models of the last evolutionary phase of low-mass stars. According to current ideas, red giants,† planetary nebulae, and white dwarfs are all genetically related. In the last stages of the red giant phase, the swollen star is ready to shed its hydrogen-rich envelope and to give birth to a planetary nebula and its nucleus. The young nebula should have practically the same luminosity as a double-shell red giant, and would be detached from the star not far below the stellar surface such that its expansion velocity is of the order of the escape velocity from the parent star. The exposed hot central star would burn out its outer shells, eventually lose its extended envelope, and finally evolve into a *white dwarf*.

In summary, a low mass red giant acquires a degenerate core which burns violently when helium is ignited. When helium fusion comes to an end, the core now filled with dense carbon and oxygen is surrounded by a double shell of burning hydrogen and helium whose massive energy production sends the star to the supergiant status. Low surface gravity, convective instability beneath the photosphere, and strong radiative flux are among the factors causing the star to eject matter copiously in the last stages of this period.

• *From ring building to an explosive climax*

Let us now turn to stars with masses greater than three solar units. The

---

† Here we mean strictly the red giants descending from low-mass stars.

evolution of these high-mass stars differs from that of the low-mass stars we just described on two points. First, they evolve much more rapidly because they lose energy through radiation at a much higher rate (remember the luminosity of high-mass stars varies with mass as $L \sim M^4$). Second, the large mass of the very massive stars can keep the core at a low enough density for the core to remain largely nondegenerate (recall $\rho \sim M^{-2}$), but at the same time can produce a pressure and temperature high enough to burn elements all the way to iron.

In high-mass stars, helium burning occurs under nondegenerate conditions and there is no helium flash. While helium fusion takes place in the core, hydrogen burns in a thin shell just outside. On the HR diagram, the star track moves leftward, toward higher temperatures. After the core helium exhaustion, the newly formed carbon core contracts and ignites a ring of helium outside. There is little systematic increase in the surface luminosity, which is in fact already quite high, because the new inflow of energy mostly goes into inflating the star. On the HR diagram, the star generally drifts to the right, almost horizontally (Fig. V.14). When the temperature in the core reaches $\sim 10^9$ K, the carbon will also ignite, starting a new set of reactions,

$$^{12}C + {}^{12}C \rightarrow {}^{24}Mg$$
$$^{12}C + {}^{12}C \rightarrow {}^{20}Ne + \alpha.$$

Since these reactions are *extremely* sensitive to temperature, the next sequence of events depends crucially on whether ignition occurs under degenerate conditions or not, and that depends on whether the mass of the star is lower or higher than $8M_\odot$. In stars with mass less than $8M_\odot$, the electrons in the core now become degenerate and the reaction rate is uncontrollable: the star bursts in a full-blown *supernova explosion*, leaving behind perhaps *nothing* more than gas and dust that disperse into space. In more massive stars, the nondegenerate core can adapt itself to the rapidly changing conditions, undergoing contraction, raising its internal temperature and pressure, and burning carbon before the electrons have the chance to become fully degenerate. Thus, very massive stars evolve smoothly, by successively burning their present stock of fuel to exhaustion, undergoing core contraction, reaching the temperature needed to ignite the ashes of the previous reaction cycle, and starting all over again. Carbon burns into oxygen, neon, and magnesium; oxygen into silicon and sulfur; and silicon into nickel. The unstable nickel isotope (^{56}Ni) decays to cobalt, and cobalt to iron. Iron, in its form of the common isotope ^{56}Fe with twenty-six protons and thirty neutrons, is the most strongly bound of all nuclei; any further fusion of it would absorb rather than release energy. The nuclear fusion process shuts off at this point; it has played out its role in the normal evolution of the stars. When a star reaches the last, iron-forming stage it becomes a *red supergiant* a thousand times the size of the Sun. Its bloated envelope of unburned hydrogen surrounds an Earth-size core which presents a characteristic onion-like structure: an innermost iron core embedded in successive slowly-burning shells of sulfur, silicon, oxygen, neon, carbon, helium, and hydrogen.

The elements and their isotopes forged by fusion can readily absorb additional neutrons (which, of course, are electrically neutral and, therefore, know no Coulomb barrier) to form new elements below and above the iron peak. Some of the new elements may have too many neutrons to be stable and may beta-decay (converting a nuclear neutron into a proton) before capturing another neutron. In ordinary stars, this form of synthesis of heavy elements is a *slow process* because free neutrons, available as by-products of fusion reactions, are not excessively abundant.

• *Supernovae*

At about 2 a.m. on February 24, 1987 at Las Campanas Observatory, eight thousand feet high in the Chilean Andes, the young astronomer Ian Shelton spotted what appeared to be a bright, new star on a photographic plate he was developing; then, stepping out into the night, he saw with his own eyes the brightest *supernova* observed on Earth for nearly four hundred years. By the next evening, nearly all major radio and optical telescopes in the Southern hemisphere were trained on the event, now dubbed SN1987a, as the first event of this kind identified that year. Within weeks, physicists operating proton decay detectors in Japan, the U.S.A., and the Soviet Union reported recordings of simultaneous brief bursts of high-energy neutrinos occuring a few hours before the first optical sighting.

This event provides astrophysicists with a unique opportunity to test their current understanding of the most spectacular phenomenon in the heavens. Up to now, their knowledge about supernovae has been largely gained through theoretical modelings and computer simulations. It is true that over the past one thousand years at least four such events have been reported to occur in our Galaxy: SN1006, described by medieval Chinese astronomers to be as bright as the half-moon; SN1054, also recorded by them and identified now as the event leading to the formation of the Crab Nebula; SN1572, sighted by Tycho Brahe, whose book about the event, *'De Nova Stella'*, gave these objects a name; and finally SN1604, reported by Johannes Kepler, whose observations and explanations contributed to the final break with the Aristotelian traditional view that the realm of fixed stars was immutable. It is also true that since the pioneering work of Fritz Zwicky in the 1930's more than six hundred supernovae have been spotted in other galaxies. But hard information about both the historical supernovae and the extragalactic supernovae is understandably lacking. So it is fortunate that SN1987a is located only some 160 thousand light years away in one of the two nearest galaxies, the Large Magellanic Cloud; its progenitor star has even been identified as a blue supergiant with known characteristics. The evolution of SN1987a is practically unfolding under our eyes!

Supernovae fall into two general classes, distinct by their spectroscopic properties. Type I supernovae show no hydrogen lines in their spectra; they occur amid young and old stars alike and in all kinds of galaxies. In their short life, they display a very rapid rise in luminosity, followed by a gradual fall-off. The events observed by Brahe and Kepler belong to this category. In contrast, Type II supernovae display intense hydrogen lines in their emission spectra. They are located in active star-forming regions, eject matter copiously at low velocities, and become radio emitters a few months after the explosion. The events SN1054 and SN1987a are examples of this type.

It is generally believed that Type I events arise from disruptive (*thermonuclear*) explosions of low-mass stars because such stars evolve slowly and, by the end of their life, will have used up their hydrogen. Consider two stars of unequal masses forming a *close* binary system. The more massive star evolves more quickly. In its red giant phase, it transfers mass to the smaller, less evolved star and its exposed core eventually becomes a white dwarf. Now we have a system composed of a white dwarf and a more massive star. As the latter quickens its pace of evolution, it becomes in due time a red giant and transfers mass back to the white dwarf. As we will see later, white dwarfs can exist only below a certain critical mass. With the added mass, our white dwarf becomes more massive than can be supported by its internal (degenerate-electron) pressure. The ensuing collapse of its carbon-oxygen core rekindles the nuclear fire, igniting the carbon at the core center, burning it at a tremendous rate ($\sim T^{120}$), which in turn *rapidly* raises the central temperature up to 500 billion degrees. The envelope, left unsupported, will begin to fall toward the center. There is a sudden enormous increase of luminosity throughout the star, setting off a deflagration, or even a detonation, which could completely destroy the star. Other theories suggest however that enough energy is carried away by weakly interacting neutrinos (produced throughout the whole core in reactions like $e^- + p \rightarrow n + \nu$ and $n \rightarrow p + e^- + \bar{\nu}$) to moderate the explosion. Then the event will not be entirely destructive and should leave behind a remnant *neutron star*.

Type II supernovae, on the other hand, descend from stars originally at least eight times more massive than the Sun. In such a massive star, the evolution proceeds very rapidly once the silicon fusion phase reaches its final stage. The iron core is built in a day, collapses and rebounds in seconds.

When iron is formed at the star center and the nuclear fire has burned out, the core has no choice but to contract and heat up, raising its temperature to several billion degrees. It is important to realize that if no substantial heat were lost, the collapse would not necessarily occur at a catastrophic speed. The star could still reorganize itself and avoid a disastrous end. Unfortunately, such is not the case; there are powerful cooling mechanisms at work in the core that rob from it much of the energy released by contraction. First, there is an intensive absorption of energy through the break-up of iron nuclei into alpha particles and neutrons. Since the average binding energy in an iron nucleus is about 2 MeV greater than in a mixture of α-particles and neutrons resulting from the iron disassociation, about 110 MeV of energy are used up in the disassociation of a single iron nucleus. The energy released by gravitational contraction is thus largely diverted to external work. The rise in temperature is halted at the present level until most of the iron near the center of the star has been split up. This pause in the heating of the contracting core is enough to cause the core to collapse catastrophically.

The implosion proper involves only a tiny inner core while the bulk of the star shrinks at a relatively slower pace. As the central density shoots up, the degenerate free electrons get squeezed into protons to form neutrons and neutrinos: $e^- + p \rightarrow n + \nu$. This reaction lowers drastically the number of electrons in the core and hence also the degeneracy pressure, which in turn speeds up the rate of collapse. Neutrons and neutrinos are produced in increasingly copious amounts. At this stage, nuclei can successively capture many neutrons before having a chance to beta-decay. It is through this *rapid process* of nucleosynthesis that heavy elements are formed. Neutrons stay bound in neutron-rich nuclei until all allowed, stable nuclear orbits are filled up, at which point the excess neutrons leak out as free particles (recall that neutrons have spin one-half, and so obey the Pauli principle). The nuclear component of the internal pressure is thus creeping up but, at this point, is still insufficient to compensate for the large loss of the electron pressure. The neutrinos, which rarely interact with other forms of matter, at first stream freely out of the core within a few milliseconds, carrying off a tremendous amount of energy — all derived from the gravitational energy set free by the collapse — and cooling down the system. However, when the central density exceeds one hundred billion grams per cubic centimeter, matter becomes so dense that even the normally elusive neutrino has difficulty finding its way out: it is scattered, absorbed, and re-emitted many times before it can escape. The neutrinos are effectively trapped for the duration of the core collapse.

Eventually, the inner core is compressed to a density comparable to that found in a large atomic nucleus; all nuclei are crushed together into a gigantic nucleus held together, not by the nuclear force, but by the gravitational force. The collapse overwhelms even the powerful pressure of nuclear origin that is being built up by nucleons. Carried by momentum beyond the point of equilibrium, the implosion squeezes matter to a density greater than that found in an atomic nucleus before it finally grinds to a halt and vigorously bounces back, rather like a rubber ball that has been sharply compressed. As the bulk of the star is still crashing down, the bounce is quickly smothered — but not before it has set off an intense shock wave that rushes outward through the iron core, collides with the infalling outer layers composed of lighter elements, reversing their motion, and touches off explosive reactions with the light elements. The shock may not be powerful enough to blow off by itself the outer envelope: it may have wasted too much energy breaking up nuclei into individual nucleons on its way out, or the tamp is just too massive. But the stalled wave is possibly heated up and revived by the intense fluxes of neutrinos that have finally percolated upward into strata of lower densities. Enough energy is deposited in the envelope through the interactions of neutrinos with matter for the shock wave to overcome gravity. The shock wave reaches the surface of the star within a few hours, and the explosion finally becomes visible. At the 1987 event, detectors silently recorded bursts of energetic neutrinos about two hours before the first photograph of the supernova was taken.

What becomes of the remaining core? The answer was given by Walter Baade and Fritz Zwicky in a 1934 paper which has become a classic in the field:

> With all reserve we advance the view that supernovae represent
> the transitions from ordinary stars into *neutron stars.*

The explosion blows off all the material of the parent star beyond a certain radius, enriching the interstellar medium with heavy elements, while the rest of the star settles down as a hot, neutron-rich ball, two hundred kilometers across, that quickly shrinks to one tenth of this size and slowly cools off. During this final period of contraction, most of the electrons are annihilated, and the matter that is left is primarily composed of tightly packed, degenerate neutrons: we have a *neutron star*. If the original star has a mass high enough ($M > 20M_\odot$), the imploding core might exceed $3M_\odot$. Then, gravity will overcome the degenerate-neutron pressure, and a *black hole* will form.

Summary

When a star heats up to ten million degrees at its center, the nuclear fusion of hydrogen (the most abundant of all elements) is ignited. Thermonuclear fusion is the only energy source that can sustain the radiation of stars for very long timespans. This process provides the most coherent explanation for a vast body of data, from the basic properties of stars on the main sequence to various phenomena marking the late stages of the stellar evolution, in passing by the relative abundances of chemical elements observed in the Universe.

According to the current stellar models, after ten percent or so of hydrogen have been burned into helium, stars become structurally unstable. The relationships between their basic properties change drastically: the stars move away from their positions on the main sequence. The subsequent collapse of the core releases enough energy to heat up the layers surrounding the core, igniting a shell of hydrogen. The luminosity generated by the shell source goes into layers lying above, causing them to expand. The bloated envelope, fed in radiation by convective currents, gives the star the appearance of a red giant.

In a low-mass star, the next cycle of fusion involving helium burning is accompanied by a large and rapid production of energy, which might force the swollen star to shed its hydrogen-rich envelope and give birth to a planetary nebula together with a remnant core. This exposed hot core goes on to become a white dwarf. High-mass stars evolve much more quickly because they radiate at vastly higher rates, yet can manage to adapt themselves to the rapid changes. They evolve smoothly, undergoing contraction, raising the temperature to the level needed to ignite the ashes of the previous reaction cycle, and starting a new cycle. But when iron is finally formed at the center, the nuclear energy source dries up, and the core has no recourse but to contract. The supernova explosion that ensues would leave behind an imploded core, which may evolve into a neutron star or a black hole.

---

5. THE FINAL STATES OF STARS

The story of the evolution of stars is the story of a death foretold: pressed by the relentless pull of gravity and surrounded on all sides by a cold and dark space, stars are fighting a battle they cannot win. Drained of their energies, they will inevitablly lose their struggle and die, either in a destructive explosion in which their ashes turn to dust, or in a lingering death in which, as a white dwarf, a neutron star, or a black hole, they finally come to terms with both gravity and the universe.

5.1 White Dwarfs

In 1844 astronomer Friedrich Bessel noticed, while examining astronomical records over many years, that Sirius, the brightest star in the Northern sky, showed a wobbly motion. He interpreted the irregular motion of Sirius as being caused by the pull of an unseen, dark and massive star that gravitated around it. This invisible companion star turned out to be the first white dwarf star ever discovered, now known as Sirius B. Although it is the brightest white dwarf detected in our Galaxy, its faint glow is overwhelmed by the light of Sirius, a star ten thousand times more luminous. Just as for many other stars, the radius of Sirius B is deduced from its light spectrum. From spectral analysis, astronomers first obtain its surface temperature, and from its apparent brightness and an independent measurement of its distance from Earth, they next find its true luminosity. With the temperature and luminosity known, the star radius is derived from the familiar radiation law.

The Galaxy contains about 10 billion such objects, or some 10 percent of all the stars. All the white dwarfs that have been identified range in radii from 5000 to 20000 kilometers and, correspondingly, in temperatures from 50000 to 4000 degrees Kelvin. However, contrary to the situation prevailing in normal stars, there is no systematic dependence of the radius on the temperature. This suggests that the white dwarfs, born in fairly similar sizes and at relatively high temperatures, cool off without further contraction. Their masses are very difficult to determine except when they belong to multiple star systems, in which case the Newton gravitational law, applied to the observed orbital motions, can give the stellar masses. For three such white dwarf stars — Sirius B, Stein 2051, and 40 Eridani B — with respective radii of 5100, 8100, and 8600 kilometers, the masses are reliably established at 1.05, 0.50, and 0.48 solar masses. White dwarfs therefore have densities of about $10^6 \, \mathrm{g\,cm^{-3}}$, or a million times greater than the mean solar density and fifty-thousand times greater than the density of the heaviest metal. One cm-cube of white-dwarf stuff brought to Earth would weigh more than 10 persons.

• Degenerate-electron pressure

A mass density of $\rho = 10^6 \, \mathrm{g\,cm^{-3}}$ implies an electron density n_e of about 10^{30} electrons per $\mathrm{cm^3}$ (because $n_e \approx 0.5\rho$/proton mass). A density of this magnitude means that electrons in a white dwarf are considerably closer than 10^{-8} cm, the atomic size, but still farther than 10^{-12} cm, the nuclear size. The electron shells are effectively crushed, whereas the atomic nuclei remain intact but stripped of their electrons. White dwarf material is a dense plasma of electrons and ions.

Using formulas obtained in Sec. 4.3, we can estimate the electron Fermi momentum to be about 0.5 MeV and the corresponding Fermi energy 0.25 MeV. From our current understanding of the final phase of the low-mass star evolution, namely, that white dwarfs evolve from the remnant central stars

TABLE 4

End states of stars

- **White Dwarf**: *Exposed core ($M < 1.4M_\odot$) of low mass stars ($M < 4M_\odot$); radius $\sim 10^{-2}R_\odot$; supported by the pressure of degenerate electrons.*
- **Neutron Star**: *Imploded core ($M < 3M_\odot$) of high mass stars; radius $\sim 10^{-5}R_\odot$; supported by the pressure of degenerate neutrons.*
- **Black Hole**: *Totally collapsed core of high mass stars; radius of event horizon $R_s = 2GM/c^2$.*

---

of planetary nebulae, we may reasonably assume the white dwarf internal temperature not to exceed 10^8 K (and is probably closer to 10^7 K, as in the Sun), which would be equivalent to a thermal energy of $kT \approx 0.01$ MeV. This is much smaller than the Fermi energy: the electrons in a white dwarf have become undoubtedly *degenerate* (although the more massive ions still remain nondegenerate). In this example the electrons are almost relativistic. But in white dwarfs at large, where mass densities range from 10^5 g cm^{-3} to 10^8 g cm^{-3}, they may be nonrelativistic or ultrarelativistic, or anything in between.

For an ideal classical gas the pressure is given by the familiar expression $P = nkT$. But for a degenerate gas, it is replaced by $P \sim n^{5/3}$ in the non-relativistic limit, and $P \sim n^{4/3}$ in the ultra-relativistic limit (see Sec. 4.3). In either case, the pressure of a degenerate gas is not influenced by the temperature at all; it depends only on the particle number n, but this dependence is stronger here than in a nondegenerate gas. Since the electron degeneracy pressure is the major component of the internal pressure of white dwarfs, we do not expect the structure of these stars to depend on their temperature. A white dwarf could in principle evolve as a stable configuration down to temperatures close to absolute zero. Nor can we expect any relationship between the mass and the luminosity of white dwarfs since the luminosity of an object is determined directly by its temperature. There exists, however, a relation between their mass and radius, as we shall presently see.

• *Mass and size of white dwarfs*

A white dwarf is in a state of equilibrium between gravity and the degenerate electron pressure, so according to the virial theorem, its total kinetic energy equals half the magnitude of its total potential energy. Here, 'kinetic energy' means the (nonrelativistic) degenerate electron energy and 'potential energy', the gravitational energy. Since both can be expressed in terms of the star's mass and radius only, we find a *mass-radius relation*† for the white dwarfs:

$$M^{1/3} R = \text{constant.}$$

It is a startling result. In our ordinary experience, we expect a solid ball twice as massive as a second made of the same material should have twice the volume. Not so with white dwarfs. A white dwarf of $1\,M_\odot$ would occupy only half the volume of a white dwarf of $0.5\,M_\odot$ and would be four times as dense.

If now more matter is added to the white dwarf (for example by accretion) so that its mass becomes greater than one solar mass, the central density will increase with the increasing gravity. More and more electrons will become relativistic. But since their velocities cannot exceed the speed of light, their kinetic energy increases less quickly with the increasing density. In fact, we have found that the mean kinetic energy of relativistic degenerate electrons depends on the particle number as $n^{1/3}$, rather than as $n^{2/3}$. The total kinetic energy is then $\sim Nn^{1/3} \sim M^{4/3}/R$, where M, R, and $N \sim nR^3$ refer, respectively, to the mass, radius, and electron number of the white dwarf. Now, by the relativistic virial theorem,‡ this kinetic energy must be balanced against the gravitational energy, which is $\sim GM^2/R$. Note that both energies scale as $1/R$ but depend differently on mass. Thus, there exists some definite value of the mass for the two forces, gravity and the pressure

---

† The constant in this relation is the product of two factors: (1) a purely numerical factor coming from a combination of fundamental physical constants given by 2×10^{17}, if M in g and R in cm; and (2) $(m_p/m_e)(Z/A)^{5/3}$, where Z is the number of degenerate particles and A the average atomic weight of the atoms in the star; in most white dwarfs $A \approx 2Z$, and the numerical value of this factor is 650.

‡ The relativistic form of the virial theorem is $KE + PE = 0$.

gradient, to be in balance. After some simple arithmetic, that value comes out to be

$$M_{max} = 5.64(Z/A)^2 M_\odot$$

where Z is the electron number, A the average atomic weight of the ions in the white dwarf, and the number 5.64 arises from a combination of some fundamental physical constants. This equation is remarkable for its simplicity: apart from factors sensitive to the composition of the star, it depends only on fundamental constants. But it is one of the most important formulas in astrophysics. It gives the maximum mass of white dwarfs; this upper mass limit is referred to as the *Chandrasekhar mass limit*, after Subrahmanyan Chandrasekhar. Since white dwarfs are an end state of stellar evolution, all the hydrogen in their parent stars should have been converted to heavier elements: $Z/A = 0.5$ for most of them. The numerical value of Chandrasekhar's limit is then $1.4\,M_\odot$. (A number of refinements lower this limit to $1.2\,M_\odot$, but, following a well-established custom, we continue to use the uncorrected value in our discussion). White dwarfs cannot exist in nature if their masses exceed this value. If the remnant star of a planetary nebula has a mass smaller than the critical mass, it will stabilize as a white dwarf. But should its mass exceed the maximum value, gravity would overwhelm the electron-degeneracy force, and the star would keep on contracting. In fact, all white dwarfs whose masses have been observationally determined obey the Chandrasekhar restriction.

• Luminosity of white dwarfs

White dwarfs do radiate energy, though weakly: their luminosity is of the order of $10^{-3}L_\odot$. But with a much smaller radius, they have a generally higher surface temperature (and so are 'whiter') than normal stars. Do white dwarfs radiate as black body radiators just like ordinary stars? If indeed they do, white dwarfs of a definite mass and of a definite radius should lie along a definite line in the HR diagram. Since white dwarfs are observed to have masses spread around $1\,M_\odot$, we expect them to occupy a narrow strip in the HR diagram to the left and below the main sequence. This prediction is borne out by data. White dwarfs, in spite of their unusual properties, are indeed black body radiators.

The surface layers of a white dwarf are nondegenrate, radiating energy by the diffusion of photons. But its interior is completely degenerate, and hence highly transparent and an excellent heat conductor. The basic reason for both properties lies in the Pauli principle. Absorption of light by matter takes place when an electron can make a transition to a higher-energy state. In a degenerate gas, where most particle states are occupied, such a possibility is not open to the majority of the electrons below the Fermi level; only the few near the Fermi level find themselves in a position where they can absorb photons. By the same token, the electrons must have a large mean free path; they cannot shake off excess energy, they cannot make collisions. It follows that the interior of white dwarfs must be isothermal, and heat is readily deposited at the base of a shell near the surface. The temperature must drop from ten million degrees at the base to a few thousand degrees at the surface proper across a distance measuring about one percent of the star radius.

With the electron thermal energy locked up, the only significant source of heat to be radiated is the residual *ion* thermal energy. This energy store is in fact quite substantial ($\sim 10^{48}$ ergs for $T = 10^7$ K), comparable to the energy output of some supernovae in the visual part of the spectrum. White dwarfs emit radiation at the expense of their heat reserves, glow ever more faintly, cool down, and finally come into thermodynamic equilibrium with the cold universe. In the process, as the internal temperature drops, the electric forces between the nuclei will eventually dominate the thermal motions. The nuclei will then begin to 'stiffen up' under their mutual net Coulomb repulsion and form a crystal lattice. The heat store of the white dwarf is somewhat reduced by the crystallization process. Still, the white dwarf will take some hundreds of millions of years to cool off.

5.2 Neutron Stars

White dwarfs were observationally discovered well before astronomers understood what these objects really are, how they hold together, how they radiate. Neutron stars, in contrast, were conceptually created by theorists as one of the three possible non-destructive outcomes in the life of stars thirty years before they were actually identified in nature.

The first possible outcome is the formation of a white dwarf when a star leaves a residual object with mass under $1.4\,M_\odot$. But if the remnant star has a mass between $1.4\,M_\odot$ and $3\,M_\odot$, it will have a completely different fate. After the nuclear fuel in the core is exhausted, the inner parts collapse catastrophically. There is a rapid heating, followed by a tremendous release of energy throughout the star. Enough energy is deposited in the star's envelope to blow it off, which we may observe as a supernova explosion. What is left of the star will implode, contracting by a hundred thousand times in a few seconds, until it shrinks to a ball some 30 km across with a density comparable to that of an atomic nucleus. This self-gravitating object, rich in neutrons, is a *neutron star*.

• *The path to neutron degeneracy*

Let us now follow more leisurely the path taken by a residual star having a mass of, say, $1.5\,M_\odot$ as it evolves from its compressed state some time before the supernova event, through its formation as the hot central remnant of a supernova explosion, and on to its final incarnation as a neutron star.

For such a massive object, the internal pressure provided by its component particles — initially, degenerate free electrons and nondegenerate nuclei — is inadequate in the face of the gravitational force. As the star (*i.e.*, the core of the parent star) keeps on condensing, matter is compressed and, at some point, the free electrons will be forced to combine with the nucleons in the atomic nuclei. They cannot combine with neutrons (because of various conservation laws) but will readily react to protons, leading to a production of neutrons according to

$$e^- + p \rightarrow n + \nu.$$

The electron has rest mass $m_e = 0.511$ MeV/c^2; the proton, $m_p = 938.28$ MeV/c^2; and the neutron, $m_n = 939.57$ MeV/c^2. Thus, for the reaction to start the electron must have at least 0.8 MeV in kinetic energy. This energy is that of an electron under degenerate conditions, and so depends directly on the high particle density now prevailing in the star. From a relation between energy and density we are already familiar with, we can estimate the corresponding minimum electron density, n_e, and hence the minimum mass density, $\rho = 2m_p n_e$. We get $\rho \approx 10^7$ g cm^{-3}. It is 10 times greater than the average white dwarf density. A star must have at least this mass density before the *neutronization* reaction, $e^- + p \rightarrow n + \nu$, can proceed.

When the density in the star exceeds the above critical value, the nuclei in the star will have more neutrons than they normally need for their stability: they become *neutron-rich*. Under laboratory conditions, neutron-rich nuclei are unstable and tend to beta-decay, converting a (bound) neutron into a (bound) proton and emitting an electron and an anti-neutrino ($\bar{\nu}$) via the reaction $n \to p + e^- + \bar{\nu}$, all taking place inside nuclei. But in a degenerate star, such reactions are in effect forbidden because the electrons produced would have energies smaller than 0.5 MeV, and would have nowhere to go, all available states being occupied by electrons already there. So the nuclei cannot beta-decay, and we have a star filled with degenerate electrons and nondegenerate nuclei accumulating more and more neutrons in their folds.

As protons are converted into neutrons in a nucleus, the newly formed particles must go to the next higher unoccupied nucleon levels (because neutrons, having spin one-half, obey the Pauli principle). For this reason, they are less tightly bound than the neutrons already there. As the neutron-ization reaction keeps on adding more and more neutrons to the nucleus, they will go to fill up states closer and closer to the continuum. Eventually, the most weakly-bound neutrons will begin to 'drip out' of the nucleus, although some of them may be recaptured by nuclei. Above a density of about $\rho \approx 4 \times 10^{11}\,\mathrm{g\,cm^{-3}}$, the star becomes a mixture of degenerate electrons and nondegenerate neutrons and nuclei.

The unbound neutrons begin to supply an increasingly larger fraction of density and pressure. At the density of $\rho \approx 5 \times 10^{13}\,\mathrm{g\,cm^{-3}}$, the neutrons will have an average kinetic energy of about 8 MeV (the average nuclear binding energy), and can no longer be recaptured by nuclei. Matter is then composed of increasingly abundant free neutrons in addition to electrons and neutron-rich nuclei embedded in a Coulomb lattice. The neutrons have evidently become almost fully degenerate because their Fermi energy, $E_F \sim 13\,\mathrm{MeV}$, now amply exceeds the thermal energy, $kT \sim 0.1\,\mathrm{MeV}$, even at the high temperatures ($T \sim 10^9$ K) now prevailing in the star. When matter in the star becomes as dense as in a typical nucleus, or $\rho \sim 2.3 \times 10^{14}\,\mathrm{g\,cm^{-3}}$, the nuclei have already merged together, creating a unique state of nuclear matter composed mainly of neutrons, with only a few percent as charged particles (protons and electrons). Each cm-cube of matter now contains 10^{41} neutrons, and the neutron-degeneracy pressure becomes powerful enough to support the object against gravitational collapse: a neutron star is formed.

• *Mass-radius relation of neutron stars*

If we assume, quite correctly as a first-order approximation, that neutron stars are made up of a nonrelativistic degenerate neutron gas, their mass and radius are related by the equation $M^{1/3}R = $ constant, just as for the white dwarfs, with the difference that the constant is now simply $2 \times 10^{17}\,\mathrm{g}^{1/3}$ cm. It follows that the radius of a neutron star is smaller than that of a white dwarf of comparable mass by a factor of 650. For our typical neutron star of mass $M = 1.5\,M_\odot$, the radius would be $R = 15$ km.

It is reasonable to expect in analogy with white dwarfs *a mass limit* above which neutron stars cannot exist. No matter how strong the nuclear pressure might turn out to be at high densities, it cannot resist gravity if the mass of the star is large enough. Indeed, to create strong force fields, energy is needed. Since energy acts like mass as a source of gravitation, the neutron gas resisting the compression also contributes to strengthen the gravitational pull that is squeezing it. Gravity will again be the winner. To estimate the maximum mass for neutron stars, we may borrow the mass-limit formula derived for white dwarfs, using however $Z = A$ (because now practically all particles in the system are degenerate). This gives $M_{ns} = 5.64\,M_\odot$ as the maximum mass of the equilibrium configuration. It is exactly four times as large as the mass limit for white dwarfs. However, the estimation is less reliable here than for white dwarfs because it does not take into account the presence of strong gravitational effects and because there are many unknowns in the structure of a collapsed object at very high densities. At present, all microscopic calculations point to a *mass limit of neutron stars* somewhat under $3M_\odot$. For a collapsed star with mass above this value, its self-gravity would crush all its content into a state of infinite density.

• *Structure of neutron stars*

We may take some time out and amuse ourselves playing with the unusual properties of neutron stars. Let us again consider our typical 1.5 $M_\odot$ neutron star. With a radius of 1.5×10^6 cm, its mean density should be $2 \times 10^{14}\,\mathrm{g\,cm^{-3}}$. A cm-cube of this stuff (if we could get it to Earth) would weigh as much as all of humanity. Or we can compare it to a cm-cube of nuclear matter. Experiments give the radius of any atomic nucleus in terms of its nucleon number, A, as $R = 1.2 \times 10^{-13} A^{1/3}$ cm. Hence the number of nucleons per unit volume is the same for all nuclei, 1.4×10^{38} nucleons cm^{-3}, which yields a mass density of $2.3 \times 10^{14}\,\mathrm{g\,cm^{-3}}$ (the mass of each nucleon is 1.7×10^{-24} g). Evidently, an atomic nucleus and a neutron star have almost identical mean densities. In this sense, a neutron star is just another atomic nucleus. But what a nucleus! It has 10^{57} nucleons compared to a paltry 10^2 for the heaviest

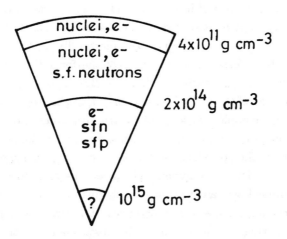

FIG. V.16 Internal structure of a neutron star.

elements on Earth. It is the only 'nucleus' known to contain vastly more neutrons than protons. And, in contrast to ordinary atomic nuclei, the force that holds its components together is gravitational, not nuclear.

We can also calculate the force of gravity at the surface of a neutron star. It is some 10^{11} times greater than the force of gravity on the surface of the Earth. To escape this field, a rocket would need a minimum velocity of more than half the speed of light.

The very strong gravitational field and the very high density of neutron stars imply that a proper treatment of their structure should take into account strong gravitational fields and the forces acting between nucleons and, possibly, between other more exotic particles. This is a difficult problem. Nevertheless, a plausible semiqualitative neutron star model has been constructed. The picture that has emerged is more complex than naively expected. It shows a system composed not simply of neutrons but of other species of particles as well, in a proportion that continuously changes from the surface of the star to its center (Fig. V.16).

The *surface* and the *outer crust* consist of very densely packed heavy nuclei with a small number of helium nuclei and lighter elements, as well as a dense degenerate electron gas like that found in the interior of white dwarfs.

As deeper we go in, the density rises from 4×10^{11} to 2×10^{14} g cm^{-3}

and the electrons are pressed into nuclei. This region, called the *inner crust*, consists of a lattice of neutron-rich nuclei bathed in an electron gas and a *superfluid* neutron gas. As it is discussed more fully in Chapter II, fermion systems driven by a force that can promote the formation of particle pairs may undergo a phase transition to the *superfluid* state, characterized by a viscous-free flow, or, if the particles are charged, to the *superconductive* state, characterized by a complete absence of electric resistance. For instance, some metals can display superconductivity when the attractive long-range electron-ion interaction in the metal lattice favors a stable pairing of electrons of opposite spins and opposite momenta. In neutron stars, where matter is superdense, a similar long-range pairing mechanism can arise because nuclear interactions are attractive at large distances. Transition to superfluidity occurs when the pairing interaction energy is substantially less than the thermal energy kT. From experiments on Earth, we know that both neutrons and protons undergo a pairing transition in cold nuclear matter and we can measure that pairing energy ($\sim 1 - 2$ MeV). From these results, we expect to find superfluid neutrons in the crust and the interior of the neutron stars that have reached temperatures around 10^8 K, or $kT \sim$ keV.

Just below this zone, there is a region where matter is denser than in atomic nuclei and which contains mainly superfluid neutrons with a smaller concentration of superfluid, superconducting protons and normal electrons (electrons remain normal in neutron stars because their long-range coupling is insufficient for pairing). Finally, near the center of the star, densities will approach 10^{15} g cm^{-3}, and the Fermi energy of the degenerate particles will rise toward 10^3 MeV. All sorts of subnuclear particles (*mesons, hyperons*), which we manage to produce at great cost in our high-energy particle accelerators on Earth, would then appear as products of reactions between highly energetic electrons, protons, and neutrons. At sufficiently low temperatures the *mesons* π, which are bosons with spin 0, may form a *Bose-Einstein condensate*, or a state of zero kinetic energy. Or the mesons and hyperons may dissolve into their elementary components, called *quarks*, and form a quark plasma. At the present time, the picture of the innermost region of neutron stars remains unclear.

• *Luminosity of neutron stars*

Neutron stars, just as any other stars, are continuously losing heat to the surrounding cold space. What kind of energy do they radiate, by how much, and for how long? This is the kind of questions we must consider before we could learn how a distant neutron star would manifest its presence to us.

As we already mentioned, a neutron star, newly formed from the implosion of the exposed core of a highly evolved star, should have an extremely high temperature, say, around 10^9 K. As with white dwarfs, we would expect it to radiate like a black body at that temperature. According to Wien's displacement law, $\lambda_{max} T = 0.3$ cm K, the wavelength of maximum radiation is $\lambda_{max} = 3 \times 10^{-10}$ cm. The corresponding energy, $hc/\lambda =$ 400 keV, would place this radiation in the range of very hard X-rays. According to Stefan-Boltzmann's law, our young neutron star would radiate at the rate of $L = 4\pi\sigma R^2 T^4 \approx 10^{45}$ erg/s for $R = 1.5 \times 10^6$ cm. It is an awesome power. For comparison, recall the luminosity of the Sun is $L_\odot = 3.90 \times 10^{33}$ erg/s, and the one hundred billion or so stars of all kinds that make up our entire Galaxy radiate only about 10^{44} erg/s spread over the whole spectrum.

It would seem that detection of such intense discrete sources of X-rays should pose no problems for astronomers. Unfortunately, it turns out differently. A neutron star, formed in the hot core ($\sim 10^{11}$ K) of a supernova explosion, cools down rather quickly. After a day or so, the *internal* temperature drops to $10^9 - 10^{10}$ K. At these temperatures, neutrinos will be produced in great abundance through the reactions

$$n + n \rightarrow n + p + e^- + \bar{\nu}_e$$
$$n + p + e^- \rightarrow n + n + \nu_e,$$

and will escape without hindrance into space, carrying away an enormous amount of energy, thereby cooling the star very quickly. Calculations have shown that the *surface* temperature drops to 10^8 K within just a month, then to several times 10^6 K, where it will remain for the next several hundred years. Afterwards, when the internal temperature has fallen to 10^8 K, photon emission overtakes neutrino emission as the main cooling mechanism, and the surface temperature hovers around 10^6 K for at least the next 10^4 years. This range of temperatures implies that photons are radiated in the soft (keV) X-ray band at rates substantially lower than we have naively estimated.

• *Pulsars*

How could you see a very dim object, thirty kilometers across and thousands of light years away? Nobody knew the answer — until 1967, when it came in an unexpected way. Anthony Hewish and his team of astronomers in Cambridge, England, had just set up a radio telescope and were planning to observe the *scintillations*, or the rapid, irregular variations in the flux of radio waves from distant sources. The twinkling arises from irregular diffraction of radio waves when the radiation passes through regions of inhomogeneous density in the plasma of the solar corona before reaching the Earth. It can be observed only if the angular size of the sources is small enough, less than half a second. Hewish therefore decided to use the scintillation technique to study the *quasars*, those extremely distant extragalactic radio sources whose angular sizes are measured in thousandths of a second. Up until then no detection devices had been built to measure short-time changes of radiation for wavelengths in the three-meter region of the radio spectrum. A young graduate student on his team, Jocelyn Bell, noticed something rather odd on the charts produced daily by the antenna. She found an unfamiliar point source that was scintillating even *at night*. More remarkable still, the flux did not fluctuate at random, as is the case with ordinary scintillations, but varied in a strictly periodic way. Very short pulses of radio waves, lasting about 50 milliseconds, were detected at very regular time intervals of approximately 1 second: the first pulsating emission source, or *pulsar*, was thus discovered.

Since then, about five hundred pulsars have been detected. All exhibit broadband *radio* emission, a few are *X-ray* or visible light sources. The pulse intensities vary over a wide range, but the basic pulse is periodic. *Pulsar periods* are remarkable for their stability. They lie in the range 1.558 ms to 4.308 s, that of the Crab pulsar (0.0331 s) being the second shortest observed. It is evident that pulse-recurrence periods as short as these can only be achieved by a *rotating* or *pulsating compact object* of some kind. Pulsar periods are found to *increase* slowly at a steady rate, typically $\dot{P} \sim 10^{-15}$ s/s; they never decrease except for occasional *glitches*, or sudden spinups (probably triggered by a starquake occuring in the crust). All observed pulsars must be in our Galaxy because they are concentrated in the galactic plane, in the same general regions as the known supernova remnants and their presumed progenitors. It is this remarkable coincidence that suggests pulsars must arise from supernova explosions. This interpretation is further bolstered by the discovery of two pulsar-supernova remnant associations: the pulsars PSR 0531 + 21 and PSR 0833 − 45 are definitely associated with the Crab and the

Vela supernova explosions, respectively. In addition, at least one probable and two possible associations have also been identified.

Now, theory tells us that neutron stars are also compact objects associated with supernova explosions. It is thus natural to identify pulsars with *rotating neutron stars.*

Suppose we have a rotating star. The maximum rate at which it can rotate without shedding matter is determined by the equality between the centrifugal force and the gravitational force exerted on each element of the star. This gives a simple relation, $\Omega^2 \sim \rho$, between the mean stellar density ρ and the maximum angular velocity of rotation Ω, or correspondingly a relation, $P \sim \rho^{-1/2}$, between ρ and the period $P = 2\pi/\Omega$. This equation implies that a typical solar-mass white dwarf cannot rotate faster than once every 10 seconds; for a neutron star of the same mass, the lower limit of the period is 60 ms. Thus a rapidly rotating white dwarf cannot produce a pulsar. Nor can a pulsating white dwarf or a pulsating neutron star. So, of all the possibilities considered, only a rapidly rotating neutron star is capable of producing the kind of clockwork mechanism associated with pulsars.

When a star in rotation begins to contract, it rotates faster to conserve angular momentum. Simultaneously, its surface magnetic field — which exists in all stars at various strengths — also increases to conserve the flux of magnetic field lines. The axis of rotation of the star and the direction of its magnetic field, however, need not coincide and, as a compact neutron star spins rapidly, it will drag its large magnetic field around with it. We know that by the laws of electrodynamics this combination of rapid rotation and strong magnetic fields must induce strong electric fields near the surface of the star. These electric fields in turn force the electric charges to flow from the surface and out of the magnetic poles of the neutron star into space along trajectories parallel to the magnetic field lines. These charged particles follow curved trajectories and pick up speed from the co-rotating fields as they move along. The resulting acceleration of the particles has two implications: first, the star is steadily losing rotational energy, and second, the accelerated particles radiate. Photons are emitted in a narrow forward cone, forming a beam of radiation that sweeps the sky like a powerful beam of light from a lighthouse. When the beam happens to point toward a radio antenna, its pulsed signals are recorded (Fig. V.17).

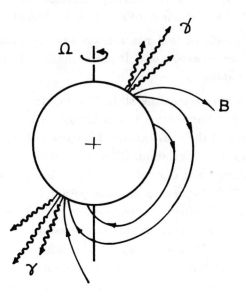

FIG. V.17 *Radiation mechanism of a pulsar. The axis of rotation of the rotating star (Ω) does not necessarily coincide with the polar axis of the magnetic field (B). Pencil-shaped beams of radiation emanating from concentrations of electrons near the magnetic poles and firmly coupled to the rotating star describe giant cones in space.*

As a pulsar radiates away its rotational energy, its rotation slows down, its magnetic field weakens, and its pulse becomes more and more irregular and occurs at longer and longer intervals. The energy emitted gradually decreases until, after about 10 million years, the pulsar disappears from the radio sky. This is the standard evolutionary scenario until 1982, when it was refined to explain a class of pulsars, newly discovered, with ages comparable to the age of the Universe. These new pulsars appear to go through a 'spin up' phase during which they accrete mass and angular momentum from a companion and are accordingly called 'spun-up pulsars'. Since the early 1970's astronomers, using space telescopes, detected X-ray signals from neutron stars. In all cases, the X-ray emission is black body radiation of matter accreted from a companion star and heated to 10^7 K as it falls into the gravitational field of the neutron star. But no emission of this kind from an isolated neutron star cooling down after formation has been identified with certainty. It is hoped that the situation will change with the operation of a new generation of space observatories having improved flux sensitivity and angular resolution.

5.3 Black Holes

In the course of its evolution and, in particular, in the late stages of its active life, a very massive star — with a mass upward of sixty solar masses during its stay on the main sequence — continually ejects matter into space. Whether it can reach a final static state or not depends crucially on the volume of the ejecta, which is absolutely unpredictable. If enough matter and radiation have been blown off so that the residual mass falls in the range of neutron star masses (up to 3 solar units), the remnant star will settle down as a neutron star, forever safe from the inward pull of gravitation — provided only that it lies far removed from any active young stars. If not, as for example when it is part of a close binary system, it may accrete enough matter from its younger companion to grow past the critical mass. Of course, if not enough matter has been ejected to begin with, the residual mass of the star easily exceeds the upper limit of stability. In either case, the star will experience the third of the three possible fates reserved for stars at the end of their evolution.

• *The ultimate victory of gravitation*

The star, or whatever it has become after its nuclear fuel has run out, collapses catastrophically because neither the neutron-degeneracy pressure nor any other force can resist gravity at this point. Once the dynamical equilibrium is broken, matter bursts inward at practically the velocity of free fall, with the free fall timescale of about 1 second. Both the gravity and the escape velocity at the star surface increase quickly.

Let us imagine a solidly built light source, placed on the surface of a spherical nonrotating star on the point of collapse and sending out signals in all directions. At the beginning, all outward rays go into space. But there is a point during the contraction when all horizontal and sub-horizontal rays are deflected back toward the star and trapped by the star gravitational field. The imaginary surface coinciding with the surface of the star at this point is called a *photon sphere*. As the star keeps shrinking, the light rays that can escape are restricted to a cone whose opening gets narrower and narrower until only a pencil of vertical light emerges free. Just a thin space below, even this light cannot escape: it is stopped and held fast by the gravitational field. A photon or any particle must move radially outward with the speed of light just to remain in place. The corresponding surface where this happens is called the *static sphere* or the *event horizon*. As the contraction presses on beyond this point, the escape velocity at the star surface exceeds the velocity of light, and so is physically unattainable. Anything, matter or light, emitted from the star is pulled back by the gravitational field; no information of any kind can be sent outside from inside (provided quantum effects are neglected). The collapse proceeds inexorably to the very end when matter is squeezed to infinite density at some space-time point: a *black hole* is formed.

The escape velocity of a particle from the surface of an ordinary star is calculated by equating its kinetic energy to its gravitational energy. If we apply the same derivation to a particle just outside the event horizon of a spherical nonrotating black hole of mass M, setting the escape speed equal to c, the speed of light, we obtain the radius of the event sphere, $R_s = 2GM/c^2$. Although the result itself is correct, the derivation is based on two invalid assumptions (a nonrelativistic kinetic energy and a Newtonian gravitational energy). The correct analysis of a spherical nonrotating black hole based on Einstein's theory of gravitation was first given by Karl Schwarzschild. For this reason, such a black hole is called a *Schwarzschild black hole*, and $R_s = 2GM/c^2$, the *Schwarzschild radius*. The Sun has $R_s = 2.95$ km; the Earth, $R_s = 0.89$ cm.

In the immediate vicinity of a strong gravitational field, space and time behave in a way that contradicts our intuition. In our part of the Universe, when we have two concentric spheres of surface areas $4\pi r_1^2$ and $4\pi r_2^2$, where r_1 and r_2 are their radii, we know that their surfaces are separated by a distance of $d = r_2 - r_1$. This is one of the many characteristics of our world, the world of *flat* space. Consider now a Schwarzschild black hole with $R_s = 3$ km. Let us define a coordinate r such that the area of a sphere of constant r and constant time is precisely $4\pi r^2$. Thus, the area of the event horizon is $4\pi R_s^2$, or about 113 km^2 in our case. But in contrast to flat space, here r does not measure directly the distances between two such spheres. For example, the distance between a sphere of 100 km in radius and a sphere of 90 km in radius, both centered at the black hole, is 10.17 km, rather than $100 - 90 = 10$ km. The deviation here is rather small, but if we consider now concentric spheres closer to the center of the black hole, we will find increasingly larger deviations: 11.25 km as the distance between spheres 20 and 10 km in radii, and similarly, 17.85 km between a 20 km sphere and a 5 km sphere. Distances seem stretched out; space is distorted, or *curved*. This effect can be visualized by imagining a ball of some kind placed on a soft mattress; the heavier the ball, the more the surface of the mattress is deformed.

The notion of time also undergoes disconcerting changes. An imaginary explorer, sitting on the collapsing star and falling along with it, would see his clock tick normally as he crosses the Schwarzschild sphere, which he does at a perfectly finite speed. He would feel the gravitational forces pulling different parts of his body at vastly different strengths; these *tidal forces* would increase rapidly and would reach infinity within milliseconds. However, an observer outside the event horizon sees the star shrinking toward R_s but never actually reaching this radius. According to her own clock, the aging rate of her explorer friend and the contraction rate of the star slow down, then virtually stop. She observes the luminosity of the star drops dramatically and the frequency of the radiation it emits decreases quickly. It is as if the outward bound photons spent more and more energy climbing out of the deepening gravitational potential well, so that the signal received at infinity is weakened by such a loss of energy. Finally, the luminosity dwindles to insignificance as light is red-shifted to infinity. The final collapse is hidden under the event horizon to all outside observers.

We already know that all stars rotate about their own axes; they all have a non-vanishing angular momentum. A rotating black hole (also called a *Kerr black hole*) conserves most of the angular momentum of its parent star and rotates at or near the maximum rate. It may be rotating so fast that it can drag anything in its immediate vicinity around to the speed of light even before the gravitational trapping makes itself felt. Therefore, in this case, the static limit — a surface on which an astronaut must keep her rocket propelled vertically outward with the local speed of light just to stay where she is — does not coincide with the event horizon; rather, it is a slightly flattened sphere, touching the horizon at the two poles on the rotation axis and intersecting the equatorial plane on a circle whose radius is larger than that of the horizon.

In principle, a black hole can carry a net electric charge, the same charge that the parent star would have had before it collapsed. In practice, it is an unlikely prospect. If a black hole is electrically charged, it exerts an electric force far stronger than the gravitational force on real or virtual (non-observable) electric charges in its neighborhood, pulling toward itself particles of charges opposite to its own, and quickly loses its charge.

A black hole can have but *three attributes*: the mass, the electric charge, and the angular momentum. These are the three properties of the original star that are coupled to long-range fields (charge to the electromagnetic field, mass and angular momentum to the gravitational field) and that are preserved throughout the catastrophic final transition. Together they define the final state of every object that has completely collapsed. Two black holes made out of the most different kinds of matter or antimatter, in the most unlikely shapes or sizes are absolutely indistinguishable, provided they have the same mass, electric charge, and angular momentum. If the collapsing star has unevenly distributed electromagnetic fields, electromagnetic radiation is emitted. If it is nonspherical, all shape distortions will be washed away during the collapse in the form of a gravitational field that propagates, at large distances, as a wave traveling with the speed of light. Whatever strangeness or beauty, whatever color or flavor the star may initially be endowed with, all will be lost after the collapse: a black hole has no 'hair'.

• *Observational evidence*

The theory of general relativity, the modern theory of gravitation formulated by Albert Einstein in 1915, predicts that total collapse is the inescapable fate of every massive star. In the limit of weak gravitational fields, as found on Earth or near any small star, this theory agrees with the ideas on universal gravity that Isaac Newton developed in 1687 in his masterwork, the *Principia*. Since its inception, Einstein's theory has been tested many times in situations involving moderately strong fields, all with unqualified success. Does the theory also hold in the extreme conditions of very strong fields, of the kind encountered in black holes? If it does, black holes must exist. Do black holes exist?

The best existing black hole candidate is a compact X-ray source, called Cygnus X-1 (Cyg X-1), located in an optical binary system in the constellation Cygnus. Its radiation is aperiodic and highly variable on all timescales, varying from months to milliseconds. The shortest 1 ms bursts set a maximum size for the emission zone of $ct \sim 300 \, \mathrm{km}$. Cyg X-1 is the unseen companion of a supergiant star whose spectral lines display variations with a period of 5.6 days, a fact used to confirm the identification of the binary system. The distance to the system is reliably determined to be about 2.5 kpc. The conclusion of careful analyses is: the mass of Cyg X-1 is greater than $3.4 \, M_\odot$ and likely to be in the range $9 - 15 \, M_\odot$. This value exceeds any upper limit we know for white dwarfs and neutron stars. So the existence of a black hole in Cyg X-1 is the most plausible, the most conservative hypothesis. In this interpretation, the radiation would be produced when the gas, which has been pulled out from the visible supergiant to form a doughnut-like accretion disk around the compact object, is funneled into the compact object, heats up by compression, converting gravitational energy into intense radiation before vanishing below the event horizon.

It has also been suggested that *supermassive black holes* ($M \approx 10^3 - 10^9 \, M_\odot$) exist at the center of quasi-stellar objects (*quasars*) and, generally, in galactic nuclei. Their immense gravitational field strongly pulls in surrounding matter, heating it up and powering an intense radiation which could be detected at all wavelengths.

• *Thermodynamics and quantum mechanics of black holes*

As one might expect from the fact that a black hole takes in everything and gives out nothing, the mass of a nonrotating black hole can only increase, never decrease. Stephen Hawking has proved a more general result: namely, the surface area of the event horizon of a (rotating or nonrotating) black hole either increases or remains constant; it never decreases. As an application of this result, imagine a simple situation in which two Schwarzschild black holes, each of mass M, coalesce to form a single Schwarzschild black hole of mass M'. Then Hawking's theorem requires that M' be greater than $\sqrt{2}M$. Hence, the maximum amount of energy that can be released in such a process is about 29% of the initial mass-energy. The actual efficiency may be much less in practice but is still far better than in nuclear processes.

The above result, known as the *second law of black holes*, should bring to mind the second law of thermodynamics which says that the *entropy* of a closed system, or its degree of disorder, always increases (or, at best, re-

mains constant); it never decreases. Our knowledge about the precise state
of a physical system tends to deteriorate with time, and the loss of such
information simply reflects the inevitable increase in entropy of the system.
Such is certainly the case with black holes: an object sinking under its own
event horizon is definitely lost to the ken of our physical world. It is then
reasonable to define, in apppropriate units, the entropy (S) of a black hole
as the surface area $(A = 4\pi R_s^2)$ of its event horizon: $S \propto A$.

This analogy between a black hole and a thermodynamic system led
Jacob Bekenstein to suggest that a black hole is physically a *thermal body*.
How to define the temperature of a black hole? Recall the *first law* of thermodynamics: it says that a small *change* in the heat (or energy) of a closed
system (dE) is given by a corresponding change in the entropy of the system
(dS) multiplied by its temperature (T): $TdS = dE$. Since the surface area
of a black hole is proportional to its squared radius, or its squared mass,
a small variation in the area (dA) is then proportional to a corresponding
variation in the mass multiplied by the mass itself, $dA \propto MdM$. With mass
identified with energy $(dM = dE)$ and surface area with entropy $(dA = dS)$,
it is sensible to define the temperature of a black hole as the inverse of its
mass: $T \propto M^{-1}$.

But now we have a problem. According to the laws of thermal physics,
we know that any object with a finite temperature must emit radiation. On
the other hand, a classical black hole is by definition a perfect absorber;
nothing could ever escape its hold. Thus, a thermal black hole could not be
black; on the contrary, it would be capable of transmitting information to the
outside world. The second part of this paradox, by itself, could be resolved
by extending the concept of a 'closed sytem', mentioned in the first and
second laws of thermodynamics, to include both the black hole and all space
and matter surrounding it. But that black holes can radiate thermally is a
logically inescapable conclusion, however embarassing it is from the classical
viewpoint. This surprising result, discovered by Stephen Hawking, can only
be understood in terms of a quantum mechanical phenomenon. According to
quantum theory, a vacuum is not an absolute void but a sort of reservoir of
pairs of *virtual* particles and antiparticles of every kind. These particles and
antiparticles are called 'virtual' because they can only appear transiently,
for a time interval too short to measure. A pair of particle-antiparticle with

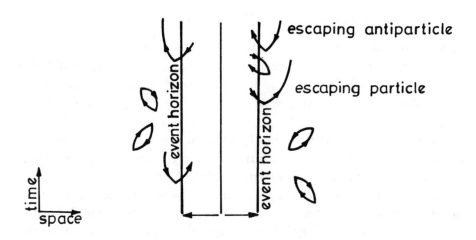

FIG. V.18 *Radiation from a black hole. In this space-time representation, the black hole is propagating vertically in time at a fixed point in space. In the course of pair creation processes near the event horizon, some pairs are split up, one member falling into the black hole, the other escaping to infinity.*

energy ΔE may be created out of the vacuum, provided they exist for a short enough lapse of time, $\Delta t < h/4\pi\Delta E$, in accordance with Heisenberg's uncertainty relation.

Imagine such virtual processes taking place near the event horizon of a black hole (Fig. V.18). One member of a pair may fall into the hole; the other may then either follow its partner to oblivion or escape to infinity. Thus, the hole appears to be emitting particles and antiparticles. According to quantum mechanics, we may view an antiparticle falling into the hole as a particle coming out of it but traveling backward in time. At the point where the pair was originally created, the particle is deflected by the gravitational field and moves forward in time. The particle appears to a distant observer to have escaped from inside the black hole. We may say that it has tunneled through the gravitational barrier on quantum mechanical principles. Hawking found that the particles emitted by such processes have a *thermal distribution* of energies. The temperature T associated with the thermal distribution for a nonrotating black hole has the form

$$T = hc/8\pi^2 kR_s \approx 0.6 \times 10^{-7} M_\odot/M \text{ (degrees Kelvin)}.$$

This relation can be interpreted as the Heisenberg's requirement for the appearance of a virtual pair with energy kT for time interval $2\pi R_s/c$. Were the Sun reduced to a black hole, its temperature would be just about 10^{-7} K; as for the Moon, it would be very nearly in thermal equilibrium with the cosmic background.

The emission of particles by a black hole results in a gain of mass-energy by the outside world. By conservation of energy, there is a corresponding decrease in mass-energy (or *evaporation*) of the black hole, and hence a loss of its surface area, or entropy. Still, there is a net increase in the total entropy of the whole system, the black hole plus the outside world, because the entropy loss by the black hole is more than compensated for by the entropy produced in the thermal radiation. We may readily estimate the power for this radiation as $L \sim R_s^2 T^4 \sim 1/R_s^2$. A solar mass black hole would radiate at the rate $L \approx 10^{-16}$ erg/s in the range of very long (km) radio waves.

The lifetime of a black hole is thus limited by evaporation roughly to $\tau \approx M/L \approx 10^{-35} M^3$ yr, where M is in g. This makes an ordinary mass black hole live for a long time and its radiation unobservable. But black holes with much smaller masses could evaporate at a more perceptible rate. In particular, if small black holes were formed at the beginning of the Universe, those of them with masses greater than 10^{15} g could have survived to the present day. Among these primordial black holes, the *mini black holes* with a mass of 10^{15} g would have the size of a proton ($R_s \approx 10^{-13}$ cm) and a temperature of 10^{11} K. At this temperature they would emit photons, neutrinos, and gravitons in profusion; they would radiate thermally at an ever increasing rate ($L \sim M^{-2}$) until they evaporate violently out of existence. Mini black holes could last for some ten billion years, about the present age of the Universe; so they could still be out there somewhere in space, sending out strong beams of very energetic gamma rays and waiting to be discovered.

Summary

Stars that have burned up most of their nuclear fuel normally die; they may be *completely* destroyed in a catastrophic explosion, with the stellar material dispersed in interstellar space, or may collapse into *compact objects* — white dwarfs, neutron stars, or black holes. Such compact objects have very small size relative to normal stars of comparable mass, and hence a correspondingly high density and strong surface gravity.

A star with a residual mass of the order of one solar unit or less, evolves into a white dwarf, an object of radius $10^{-2} R_\odot$ and average density less than $10^7 \mathrm{g\,cm}^{-3}$. It radiates away its residual thermal energy and, because of its small radius, is characterized by a higher effective temperature (and hence is whiter) than a normal star of comparable mass. It can be described as a globe of dense gas of nuclei and degenerate electrons, surrounded by a comparatively thin envelope of ordinary gas.

A star with a residual mass between 1 solar unit and 3 solar units will collapse into a neutron star (of radius $\sim 10^{-5} R_\odot$) after a short fling as a pulsating radio source (pulsar). It has a density comparable to nuclear values and a complex internal structure marked by a preponderance in degenerate neutrons.

A star with a residual mass greater than 3 solar masses collapses to infinite density in a zero volume, a phenomenon hidden from our physical world by the event horizon. No light or matter can escape from inside the event horizon; a classical black hole looks black to a distant observer. But it may emit particulate radiation if quantum effects are important.

Compact objects are supported against gravitational collapse by other means than thermal pressure: the electron degeneracy pressure in white dwarfs, the neutron degeneracy pressure in neutron stars; but black holes have completely collapsed to singularities. Considered as classical and isolated objects, compact stars are static over the lifetime of the Universe; they are the *end states* of stars.

White dwarfs can be seen as optical objects. Neutron stars can be observed directly during their initial phase as pulsating radio sources, or indirectly as periodic X-ray sources if they are parts of close binary systems. Black holes can be detected indirectly, for example as gas-accreting aperiodic X-ray sources.

CHAPTER VI

THE INNERMOST SECRETS

OF THE UNIVERSE

CONTENTS

1. ELEMENTARY CONSTITUENTS OF MATTER

Ever since the dawn of civilization, one of the most cherished ambitions of mankind has been to establish order and regularity in our complex surroundings. The ancient Greeks thought that all matter was made up of four elements: air, fire, water, and earth. The ancient Chinese had a similar thought, though they insisted that there were five elements: metal, wood, water, fire, and earth, and they named the five classical planets accordingly. Years of work in chemistry showed that both of these views are, to say the least, somewhat oversimplified. Matter is now known to consist of some one hundred different elements, instead of just the four or five. In 1808, John Dalton proposed that there is a most elementary constituent within each of these elements, which is itself unalterable and indestructible. He called it the *atom*. Further progress came in the middle of the 19th century when Dmitry Mendeleyev discovered the regularity of chemical elements in his famous periodic table, though the reason behind these regularities was not understood until the end of the first quarter of the 20th century when quantum mechanics was established.

Ironically, though by no means uncommon in the annals of scientific discoveries, this understanding of the periodic table began with the destruction of the cherished atomic concept of Dalton. *Electrons* were discovered by J.J. Thomson in 1897, and the atomic *nucleus* was discovered by Ernest Rutherford in 1911. These discoveries led to the planetary model of atoms, in which atoms are thought to be made out of electrons circulating about an atomic nucleus, like planets revolving around the sun. Now in this picture atoms are no longer the most elementary entities as Dalton envisaged. Electrons and nuclei now share that honour. Yet this destruction of Dalton's atomic concept actually opens up the door for an understanding of the periodic table. With electrons we can relate different chemical elements. Elements are different merely because they have different number of electrons in the atom. The similarity of chemical properties is now translated into the similarity of electronic arrangements, and the latter can be understood from the laws of quantum mechanics (see Appendix B for a discussion of quantum mechanics).

A chemical element may have several *isotopes*. They have identical chemical properties, the same number of electrons in each atom, but they differ from one another in the mass of their nuclei. This simply suggests that it is the electrons of an atom, and not the nuclei, that are responsible for its chemical properties. But why should there be different nuclei for the same element? The mystery was finally solved when James Chadwick discovered the *neutron* in 1932. This is an electrically neutral particle that resides in the nucleus. The other particle in the nucleus is the *proton*, which carries one unit of positive charge and is nothing but the nucleus of the hydrogen atom. All nuclei are made up of protons and neutrons, and these two particles are collectively referred to as *nucleons*.

The charge of the electron is equal and opposite to that of the proton, so it carries one unit of negative charge. Now a given chemical element has a definite number of electrons. Since atoms are electrically neutral, there must be an equal number of protons as electrons in each atom. However, this argument does not limit the number of neutrons in the nucleus because neutrons are electrically neutral. Atoms which differ only by the number of neutrons in their nuclei are isotopes of one another. This is why a chemical element may have several isotopes.

From this reasoning one might conclude that one could obtain an infinite number of isotopes just by adding more and more neutrons. This turns out to be false because a nucleus with too many neutrons becomes unstable by undergoing *beta decay*.

Now the laurel of elementarity is passed on to the electrons, the protons, and the neutrons.

Since the 1930's many other new particles have been discovered in cosmic rays, or produced and detected in experiments using high energy accelerators. Most of these are *hadrons*, a name reserved for particles which respond to nuclear forces. Protons and neutrons are definitely hadrons, for what holds them together in a nucleus is by definition the *nuclear force*. In contrast, electrons are not hadrons; what keep them held to the nucleus in an atom are the *electric and magnetic forces*. With the presence of tens of hadrons, the stage is set for another shakedown. There were many hadrons and they all seemed to be unrelated. It was just like the old days with many unrelated chemical elements before the discovery by Mendeleyev. For that reason the rush was on to find a regularity, or if you like, some sort of a periodic table for them. Murray Gell-Mann and Yuval Ne'eman finally succeeded in the early 1960's, and the regularity goes under the name *'Eightfold Way'*, or more technically, an $SU(3)$ symmetry. Just like Mendeleyev's periodic table, which opened the way to the discovery of electrons, protons, and neutrons, and the inner structure of an atom, Eightfold Way led to the discovery of an inner structure of the hadrons. In 1964, Murray Gell-Mann and George Zweig proposed that protons and neutrons, and all the other hadrons, were themselves made out of three fundamental particles called the *quarks*. The electron, however, is not a hadron and is still believed to be elementary.

All of this, so to speak, is ancient history. Nevertheless, it serves to illustrate how discoveries are made through an intimate interplay between theory and experiment. Puzzles lead to new ideas, and regularities lead to new structures. This is the same in all fields. What distinguishes this field of elementary particles from others, is the small size of the elementary constituents. As the substructures become smaller and smaller, we need a larger and larger 'microscope' to 'see' them. The only difference is that these large 'microscopes' are called *accelerators*, and they occupy miles and miles of land. An optical microscope yields to an electron microscope when higher resolution is called for, because the beam of electrons in the latter has a smaller wavelength than the beam of light in the former, and better resolution can only be achieved by an illuminating beam with a smaller wavelength.

It is a fundamental fact of the quantum nature of matter that all particles in the microscopic world behave like waves. Moreover, the wavelength of a particle gets shorter as its energy gets higher. This is where an accelerator comes in: to produce for us a higher energy beam with enough resolution to probe the tiny world of elementary particles.

One never sees a cell directly from a beam of light without the complicated array of lenses in a microscope. Similarly, one cannot see what is illuminated by the beam emerging from an accelerator without a complicated bank of detectors and computers. What is different between the two, however, is the destructive nature of the latter. One can illuminate a sample under an ordinary microscope without destroying it, but it is virtually impossible to have a very energetic beam impinging on a target particle without shattering it to pieces. This complication adds substantially to the difficulty of reconstructing the original image before it breaks apart, but the other side of the coin is that such pulverization and disassembly allow us to look into the guts of the target particle. With the help of theory, the reconstruction can be achieved.

In what follows, we will adopt a pedagogical approach. We will skip all these reconstructions, and hence most of the direct experimental data, to concentrate on the physics that finally emerges. We will make free use of the licence of hindsight to discuss the beauty and the intricacies of the physics in as logical a manner as possible without mathematics. Let us be forewarned however that this approach also contains an inherent danger of the misconception that physics is purely a logical and theoretical science. It is definitely not; experiments are needed to confirm a conjectural theory and to guide our thinking in future developments. The final judge as to whether something is correct or not is experiment.

Accordingly, we will now skip over the intervening history in the second half of the 20th century, and jump directly to a summary of the elementary constituents of matter as we know them today.

These constituents fall into two categories: the *quarks* and the *leptons*. They are distinct because quarks experience nuclear forces but leptons do not.

There are three *generations* each of quarks and leptons. Constituents in different generations are almost clones of each other, except that those in higher generations have larger masses. Constituents of the first generation make up the usual matter that we are familiar with. Constituents of the second and third generations are unstable and cannot be found naturally on earth.

Recent experiments show that it is unlikely that there are more than three generations of matter. But why three? Why do we even have a second and a third generation anyway? This nobody really knows. When the *muon*, a second generation lepton, was discovered in 1936, it was so astonishing, and the muon was deemed to be so 'unnecessary', that I.I. Rabi was prompted to utter his famous remark: "Who ordered that?" Fifty five years later, we still do not know the answer.

We will now proceed to give a more detailed description of these constituent particles.

There are two quarks and two leptons per generation. The quarks of the first generation are called the *up quark* (u) and the *down quark* (d), and the leptons of the first generation are the *electron* ($e = e^-$) and the *electron-neutrino* (ν_e). Symbols in the parentheses are used to designate the particles; a superscript, when it is present, indicates the electric charge carried by that particle.

The quarks of the second generation are the *charmed quark* (c) and the *strange quark* (s), and the leptons are the *muon* (μ^-) and the *muon-neutrino* (ν_μ). Finally, the quarks of the third generation are the *top quark* (t) and the *bottom quark* (b) (sometimes also known respectively as the *truth* and the *beauty* quarks), and the leptons are the *tau* (τ^-) and the *tau-neutrino* (ν_τ). All these particles except the top quark have been experimentally observed. As for the top quark, there are rather strong theoretical reasons to suggest its existence, although so far its discovery has eluded us, presumably because it is too heavy to be produced effectively by the present-day accelerators.

If the electric charge carried by an electron e is counted as -1, then the electric charges of μ and τ are also -1, but those for the neutrinos ν_e, ν_μ, and ν_τ are zero (hence the name neutrino, which means 'little neutral one' in Italian). The electric charges of u, c, and t are $+\frac{2}{3}$, and the electric charges of d, s, and b are $-\frac{1}{3}$.

The proton (p), which has a unit ($+1$) of electric charge, is made up of two up quarks and a down quark. The neutron (n), which is electrically neutral, is composed of two down quarks and an up quark: $p = uud, n = udd$. Other hadrons are made up of three quarks, called *baryons*, or a quark and an antiquark, called *mesons* (see the following for a discussion of antiquarks).

There is an asymmetry between quarks and leptons in that leptons have integral units of electric charges while the electric charges of quarks are multiples of $\frac{1}{3}$. This factor of 3 is actually accompanied and 'compensated' by another factor of 3: each quark comes in three *colors*. We will explain more fully what 'color' is in Sec. 4, but whatever it is, it has nothing to do with the color we see with our own eyes. It is just a name invented to label the different varieties that exist for each of the quarks described above.

To this list of constituents of *matter particles* we must add the *anti-matter particle*, or simply *antiparticle*, for each of them. Antiparticles will be denoted by a bar on top of the symbol for the particle. For example, the antiparticle for the u quark is called the anti-u quark and is denoted by $\bar{u}$. Antiparticles have the same mass but the opposite electric charge as the particle. They also have all the opposite additive quantum numbers (see Sec. 4). When they encounter particles of the same kind, they annihilate each other. Since the earth is made out of particles (by definition), no antiparticles can live for long in our vicinity, so all the antiparticles observed to date are produced either by cosmic ray bombardments or by high energy accelerators.

The existence of antiparticles for matter was first predicted theoretically in 1929 by Paul Dirac long before any of them was found experimentally. (See the end of Sec. 7.3 for more details). This prediction was based on the theory of special relativity and quantum mechanics, and represents one of the greatest triumphs of theoretical physics in the twentieth century. Nowadays the existence of the antiparticles is very firmly established experimentally.

Beyond the quarks and leptons, present theory (called the 'Standard Model') would also like to have, for technical reasons to be discussed in Sec. 5.3, another electrically neutral particle, called the *Higgs boson* (H^0). Just as for the top quark, an intense search for the Higgs is underway, but as yet it has not been found. Unlike the top quark, whose existence we are relatively sure of, the existence of the Higgs boson does not enjoy the same degree of confidence. We are far from being sure that it has to exist as an elementary particle, and we do not know whether there should not be more than one of them. The form and shape of the future theory (which is more fundamental than the existing Standard Model) will depend quite critically on such details, which is why it is such an important matter to discover the Higgs particle or to rule it out experimentally.

The Higgs boson is certainly not a matter particle in the usual sense. Matter generally cannot disappear unless being annihilated by its own antiparticle. In contrast, H^0 can be created and absorbed singly.

The Standard Model also requires a set of *gauge particles* to transmit the forces (see Secs. 3–5). These particles are again *not* indestructible; in fact forces are transmitted only because these particles can be freely absorbed and created (see Sec. 3). There are a number of gauge particles, each responsible for transmitting a different kind of fundamental forces. As we shall see in Sec. 2, the fundamental forces are the *strong* or nuclear force, the *elec-*

troweak (electromagnetic and weak) force, and the *gravitational* force. The corresponding particles are the *gluons* (g) for the strong force, the *photons* (γ), the $W^\pm$ and the Z^0 particles for the electroweak force, and the *graviton* (G) for the gravitational force. With the exception of the graviton, all the other gauge particles have already been found.

All in all, there are several kinds of elementary particles — some responsible for giving us the material universe (quarks and leptons), some responsible for transmitting the forces between particles (the gauge particles), and some are there for more complicated technical reasons (the Higgs particle). These particles are summarized in the Table below.

The Standard Model Particles

Matter/anti-matter particles			Gauge particles		
Quarks		Leptons	Strong	Electroweak	Gravitational
(u,d) $(\bar{u},\bar{d})$		(ν_e, e^-) $(e^+, \bar{\nu}_e)$	g	γ, $W^\pm$, Z^0	G
(c,s) $(\bar{s},\bar{c})$		(ν_μ, μ^-) $(\mu^+, \bar{\nu}_\mu)$	Higgs particle		
(t,b) $(\bar{b},\bar{t})$		(ν_τ, τ^-) $(\tau^+, \bar{\nu}_\tau)$	H^0		

Summary

Matter is composed of two kinds of particles, the quarks and the leptons. Quarks experience the nuclear force but leptons do not. Each quark has three possible colors but leptons are colorless. Quarks and leptons come in three generations each. The first generation quarks are the (u,d), the second generation quarks are the (c,s), and the third generation quarks are the (t,b). The corresponding three generations of leptons are (e^-, ν_e), (μ^-, ν_μ), (τ^-, ν_τ).

To each quark or lepton there exists an antiparticle with the same mass and the opposite electric charge.

On top of these, there may also be one or several Higgs bosons.

2. FUNDAMENTAL FORCES

Of equal importance to the identity of elementary particles is the nature of fundamental forces. What holds protons and neutrons together to form a nucleus? What binds electrons and a nucleus together into an atom, and atoms together into a molecule? For that matter, what holds molecules together in ourselves, and what holds us on the surface of this planet? Can we study the forces of Nature and classify them systematically?

At first sight this seems to be quite an impossible task. There are simply too many varieties of forces. After all, you can tie things together with a rope, hold them together with your hands, or bind them together with crazy glue. All these correspond to different forces. How can we possibly classify them all?

Fortunately, things are not so bleak! We could have said the same thing about the classification of matter, for certainly there is a great variety there also. But we now know that all bulk matter is made up of electrons and quarks. Similarly, it turns out that all these complicated forces are nothing but very complex manifestations of four fundamental forces! In order of decreasing strength, these are the *nuclear force*, the *electromagnetic force*, the *weak force*, and the *gravitational force*.

Before we go into the details of these four fundamental forces, it might be interesting to reflect on what kind of a world we would live in should these forces be very different from what they are.

If the electromagnetic force holding the electrons and the nucleus together had a short range and was very weak, then there could be no atoms and no molecules, because electrons could no longer be bound to the nuclei. The whole world would be a soup of negatively charged electrons and positively charged nuclei, in what is known as a *plasma*. Needless to say we would no longer be present. Now think of the other extreme, when the electric force was much stronger than the nuclear force. Then the electrostatic repulsion between the protons in the nucleus would overcome the nuclear attractive forces between them, and a nucleus containing more than one proton would be broken up. The world would then be left with only one chemical element, the hydrogen, though hydrogen isotopes could exist and it is conceivable that very large molecules of hydrogen could be formed. In any case the world would also look very different from what it is today.

In both extremes, when the electromagnetic force is very weak or very strong, it is almost impossible to *see* anything. In the former case, when the world is a plasma, light is constantly emitted and absorbed by the charged particles in the plasma. This means that light can never travel very far, and we cannot really see clearly like we do in an electrically neutral surrounding when this does not happen. At the other extreme, when the electromagnetic force is very strong, ironically we cannot see either. The world is now neutral, but another effect takes over to block our vision. In the present world, light can travel great distances as long as there is no matter blocking its way. It does not matter that there are other light beams travelling in different directions crossing our light beams — they just go through each other. There are no traffic police in the sky, no traffic lights, but there are never collisions between the light beams in the sky. Otherwise whatever image light brings will be all shattered by the collsions from cross beams and you will never be able to see anything. Alas, this happy phenomenon will no longer be there if the electromagnetic force is very strong. In that case, it is easy for the light beam to create 'virtually' a pair of electron and positron, which are charged and can thus absorb and emit light. In this way an effective plasma would be present in the vacuum, which as we have seen, would severely limit our vision. The difference between this 'virtual' plasma and the real one of the previous case, is that none of the pairs in this virtual plasma can live very long. But as one dies, others are born, and there are always some pairs there to obstruct the passage of the light beam.

If the nuclear force holding the nucleons together were too weak, we would not have isotopes. On the other hand, if the nuclear force were very much stronger than it is, nuclei could be much larger and very heavy chemical elements could be formed. We can go on and on with different scenarios. If the gravity were too weak? Then we would all float away like an astronaut in orbit.

So the kind of world we have around us depends very sensitively on what forces are present and how strong they are. Our mere existence depends on that too. It is therefore intellectually important to understand the origin and the detailed nature of the fundamental forces. We will indeed discuss this very important and fascinating problem, but in later sections. For now, let us first get familiar with these forces.

It is ironic that the first of these four fundamental forces to be discovered is actually the weakest in strength. It is the gravitational force. This discovery was made by Isaac Newton in the seventeenth century while endeavouring to explain Kepler's three laws of planetary motion. He found out that the planetary motions could be explained by postulating a universal attractive force between any two bodies. The force has to be proportional to the product of the masses of the two objects, and inversely proportional to the square of the distance separating them, no matter what the distance is. For this reason this force law is also known as an *inverse-square law*. This force is universal and all pervasive. It exists between any two of us, but we do not feel a force pulling us towards the other person because neither of us is massive enough. In contrast, the earth is heavy and it is this force that holds us to the surface of the earth. Similarly, it is this force that keeps the planets going around the sun, and the satellites going around the earth, like a weight at the end of a string. Without this force they would fly apart in a straight line. This force also exists between two elementary particles, but it is so feeble compared to the rest of the forces present that it can be safely neglected in present-day experiments. However, it may play a very fundamental role in some future grand theories. We will come back to this aspect in the last section.

Newton's law of gravitation was modified and improved by Albert Einstein in his famous 1915 *theory of general relativity*. The result differs from Newton's law in mathematical details, but this difference produces only very small effects on planetary motions and day-to-day gravitational actions. If you should fall off a ladder, God forbid, Einstein would neither help you nor hurt you any more than Newton.

One important aspect of Einstein's gravity is that it affects light. The Newtonian gravitational force is proportional to the mass of the particle it acts on. Light consists of massless particles called *photons*. Being massless, they are not subject to gravity according to Newton's law, but they are according to Einstein's. Observations show that light passing near a massive body is indeed pulled towards that body by gravity, just the way Einstein predicted. In fact, wherever there is energy or pressure, there is gravity, according to Einstein. Since photons have energy, they can therefore be influenced by gravity.

The next fundamental force to be discovered was the *electromagnetic force*. Electricity and magnetism were first thought to be two unrelated forces, but subsequent experimentation showed that they were intimately connected, a connection that was finally codified and formulated into the famous *Maxwell's equations* by James Clerk Maxwell in 1873. Under ordinary circumstances, the electric force is more important than the magnetic force. Like gravity, the electric force also obeys an inverse-square law, called *Coulomb's law*. The difference is, instead of being proportional to the masses of the two bodies, the electric force is proportional to the product of their electric charges. Since electric charges can be either positive or negative, electric forces can be attractive (between opposite charges) or repulsive (between charges of the same sign). In contrast, masses are always positive and the gravitational forces are always attractive. Antigravity does not exist; even antiparticles have positive masses.

It is this bipolar property of the electric force that allowed the gravitational force to be discovered. Because of this propery, there is no electric force between neutral objects, for the electric force on (or from) the positively charged part of the object is cancelled by the electric force on (or from) the negatively charged part of the object. If this were not the case, the stronger electric force between the planets and the sun would have masked completely the gravitational force and prevented its discovery by Newton.

The alert reader may have noticed a peculiar feature of these two forces. The gravitational forces are *attractive* between masses of the *same* sign (because masses are necessarily non-negative), while the electrical forces are *repulsive* between charges of the *same* sign. The difference comes about because of the different nature of these two forces, in spite of the fact that both of them obey the inverse-square law. We will come back to this point later in the next section.

The electromagnetic force is the force responsible for the creation of light, radio, and television waves. It is also responsible for the working of electronics. It is the force that binds electrons to the nucleus to make an atom, the atoms together to make a molecule, and the molecules or atoms together to form a liquid or a piece of solid. It is the fundamental force, when dressed up in a complicated way by the presence of these atoms and molecules, which gives rise to the strength of a piece of rope, or the adhesion of crazy glue. In short, other than gravity, the electromagnetic force is the only force we encounter in daily lives.

The remaining two fundamental forces were discovered in the twentieth century. They are the *nuclear* or the *strong force*, and the *weak force*. They are not detectable in daily lives because of their incredibly short ranges. Nuclear forces between two nucleons are effective only when they are within 10^{-15} m (one million-billionth of a meter) of each other. Weak forces have a shorter range still: about 10^{-17} to 10^{-18} m. At distances short compared to their ranges, these two forces also obey the inverse-square law. Beyond these distances, as is the case in daily lives, these forces can become very small.

The great strength of the nuclear force can be seen from the fact that an enormous amount of energy can be derived from a small quantity of fissionable material used in a nuclear power plant or an atomic bomb.

The weak force is responsible for the instability of neutrons when left alone. Under the influence of this force, an *isolated* neutron will be torn into a proton, an electron, an antielectron-neutrino ($n \rightarrow p + e + \bar{\nu}_e$) within a time of some fifteen minutes or so, in a process known as *radioactive beta decay*, (or β-*decay*). Under the influence of the strong or the electromagnetic force, some hadrons also become unstable, but their lifetimes are typically many orders of magnitude shorter than the neutron lifetime. This shows that the strength of the force responsible for the beta decay is weak compared to the other two—hence the name 'weak force'.

Neutrons in a nucleus are subject to the protection of the nuclear and the electromagnetic forces from the other nucleons, and they will remain stable provided there are not too many of them. If there are too many, such protection would not be sufficient for all of them to remain secure, and the nucleus would undergo a radioactive beta decay. If you have seen a watch with glowing dials, you have probably seen beta decay at work. A radioactive material is painted on the dial, whose emitted electron excites the nearby fluorescent material to produce light.

As a result of the beta decay, a neutron inside the nucleus is changed into a proton, while the produced electron and the antielectron-neutrino leave the nucleus. The new nucleus has one more proton than the old nucleus, and soon it will capture an electron somewhere to neutralize it. Since the chemical nature of an element is determined by the number of electrons or protons, a chemical element undergoing a beta decay changes into another chemical element.

When such a chemical change was discovered by Rutherford at the beginning of this century, it amounted to heresy especially among the chemists. It seemed like modern alchemy; it certainly violated the cherished 'unalterable and indestructible' doctrine of Dalton's atomic theory. For this reason when Rutherford gave a talk on this discovery at the Physical Society of McGill, he was severely attacked and ridiculed by his chemistry colleague Frederick Soddy. Eventually Soddy became a convert and a close collaborator of Rutherford, and both Rutherford and Soddy went on to win a Nobel Prize in chemistry (1908 and 1921 espectively).

The strong force is the strongest of the four forces. As remarked above, the force between two hadrons has a range of the order of 10^{-15} m. The force between quarks, however, appears to be quite peculiar. At short distances, the force obeys the inverse-square law like everything else. From a distances of the order of 10^{-15} m on, the force is constant, and quite independent of the separation of the two quarks. That means no matter how far apart these two quarks are, there is always a constant force to pull them back. It also means that the work done and the energy required to pull the two quarks apart is proportional to their separation, hence an infinite amount of energy is needed to separate and isolate them. As a result two quarks cannot be isolated from each other, a property that has come to be known as *confinement*. Note that hadrons such as the nucleons are made out of quarks, and there is no doubt that hadrons can be isolated. So a group of quarks is sometimes confined, sometimes not. We will come back in Sec. 5 to discuss which is which. We should also note here that the idea for quark confinement is actually forced upon us by the failure to find *isolated* quarks in spite of an intense search.

Summary

All forces are complex manifestations of four fundamental forces. In order of decreasing strength, these are the nuclear (or strong) force, the electromagnetic force, the weak force, and the gravitational force.

3. THEORY OF FORCES

We saw in the previous section that our mere existence, as well as the character of the whole universe, depends critically on the range, the strength, and various other details of the fundamental forces. The electromagnetic and gravitational forces have long ranges and obey the inverse-square law, but the weak and the nuclear forces have only short ranges. Why are their ranges different? In spite of that why do they all obey the inverse-square law within their respective ranges? What makes the strengths of the various forces so different? What generates the forces anyway? Why do some particles experience nuclear forces and others do not? Why are there four fundamental forces? Why are some forces attractive and others repulsive? There is a whole series of interesting questions of this type. We know the answers to some of them, but unfortunately not to many others at this moment.

The questions above are unanswerable, and perhaps meaningless, in the realms of classical physics and the quantum mechanics of atoms and molecules. A force is just there. One can measure to determine what it is, and one can ask how it affects the motion of particles. But one has absolutely no idea where it comes from.

The theory of elementary particles belongs to a class of theories known as *local quantum field theories*. In such theories, some of the questions asked above are answerable. What characterizes these theories is the fact that all interactions take place *locally* in the form of creation or annihilation of particles; forces across a distance are transmitted by the propagation of particles (known as *exchange particles*) so created. Forces differ because the exchange particles are different, or else because the manner they are created and absorbed is different. Let us see how that works for the electromagnetic forces.

To do that we go back to the explanation of the *photoelectric effect* by Einstein in 1905.

Electric currents are known to flow in some metals when a beam of light is shone on it. This is the photoelectric effect discovered by Heinrich Hertz in 1887. Nowadays phototubes are constructed to make use of this effect to turn light into electricity, a device which enables an alarm signal to be sounded or a door to be open when a beam of light is interrupted.

What Einstein discovered, by studying this effect in detail, was that electromagnetic waves are made out of many particles called *photons* (γ). When a photon hits an electron in the metal and knocks it into motion, a photoelectric current is generated.

According to the quantum field theory of electrodynamics (*quantum elec-trodynamics*, or QED for short), electromagnetic forces between two charged particles are generated by the transmission of a photon between them. This photon brings with it a 'command' from one particle to another, according to which the recipient particle reacts. We interpret this reaction as the result of a force. The photon passes on a (force) message just like a radio signal to a spacecraft does. The difference is, a radio signal contains many photons, and these can be coded into a great variety of different commands: to order it to change orbits, to order it to report to earth, or to order it to take a picture. If the signal is reduced to a single photon, then clearly only a very limited amount of information can be passed on, but nevertheless enough to generate what is required for a force.

There is however one more point to add. One needs energy to send out a radio wave. Such energy is not always available in the microscopic world of motions of two charged particles. This is where the *quantum* nature of the microscopic world comes to the rescue. According to Heisenberg's famous *uncertainty principle* (see Appendix B for a discussion of this principle), en-ergy can be borrowed if necessary for a short time, free of interest, to enable that command signal carried by the *exchange particle* to be sent out.

In the following subsections, this force mechanism will be studied in greater detail.

3.1 Range and Mass

Let m be the mass of the exchange particle. To exchange it between two particles A and B, it must be first created by A and later on absorbed by B. According to the most famous equation of Einstein, this requires an amount of energy $E = mc^2$, where $c = 3 \times 10^8$ m/s is the velocity of light. Since this amount of energy E is not readily available, it must be borrowed. This financing is permitted under the rules of quantum mechanics, but the uncer-tainty principle allows this borrowed energy to be kept only for an amount of time Δt of the order of $\hbar/\Delta E$, where $\hbar = 6.6 \times 10^{-25}$ GeV·s is the *Planck's constant*. During this short interval of time, the exchange particle created at A must travel far enough to reach B if a force is to be transmitted between them. Even travelling at the speed of light (and nothing can travel faster than that), the distance it can cover in time Δt is only $R = c\Delta t \sim \hbar c/mc^2$, so B cannot be further away in order to feel the force effectively. This formula which relates the *range R* of the force and the mass m of the exchange particle is sometimes called the *Yukawa formula*.

To have a proper appreciation of this formula, we should really develop some feeling about the magnitude of the quantities $\hbar c$ and mc^2, as well as the units used to express them. We shall defer such a discussion until the end of this subsection.

In the case of the *electromagnetic force*, the photon has mass $m = 0$, so the range R is infinite, which agrees with the inverse-square law. The *gravitational force* also obeys the inverse-square law, so its exchange particle, called a *graviton* (G), should also be massless. Unfortunately, the gravitational interaction is so feeble that in spite of intense efforts around the world, even a large aggregate of gravitons in the form of a gravitational wave has not yet been found, let alone the detection of single gravitons.

In the case of the *strong interaction* between two hadrons, the range R is of the order of 10^{-15} m. Using Yukawa's formula, the mass of the exchange particle comes out to be something of the order of $m \approx 0.2$ GeV/c^2. It was Hideki Yukawa who first recognized this connection in 1935, at which time the range of the nuclear force was already known. Using that, he predicted the existence of a particle with this mass which carries the nuclear force. This particle was later found, with a mass $m = 0.14$ GeV/c^2, and is nowadays known as the *pion* (π). It actually comes in with three possible electric charges: π^+, π^0, and π^-. Their quark contents are respectively $\pi^+ = u\bar{d}$, $\pi^0 =$ a combination of $u\bar{u}$ and $d\bar{d}$, and $\pi^- = d\bar{u}$.

In the case of *weak interactions*, the range $\sim 0.2 \times 10^{-17}$ m implies an exchange particle with mass $m \approx 100$ GeV/c^2. There are actually three particles found. They are W^+ and W^-, with the same mass 80 GeV/c^2, and Z^0, with a mass of 91 GeV/c^2.

The force that leads to the confinement between quarks is constant at all large distances, so it has an infinite range. The exchange particle must be massless, and it is called the *gluon* $(g = g^0)$. As we shall see later, there are actually eight such gluons.

The alert readers may have noticed a very curious fact. The force range between quarks is infinite. However, hadrons are made out of quarks but the force range between hadrons is short. What is going on? A look at the more familiar atomic world would explain why.

The force between neutral atoms, the so-called *van der Waals force*, also has a range much shorter than the range of the Coulomb forces between electrons and/or protons. This is because an electron in one atom sees both a repulsive force from the electrons of another atom, and an attractive force from the protons of the other atom. These two forces are almost equal and opposite and they tend to cancel each other, especially when the separation between the two atoms is much greater than the size of the atoms. This is why the van der Waals force has a range much shorter than the Coulomb force. In the same way, a cancellation of the forces between quarks in different hadrons must be responsible for the short-range nuclear force between two hadrons.

We return now to the Yukawa formula $R \sim \hbar c/mc^2$ in an attempt to develop some intuition by studying the magnitude of the various quantities and units. Using the numbers for $\hbar$ and c quoted before, one obtains $\hbar c = 0.197$ GeV·fm, where 1 fm$=10^{-15}$ m (fm is called a 'femto-meter or a 'fermi'). If the rest energy mc^2 is 0.14 GeV, like the pion, then the range R is a bit over 1 fm. In this way we see that the length unit fm is quite appropriate for the study of elementary particle physics. In fact, the size of a nucleon is also roughly 1 fm.

The energy unit GeV is also quite suitable; in terms of this unit the rest energy mc^2 of an elementary particle is not exceedingly large or exceedingly small. The notation GeV stands for 'Giga electron Volt', or 10^9 eV; an electron Volt (eV) being the amount of energy an electron gains when it slides down a one Volt potential. The unit eV is a bit small in the study of elementary particle physics, but it is quite a suitable unit for atomic physics. The energy required to tear away the electron bound to a hydrogen atom is, for example, 13.6 eV.

In terms of units closer to daily usage, an eV is an awfully small unit. 1 eV is equal to 1.6×10^{-12} ergs, or 1.6×10^{-19} Joules, or 3.8×10^{-23} Calories. Remembering that one Calorie is the energy required to raise the temperature of a kilogram of water by 1° C, and that a slice of white bread contains some 80 Calories, we gain a proper appreciation as to how small

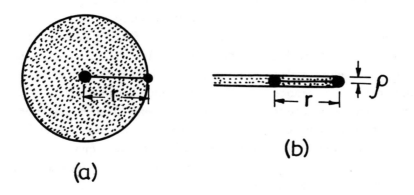

FIG. VI.1 *(a) A sphere with radius r and surface area proportional to r^2, illustrating the origin of the inverse-square law; (b) a cylinder of radius ρ, illustrating the origin of confinement.*

an energy unit eV is in daily lives. However, there are some 3×10^{25} H_2O molecules in a kilogram of water, so one Calorie per kilogram works out to be about a thousandth of an eV per water molecule. In this way we see once again that an eV is actually not a small unit in the realm of atomic and molecular physics.

3.2 Inverse-Square Law vs Confinement

A force will taper off to zero beyond its range R. What is it going to be at a distance r small compared to its range R? We know from the previous discussions that inverse-square law holds at these short distances. But why?

At such a short distance, the exchange particle simply spreads out from its creator like a ripple. At a distance r from the emitting particle, the surface area of a sphere is $4\pi r^2$ (see Fig. VI.1a), and the exchange particle can be anywhere on this surface. The probability of its being along some fixed direction is therefore proportional to $1/r^2$, which is then the origin of the inverse-square law.

This argument may still be valid even if the exchange particle is not emitted isotropically. For example, if it is emitted uniformly within a 30°

angle, then the area of the surface within this 30° cone would still increase like r^2, and the force would still decrease like $1/r^2$. If the exchange particle carries a spin (see Sec. 3.3), anisotropy can indeed occur.

The argument, however, would fail if the emission were confined to a cylinder of constant radius ρ, as in Fig. VI.1b. For the sake of definiteness let us assume $\rho \ll R$, the range of the force. At a distance $r \ll \rho$, the same argument as before leads to an inverse-square law. For $\rho \ll r \ll R$, the story is different. The exchange particle is now guided down the cylinder, without even having the chance to spread out into a cone. The cross-sectional area down the cylinder is constant, so the force also remains independent of r. This is what happens to the confinement force between quarks, where $\rho \sim 1$ fm. What is still to be understood is why the exchange particle is confined to the cylinder. We will return to this in Sec. 5.4.

3.3 Spin and the Nature of Forces

The spin of the exchange particle also conveys important information about the force it transmits.

Spin is the angular momentum of a particle at rest. It is an intrinsic and unchangeable attribute of the particle, just like its mass. We might think of a particle as a miniscule earth, and its spin is like the rotation of earth about its own axis. While this picture has some merit, it is unable to explain how a very small object like the electron can rotate that fast to yield a finite amount of spin that it actually carries.

The spin of a particle is specified by a number s, which may take on non-negative integer values ($s = 0, 1, 2, \cdots$), or half-integer values ($s = \frac{1}{2}, \frac{3}{2}, \frac{5}{2}, \cdots$). The meaning of this number will be explained below. The spin of an electron is $s = \frac{1}{2}$.

Angular momentum is described classically by a three-dimensional vector or, equivalently, by three real numbers representing the three components of the vector (see Appendix A). In the microscopic world, on account of the uncertainty principle, it is impossible to get accurate values of the three components simultaneously. The best we can do is to specify the magnitude of the vector, as well as one of the three components. It is traditional to call the direction along which the component is specified the z-direction. For a given spin s, the magnitude of the spin angular momentum vector is $\sqrt{s(s+1)}\hbar$. This vector has $2s + 1$ possible orientations, each given by

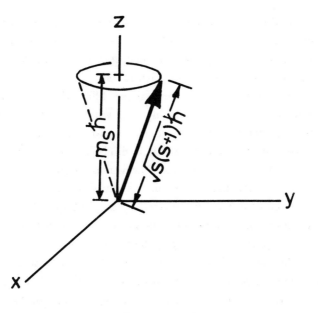

FIG. VI.2 *The thick arrow represents a quantum-mechanical angular momentum vector, which can be pictured to be precessing about the z-axis on the surface of the cone.*

a different z-component $m_s\hbar$ which may take on any one of the following values: $s\hbar, (s-1)\hbar, \cdots, (-s+1)\hbar, -s\hbar$. Since the x- and the y- components cannot be fixed once the z-component is specified, we might think of the spin vector to be precessing in a cone as shown in Fig. VI.2.

The nature of the force carried by an exchange particle depends on the spin of that exchange particle, and this spin is always an integer rather than a half-integer.

Quantum field theories dictate that if s is an even integer, then the force between identical particles is attractive. If s is an odd integer, then that force between identical particles is repulsive, and the force between a particle and its own antiparticle is attractive.*

Pions have spin $s = 0$ and gravitons have spin 2. This is the reason why

---

* Strictly speaking, this rule is true only for the spin-indpendent forces. Since these are usually the dominant forces, the rule stated in the text does give a good indication.

the nuclear and the gravitational forces of two protons are both attractive. Photons have spin 1, so the Coulomb force between two protons is repulsive.

The spin of the exchange particle also determines the complexity of the force. An exchange particle carries with it $2s + 1$ bits of information on its orientation. These different bits can trigger different forces, so the number and complexity of the forces would increase with the value of the spin. For example, the Yukawa nuclear force between two nucleons is carried by the pion, which has spin 0. So the force is a single attractive one. The electromagnetic force between two protons, however, is carried by a photon, which has spin 1. As a result, there is in this case an electric force *and* a magnetic force. Gravitational force is transmitted by a spin 2 graviton, so the force is even more complex than the electromagnetic one. Nevertheless, in all these cases, unless the particles move very fast, the magnetic-like force is much smaller than the electric-like force, so the latter dominates. In the case of electromagnetism, this is the Coulomb force. In the case of gravity, this is the Newtonian force. When we talked about the association between the spin of the exchange particle and the type of force it transmits, we were tacitly referring to these dominant forces.

3.4 Feynman Diagrams

We have so far concentrated on the exchange particle, and how its mass and spin affect the range and the complexity of the force it propagates. We are now in a position to consider other details of the force, as well as the effect of the force on the emitting and the absorbing particles.

The best way of doing so is to picture the physical process graphically in a set of diagrams like those shown in Fig. VI.3.

Each of these diagrams can be interpreted both as a *Feynman diagram*, or an *old-fashioned diagram*. We shall make the distinction much clearer later on, but for now, all that we have to know is that a Feynman diagram can be considered as a representative of several old-fashioned diagrams; different old-fashioned diagrams within this class actually depict different *physical mechanisms*.

In an old-fashioned diagram, *time* runs generally upwards, so we should read them from the bottom up. Consider for example the old-fashioned diagram in Fig. VI.3a. What it depicts is an $e^- + e^+ \rightarrow e^- + e^+$ collision process through the following mechanism. An electron e^- (represented by the solid

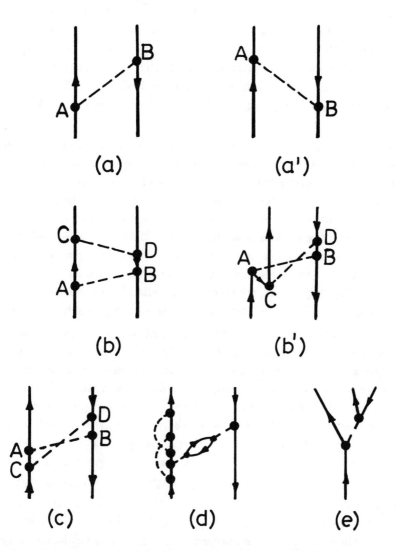

FIG. VI.3 *Feynman and old-fashioned diagrams. A solid line with arrow pointing upwards is a fermion, downwards is an anti-fermion, and a broken line is a boson. Diagrams (a)–(d) depict a number of mechanisms for electron-positron elastic scattering; (e) is the lowest-order diagram for the β-decay of the down quark (d), which in turn causes the neutron (n) to decay.*

line with an arrow pointing up) comes in from time $-\infty$ and emits a photon γ at point A. The emitted photon (represented by the broken line) propagates on and is absorbed by a *positron* (*i.e.*, antielectron) e^+ (represented by a solid line with an arrow pointing down) at point B. A force between the electron and the positron is thereby transmitted by the photon. As a result the electron and the positron will be bounced off at a different direction, although this aspect of going in a different direction is not really shown in the diagram.

It is important to realize that as a result of this exchange of the photon, the positron is not being pushed away by the impact of the photon. After all, the Coulomb force between the electron and the positron is attractive, so the positron should be pulled towards the electron rather than being pushed away. As explained at the beginning of this section, the point is that one should not view the photon as a bullet or a billiard ball, whose sole function may be to impart a momentum on the absorbing particle. Rather, one should regard the photon to be a signal carrying a message encoded in its mass and spin. Just like a spacecraft, which can be recalled by a command signal, it is possible for the recipient positron to be ordered to approach the electron.

The other point to note is that the emission of the photon at A is a process forbidden by classical mechanics, for it does not conserve energy and momentum. It is however allowed by quantum mechanics because of the possibility of borrowing or lending energy for a short time. This photon is called a *virtual particle*. Since its energy is borrowed, and is not derived from the momentum and the mass of the particle in the usual way, usual kinematics generally fails. For example, the speed of the virtual photon propagation between A and B may no longer be the speed of light.

Let us now return to the diagrams of Fig. VI.3. Diagram (a') is very much like diagram (a), except in this case the virtual photon is emitted by the positron rather than the electron. Since the only thing different between these two diagrams is the time ordering, A before B in (a) but A after B in (a'), there seems to be no point in duplicating the work to draw two of them. We will therefore combine the two and draw only one diagram — it does not matter which of the two we draw. The result is a *Feynman diagram.*

So a Feynman diagram differs from an old-fashioned diagram in that the former is actually a representative of a class of the latter; diagrams within the class differ from each other only by the time-orderings of the emission and absorption events. These emission or absorption events, indicated by a heavy dot in Fig. VI.3, are called *vertices*, and the lines between two of these heavy dots, indicating the propagation of virtual particles (solid or broken lines), are called *propagators*. A Feynman diagram with n vertices therefore is a representative of a set of $n! = n \cdot (n-1) \cdot (n-2) \cdots 3 \cdot 2 \cdot 1$ old-fashioned diagrams — $n!$ being the number of possible time-orderings of n vertices. Again, we may use any of the $n!$ as the Feynman diagram for the whole class.

We should elaborate a bit more on the connection between the Feynman and the old-fashioned diagrams. The former clearly loses all time orderings, for it actually represents the *sum* of the old-fashioned diagrams with all possible time orderings. It should also be noted that these diagrams are more than pictorial; mathematical expressions can actually be written down for each vertex and each propagator, and hence for each old-fashioned diagram. In this way we can actually *compute* the *T-matrix elements*, which are mathematical expressions from which physical quantities like cross sections and decay rates can be calculated. Moreover, and this is the important contribution of Feynman, one can write down rules for vertices and propagators for *Feynman diagrams* as well. These rules are necessarily different than the rules for old-fashioned diagrams. With these new rules, we can compute the T-matrix for the *sum* of $n!$ *old-fashioned diagrams* all at once by applying the new rules to a *single Feynman diagram*. An enormous amount of labor is thus saved. That this is possible is a consequence of the relativistic nature of elementary particle interactions.

As convenient mathematically and graphically as the Feynman diagram is compared to the old-fashioned diagrams, it has its drawbacks. It is the old-fashioned diagrams which actually represent the *physical mechanisms*. For example, take the old-fashioned diagrams (b) and (b'); they differ by time orderings for the events A,B,C,D, being A<B<C<D for the former, and C<A<B<D for the latter. When we add them, together with 22 other old-fashioned diagrams, we get a single Feynman diagram. If we write this Feynman diagram in the form (b), say, we might tend to think of it to represent the emission of virtual photons at A and D, and their subsequent absorption at B and C. We might fail to appreciate that time orderings are lost in Feynman diagrams, or that it also includes the physical mechanism

(b'), which involves a creation of a pair and a photon at the point C, followed by subsequent annihilation between the created positron and the incoming electron at point A, at which time a photon is emitted and later absorbed at B, then finally the photon created at C is absorbed at D.

Note in this connection that (c) belongs to a separate Feynman diagram than (b) and (b'). Diagram (b') can be 'straightened out' to give (b), but we cannot do so for (c) to get (b).

To summarize, Feynman diagrams are much more compact than the old-fashioned diagrams, but we must realize that a Feynman diagram with n vertices actually contains $n!$ different physical mechanisms described by the individual old-fashioned diagrams.

There is an infinite number of Feynman diagrams that can be drawn for each physical process, and all of them have to be taken into account. For example, diagrams (a), (b), (c), (d) and many others all describe different ways an electron and a positron can collide, and in principle all of them should be included in our considerations. In practice, as we shall see, diagram (a) is more important than diagrams (b), (c), and (d).

Feynman diagrams can be used to describe any other physical process as well. For example, Fig. VI.3e can be taken to represent the decay $d \to u + e^- + \bar{\nu}_e$ mediated by the exchange of a W^- particle. In so doing, the solid lines now represent quarks rather than electrons and positrons, and the broken line is a W^- and not a γ. The mathematical factors corresponding to the propagators and vertices also differ, but these are details which do not worry us here.

Before we close this subsection, it is convenient to introduce some terminology for later use. Diagrams without any closed loops are called *tree diagrams*. Thus (a), (a') and (e) are tree diagrams. In contrast, (b), (b') and (c) are one-loop diagrams, and (d) is a three-loop diagram.

Among other things the mathematical expression for a vertex contains a numerical factor g called the *coupling constant*. The magnitude of this factor conveys the strength of the interaction. For electromagnetic and weak processes, this coupling constant is small. As a result only a small number of simple (often tree) diagrams has to be taken into account, because more complicated (loop) diagrams contain more vertices and therefore more factors of g. If g is small and whatever multiplies into it is not too big, then the contribution to the T-matrix from these diagrams with a large number of g factors would be relatively small, and often can be neglected. This is the case for diagrams (b), (c), and (d) compared to (a).

The number of factors of g in a Feynman diagram is called the *order of the perturbation theory* (the computation of the T-matrix through diagrams is called *perturbation theory*). Thus diagram (a) is of second order, diagrams (b, c) are of the fourth order, and diagram (d) is of eighth order. If g is small, higher order diagrams are expected to give relatively unimportant results and can thus be neglected, unless very accurate outputs are needed to compare with high precision experiments.

Summary

Forces are transmitted by the exchange of particles. The mass m of the exchange particle determines the range R, and its spin s determines the nature of the resulting forces. For distances small compared to R, all forces obey the inverse-square law. For distance large compared to R, they taper off rather sharply, unless confinement is involved. In that case the force is independent of the distance.

Physical results from collision, production, and decay processes can be computed from Feynman diagrams. The important ingredients of a diagram are the vertices, propagators, and external lines. All distinct diagrams should in principle be drawn and summed. A Feynman diagram with n vertices contain $n!$ old-fashioned diagrams differing from one another by the time-ordering of their vertices. It is the old-fashioned diagrams that describe the actual physical mechanisms.

---

4. QUANTUM NUMBERS,
CONSERVATION LAWS, AND SYMMETRY

We have now met a large number of particles: the quarks, the leptons, and the various exchange particles carrying forces. How do we tell them apart?

How do we tell any two persons apart? By their individual characteristics, of course. John Doe might be 2 meters in height and 100 kg in weight, is male, and has black hair and brown eyes. Jane Smith is, say, 1.8 meters high and weighs 70 kg, is female, and has blond hair and blue eyes. They are different and they look different. There is no problem in distinguishing them.

Similarly, we distinguish the different particles also by their special characteristics, except that these characteristics in physics are usually expressed as numbers. We shall refer to these numbers as *quantum numbers*.

The mass of a particle is such a number, and so is its size. Other examples are the electric charge and the spin s of a particle.

There are also numbers that describe the state of motion of a particle, rather than its intrinsic characteristics. The energy of a particle, its momentum, and its angular momentum are numbers of this kind. We will not be discussing systematically these numbers related to motions in space-time, though we will use them for illustrations and comparisons.

What we will concentrate on in this section are the *internal quantum numbers*, as well as the quantum numbers that have to do with *discrete symmetry* operations. The latter will be discussed at the end, in Sec. 4.5. Internal quantum numbers are numbers specifying characteristics of a particle which have nothing, directly or indirectly, to do with our space-time. Thus neither the mass, the spin, nor the size of a particle is an internal quantum number, but the electric charge of a particle is.

It is interesting to ponder on what one means by an 'internal quantum number'. For cataloging purposes, we may certainly assign the catalog number 1423AZ, for example, to an electron, and the catalog number $\beta\alpha$54A to a proton. But these numbers are artificial. They have no intrinsic significance, and we will not consider them. The only kind of internal quantum numbers that are significant are those which you cannot change. This is certainly not so for 1423AZ — in another catalog this could very well have been 3241BY. But the charge of the electron, for example, is -1 once the unit of charge is fixed, and it can never be $+2$. Why? This is because genuine intrinsic quantum numbers satisfy *conservation laws*. This means that the 'sum' of these numbers over every particle in an isolated system is (exactly or at least approximately) the same at all times. The electric charge of a particle is an exactly conserved quantum number. As we shall see later, the *'strong isospin'* and the *'weak isospin'* are approximately conserved quantum numbers.

In stating what is meant by a *conservation law* above, we put the word 'sum' in quotation. This is because the quantum numbers sometimes have to be added in a funny way, not just arithmetically. We will call those that add arithmetically *additive quantum numbers*, and those that do not *non-additive quantum numbers*. The electric charge is an additive quantum number; the strong and weak isospins contain both additive and non-additive quantum numbers. See Secs. 4.2 and 4.4 for discussions of these topics.

The importance and the significance of internal quantum numbers go far beyond what has just been discussed, as an intrinsic label of the particles. It turns out that they also govern the strength and the type of forces allowed on the particles. Theories of this type, where the force is completely governed by the internal quantum numbers, are called *gauge theories*. They are the most symmetrical and the most elegant theories possible, and they will be discussed in Sec. 5. All our fundamental forces are gauge forces, though we do not quite know why. Whatever the reason, clearly internal quantum numbers play an extremely important role in the theories of elementary particle physics. For that reason we will give a rather detailed discussion of them in this section.

4.1 Exactly Conserved Quantum Numbers

Exactly conserved numbers are particularly important for at least three reasons. First, because of their conservation, these numbers can be determined relatively easily for a new particle. Take, for example, neutron decay: $n^0 \to p^+ + e^- + \bar{\nu}_e$. Suppose the electric charges of the neutron n, the proton p, and the electron e^- are known. Then from charge conservation, the charge of $\bar{\nu}_e$ can simply be calculated arithmetically to be zero. Similarly, if the energy and momentum of these three particles are known, the energy and the momentum of $\bar{\nu}_e$ can be obtained from the energy and momentum conservation laws.

As a matter of fact, that is how the existence of (anti-)neutrinos was deduced by Wolfgang Pauli. The neutrino is a neutral lepton, so it interacts only weakly with matter. The interaction is so weak that there is no chance of detecting it unless one has a very large chunk of matter with sophisticated instrumentations, and these were not available at the time when Pauli made his deduction. What was observed then was the emergence of a proton and an electron from a neutron decay, with the total energy and momentum of the proton and the electron not adding up to the energy and momentum of the initial neutron. Either the cherished energy and momentum conservation laws were broken, or there was an invisible particle carrying away the missing energy and momentum. Pauli boldly chose the latter. He was right, though the actual detection of the neutrino did not occur until years later.

The second reason why conservation laws are important is that they tell us what processes *cannot* happen. Reactions in elementary particle physics are very complex. Before a dynamical theory and a suitable computational technique are available to enable us to calculate what processes can occur and with what probability, it would be useful to know at least what processes *cannot* occur. This is where the conservation laws come in. Processes that violate one of the conservation laws will not occur. For example, the process $n^0 \to p^+ + \nu^0$ can never happen because it violates charge conservation.

The third, and perhaps the most important aspect, is that conservation laws allow us to peek into the dynamical world, thereby helping to nail down the correct dynamical theory. There is an intimate connection between a conservation law and a dynamical symmetry. *Noether's theorem*, due to Amalie Emmy Noether, tells us that whenever there is a continuous dynamical symmetry, there is always a corresponding conservation law. A dynamical symmetry, in this context, simply means that the dynamics remains *invariant* (the same) when certain other things (known as a *symmetry transformation*) are altered. The proof of this theorem is beyond the scope of this book. See however Chapter III on Symmetry for motivations and further discussions of this topic. The relation between the symmetries and the quantum numbers will be discussed in Sec. 4.4.

4.2 Vectorial Quantum Numbers

Spin is what we shall refer to as a *vectorial quantum number*. Let us recall from Sec. 3.3 what we know about it. See Fig. VI.2.

Classically, spin angular momentum is a vector, whose three components are specified by three numbers. Quantum mechanically, only one of the three components (say the z-component) can be specified, but on top of that we can also specify the length of the vector. Both of these numbers are *quantized*: the length of the vector is $\sqrt{s(s+1)}\hbar$, where s is an integer or a half-integer; for a given length, the z-component may take on values $m_s\hbar$, where m_s can take on one of the $2s+1$ values $-s, -s+1, \cdots, s-1, s$. We shall refer to this pair of numbers s, m_s as *vectorial quantum numbers*, because classically both of them refer to some attribute of a vector.

Now let us discuss how to find the sum of the spins of two particles.

Classically, you just add the two vectors, or equivalently, add each of the three components of the vectors. Quantum mechanically, it is not possible to specify all three components, so we cannot quite do it this way. Nevertheless, the z-component can be specified so we can still add it in the usual way. Thus the quantum number m_s is an additive quantum number.

The only other quantum number we can specify is s. This is related to the length of the spin vector. Even classically, the length of the sum of two vectors is *not* equal to the sum of their lengths. So we do not expect s to be an additive quantum number. This is what we referred to as a non-additive quantum number.

Although s cannot be added arithmetically, it can be shown that s can be 'added' in a funny way. The derivation is beyond the scope of this book, but the result is easy to state. If s_1 and s_2 are the spins of the two particles, and s is the total spin quantum number so that the magnitude of the total spin vector is $\sqrt{s(s+1)}\hbar$, then s may take on any one of the values $s_1 + s_2$, $s_1 + s_2 - 1$, $\cdots$, $|s_1 - s_2| + 1$, $|s_1 - s_2|$. One also knows the probatility of s having one of these allowed values, but we will not go into that.

As we said, the proof of this is beyond the scope of this book, but some sort of understanding of this funny rule is not hard to obtain. Let $m\hbar, m_1\hbar, m_2\hbar$ be the z-component of the spin vectors associated with s, s_1, s_2 respectively. Since the z-component is additive, $m = m_1 + m_2$. On the other hand, m_1 can take on any value between s_1 and $-s_1$, in steps of 1; similarly m_2 can be between s_2 and $-s_2$, also in steps of 1. This means that the maximum $|m|$ one can get is $s_1 + s_2$. It also means that the maximum s one can get is $s_1 + s_2$, because for each s there must be an m value equal to s. Physically, this corresponds to the case when both of the two spin vectors are lined up (say along the z-axis) as much as is allowed by quantum mechanics. Given that, it is easy to guess that the minimal s is achieved when the two spin vectors are as much anti-aligned as possible, and this presumably means $s = |s_1 - s_2|$. When the two spin vectors are neither aligned nor anti-aligned, s would be somewhere in between. This is why s can take on all possible values between $s_1 + s_2$ and $|s_1 - s_2|$, in steps of 1.

There are *internal* quantum numbers which mathematically (but not physically because they are internal) behave like a spin. *Isospin* is such an internal quantum number, as we shall see in the next subsection. However, isospin is only approximately conserved.

There are also vectorial quantum numbers that do not behave like spin. One can show that these are necessarily even more complicated than the spin. We can think of them as vectors in a d $(d > 3)$ dimensional space, with r $(r < d)$ of its components simultaneously specifiable. These r quantum numbers are additive. On top of that, there are also a number of non-additive quantum numbers. In the cases we will be interested in, there will always be r non-additive quantum numbers as well. Although the non-additive quantum numbers do not add arithmetically, there are again funny additional rules for all of them.

What d and r are, and how the vectorial quantum numbers add, are determined by a branch of mathematics called *group theory*. Quantum numbers are usually classified by their associated symmetry via Noether's theorem. The associated symmetry is specified by a *symmetry group*, which we will discuss in Sec. 4.4. The symmetry group associated with an additive quantum number is known as $U(1)$, that associated with the isospin is known as $SU(2)$, and that associated with the color spin mentioned below is known as $SU(3)$. The groups for additive quantum numbers are known as *abelian groups*, and those for the vector quantum numbers, like $SU(2)$ and $SU(3)$, are known as *non-abelian groups*.

The only exactly conserved vectorial quantum number known today is the *color spin*, or simply *color*. The word 'spin' here is perhaps a misnomer, for in detail it does not behave like a spin at all. The symmetry group is $SU(3)$ rather than $SU(2)$; the number of additive as well as non-additive quantum numbers is 2 each, rather than 1 each for spin; and the dimension d of the color vector is 8 rather than 3 for the spin. Nevertheless, the word 'spin' does conjure up the image of a vector and is therefore sometimes used.

Color is the quantum number governing the strong interaction. We will come back to it in Sec. 5.4.

Particles having the same non-additive quantum numbers but differing from each other by their additive quantum numbers are said to belong to the same *multiplet*. The number of members of a multiplet is called its *multiplicity*. Since internal symmetries are by definition independent of space-time, particles in the same multiplet must have the same space-time properties. In particular, they must have the same mass and the same spin, as well as the same parity (see Sec. 4.5 for a discussion of parity).

In the case of isospin $(SU(2))$ with quantum numbers s and m_s, a multiplet consists of particles with the same s and different m_s, and the multiplicity of such a multiplet is $2s + 1$.

A multiplet with multiplicities 1,2,3, etc., is often referred to as a *singlet*, *doublet*, *triplet*, etc.

4.3 Approximate and High Temperature Conservation Laws

It is often useful to consider quantum numbers that are not exactly conserved. There are two types.

The first is such that the symmetry would have been exact, and the quantum numbers would have been exactly conserved, if we could turn off some weaker forces. Since in reality this cannot be done, the symmetry is only approximate, though often the approximation is very good.

An example of this kind of approximate symmetry is *strong isospin*. This is an $SU(2)$ internal symmetry which would have been exact if electromagnetic and weak interactions could be turned off. In that case, we would not be able to tell the difference between quarks u and d, nor between nucleons p and n. This symmetry is very useful and very successful in nuclear physics and other strong interaction problems. Under this symmetry, (u, d) and (p, n) each forms an iso-doublet, and the pions (π^+, π^0, π^-) form an iso-triplet.

The second type is what we call *high temperature symmetry*. These quantum numbers would have been conserved if the temperature of the universe were very high. So in the early hot universe these quantum numbers should have been exactly conserved, though they are only approximately conserved today.

As we shall see (Secs. 5.3 and 5.4), what happens at low temperature is that the vacuum finds it more stable to have a 'condensate' present. If this condensate breaks the symmetry present at high temperatures, then whichever process that interacts with the condensate will not preserve the symmetry. This way of breaking a symmetry is known as *spontaneous symmetry breaking*. However, processes that do not interact with the condensate, or interact with it only weakly, will still preserve the symmetry well. At high temperatures, the condensate is boiled away so the symmetry becomes exact for all processes.

These high temperature quantum numbers are inferred from the existing theory of elementary particles. To the extent that present-day theory is very successful in predicting many experimental results, it is reasonable to expect the high temperature symmetries so derived to exist. However, it is important to bear in mind that there is no *direct* experimental evidence for these symmetries, in the sense that it is not yet possible to generate and sustain a high enough temperature in the laboratory to boil away the vacuum condensate in order to test the exact symmetry.

An example of a high temperature symmetry is to be found in the electromagnetic and weak interactions. An $SU(2)$ *weak isospin* symmetry is supposed to be present. This will be discussed in Sec. 5.3. Note that the weak isospin turns out to be identical to the strong isospin for (u, d), but this is not necessarily so for other particles.

4.4 Symmetry and Conservation Laws

We discuss in this section how conserved internal quantum numbers can be classified. Using Noether's theorem, the problem of classifying conserved quantum numbers becomes the problem of classifying symmetries, and the latter can be accomplished with the help of a branch of mathematics called *group theory.*

We will call the mathematical space on which such symmetry transformations operate an *internal space.* Since we are dealing with internal quantum numbers, this space has nothing to do with the space-time we live in. We may think of it as a mathematical construct useful in our discussions. It is useful because the symmetry operations on the internal space is very analogous to rotations in our real space-time, and the internal quantum numbers are generalized spins.

We will now embark on a description of the classification of internal symmetry operations. We will not give a complete classification; we will confine ourselves only to those classes that are known to be used in Nature.

The discussion will be a bit technical, and the rest of this subsection can be skipped at a first reading. Readers who wish to do so may skip directly to Sec. 4.5.

Consider all continuous rotations in an n dimensional space. These rotations form what is known mathematically as a *group.* We really do not have to know what a group is (it is however discussed in Chapter III on Symmetry);

all that we have to know is that it is a collection of symmetry transformations. The group of all n dimensional rotations is usually denoted by $SO(n)$, and it stands for 'Special Orthogonal group of n dimensions'. The word 'special' simply means that we are considering only continuous rotations, and not reflections. The word 'orthogonal' comes in because mathematically these rotations are described by $n \times n$ orthogonal matrices. In this language, 'special' corresponds to those matrices whose determinants are $+1$.

It can be shown that such $SO(n)$ matrices can be labelled by $d(n) = n(n-1)/2$ parameters. In general, the number of parameters for a group is called the *dimension* of the group. Please do not confuse the dimension of the *group* and the dimension of the *internal space*. The former is d and the latter is n.

We will now discuss the meaning of these d parameters. Let us start with two-dimensional rotations, where $n = 2$ and $d = 1$. The single parameter in this case is simply the angle of rotation in the plane. Next, consider three-dimensional rotations, where $n = 3$ and $d = 3$. These three parameters are the so-called *'Euler angles'*, and can be understood in the following way. A three-dimensional rotation can be defined by fixing both an axis, and the angle of rotation in the two-dimensional plane perpendicular to this axis. The axis requires two angles to fix it. Together with the rotational angle, we have the three required parameters.

This interpretation of the parameters can be generalized to n dimensions. To start with, the following algebraic identity can easily be verified: $d(n) = d(n-1) + (n-1)$. This identity has the following geometrical meaning. Just like the three-dimensional rotation discussed above, an n dimensional rotation can be specified by the orientation of a rotational axis, and the amount of rotations in the $(n-1)$ dimensional hyperplane perpendicular to this axis. The axis in n dimensional space requires $(n-1)$ angles to fix, and the amount of rotations in the $(n-1)$ dimensional hyperplane is specified by $d(n-1)$ parameters. This is then the geometrical content of the algebraic identity above. By induction, we can get down eventually to three and two dimensions and understand completely the meaning of the $d(n)$ parameters.

Among the rotational groups, $SO(2)$ is special. One can make a rotation of $30°$, followed by a rotation of $40°$. The result is a rotation of $70°$, and this can also be achieved by reversing the order of rotations, by first rotating through 40°, then followed by a rotation of 30°. This property, that the order of rotation is immaterial, is called *abelian*. A group whose operations all have the abelian property is called an *abelian group*. $SO(2)$ is an abelian group.

$SO(2)$ is parametrized by one angular parameter. The corresponding conserved quantum number à la Noether, when a dynamical $SO(2)$ symmetry is present, is a single additive quantum number. The quantum number m_s for the z-component of the spin angular momentum is one of this kind. This is so because the z-component of the angular momentum is nothing but the angular momentum corresponding to rotations about the z-axis, and a rotation about the z-axis is a rotation in the xy plane, and hence a two dimensional rotation.

All the other rotational groups, $SO(n)$ for $n > 2$, are *non-abelian*. This means that different orders of rotations generally give different answers. Take for example $SO(3)$. A rotation (R_1) through 90° about the z-axis in the counterclockwise direction leaves the z-axis unchanged, moves the x-axis to the old position of the y-axis, and the y-axis to opposite the direction of the original x-axis. See Fig. VI.4.

We shall use the following notation to denote this change: $R_1 : (x, y, z) \rightarrow (y, -x, z)$. On the other hand, a rotation (R_2) through 90° about the x-axis in the counterclockwise direction changes (x, y, z) to $(x, z, -y)$, so $R_2 : (x, y, z) \rightarrow (x, z, -y)$. Doing these two rotations in succession, we get in one order $(R_2 R_1)$: $(x, y, z) \rightarrow (y, -x, z) \rightarrow (z, -x, -y)$, and in the other order $(R_1 R_2)$: $(x, y, z) \rightarrow (x, z, -y) \rightarrow (y, z, x)$. They give different results. The group is therefore non-abelian.

This rather dry mathematical demonstration of the non-abelian nature of $SO(3)$ can be given a more human dimension. See the end of Sec. 2 of Chapter III for such an explanation.

The notation $SO(m) \times SO(n)$ stands for a group of rotations in an $m + n$ dimensional space, with the first m coordinates rotated by $SO(m)$ and independently the last n coordinates rotated by $SO(n)$. It is called the *direct product* of $SO(m)$ and $SO(n)$. By definition there is never any rotation mixing up the first m coordinates and the last n coordinates. These latter rotations belong to $SO(m + n)$, but not $SO(m) \times SO(n)$. Similarly, one can define $SO(m) \times SO(p) \times SO(n)$ as the rotations in an $m + n + p$ dimensional

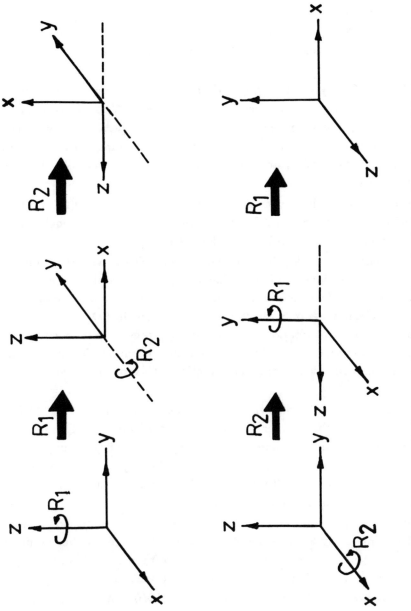

FIG. VI.4 R_1 is a counter-clockwise rotation of 90° about the (original) z-axis, and R_2 is the same about the (original) x-axis. The upper diagram shows the result of R_2R_1, and the lower diagram shows the result of R_1R_2. They are not equal.

space where the first m coordinates, the middle p coordinates, and the last coordinates are never mixed up.

The group $SO(m)$ is a *subgroup* of the group $SO(m+n)$, because the m dimensional rotations specified by $SO(m)$ are special cases of $m+n$ dimensional rotations in $SO(m+n)$.

Quite generally, given two groups G_1 and G_2, operating respectively on m and n dimensional spaces, then the *direct product* $G_1 \times G_2$ is a group operating on the $m+n$ dimensional space, with G_1 operating only on the first m dimensions, and G_2 operating only on the last n dimensions. If $n < m$ and all the operations of G_1 are covered by some operations of G_2, then G_1 is said to be a *subgroup* of G_2.

The dimension of the abelian group $SO(2) \times SO(2) \times \cdots \times SO(2)$ (product of d $SO(2)$'s) is d, for each $SO(2)$ has dimension 1. Correspondingly there are d conserved quantum numbers, one for each $SO(2)$.

In comparison, consider the non-abelian group $SO(n)$ with the same dimension d. The rotations in a non-abelian group cannot be thought of as being all *independent* of one another, for otherwise the order of rotations should not matter. The number of additive quantum numbers in $SO(n)$ depends on how many independent rotations we can find simultaneously, independent in the sense that the order of rotations between them is immaterial. In general, the number of independent symmetry operations of a group is called the *rank* of the group.

For $SO(n)$, the rank is known to be $r = n/2$ if n is even, and $(n-1)/2$ if n is odd. This gives the number of *additive quantum numbers*. This number is clearly smaller than the dimension $d = n(n-1)/2$ of the group unless $n = 2$, which is abelian. The number of *non-additive quantum numbers* for $SO(n)$ turns out to be also r. Technically, what we call non-additive quantum numbers are known as *Casimir invariants*.

Spin is the quantum number associated with three dimensional rotations. The rank of $SO(3)$, for example, is $r = 1$. The corresponding additive quantum number is just m_s, and the corresponding non-additive quantum number is just s. See below for the connection between $SO(3)$ and $SU(2)$.

We have now completed the discussions on $SO(n)$. We will discuss another class of symmetry groups: $U(1)$ and $SU(n)$.

$SU(n)$, which stands for 'Special Underline{U}nitary group of $\underline{n}$ dimensions', is in a sense the complex-number version of $SO(n)$. It is composed of all $n \times n$ unitary matrices of determinant $+1$. A more geometrical way of describing $SO(n)$ and $SU(n)$ is the following. Consider the space of real vectors in n dimensions. Linear transformations keeping the magnitudes of all such vectors fixed define $SO(n)$. In other words, the members of $SO(n)$ describe rotations. Similarly, in the space of complex vectors of n complex dimensions, the linear transformations keeping the magnitudes of all such complex vectors fixed define $U(n)$. If we confine ourselves to those with determinant $+1$, then we get $SU(n)$. So $U(n)$ and $SU(n)$ can be thought of as some sort of rotations in a complex space of n dimensions.

Quantum mechanics deals with complex numbers and complex wave functions. It is therefore not surprising that the $SU(n)$ and $U(1)$ groups would find their way into quantum field theories. The reason why only $SU(n)$, and not $U(n)$, is considered for $n > 1$, is simply that a $U(n)$ can be thought of as being made up of an independent $SU(n)$ and a $U(1)$. Consequently there is no need to consider both of them separately.

The group $U(1)$ is abelian. It is actually the same as $SO(2)$. To see that, we note that a vector in a complex one-dimensional space is nothing but a complex number $z = x + iy$, which can be thought of as a two-dimensional real vector with components x and y. The magnitude of the complex number is by definition $\sqrt{x^2 + y^2}$, which is also the magnitude of the corresponding two-dimensional real vector. Hence $U(1)$ and $SO(2)$ are the same, and $U(1)$ has also a single additive quantum number.

The group $SU(n)$ is non-abelian. It has dimension $d = n^2 - 1$ and rank $r = n - 1$. The fact that $r < n$ confirms again that it is non-abelian. Note that the dimension for $SU(2)$ is 3, the rank is 1, the same as $SO(3)$. As a matter of fact, other than some subtle mathematical (topological) differences, these two groups can be thought of as being the same. This means that $SU(2)$ is also specified by one additive and one non-additive quantum number. In general, $SU(n)$ is specified by r *additive* and r *non-additive* (Casimir invariants) quantum numbers.

Other than subtle mathematical (topological) considerations, we may therefore use either $SO(2)$ or $U(1)$, and either $SO(3)$ or $SU(2)$, to describe the same thing. In both cases it is the latter that is traditionally used.

4.5 Multiplicative Quantum Numbers and Discrete Symmetries

The quantum numbers we have spoken of so far are mainly additive, in the arithmetic or the vectorial sense. These numbers arise from continuous symmetry groups like $SU(2)$. There is another kind of symmetry, which is *discrete* and not continuous. The most notable of these are the reflection in ordinary space, otherwise known as a *parity operation* ($\mathcal{P}$), and the reflection in charge space, otherwise known as a *charge conjugation operation* ($\mathcal{C}$). These groups may or may not lead to a quantum number, but if they do, the corresponding quantum numbers are just $+1$ or -1, and they are multiplicatively conserved. This means that the total quantum number for the system should be taken to be the product of the individual quantum numbers.

Let us discuss *parity* first. Invariance under a parity operation means that if a process or a motion can occur, then the same process reflected in space, or viewed in a mirror, can equally occur as well. Before 1957, it was believed that such an invariance was universal. Then Tsung Dao Lee and Chen Ning Yang discovered that while this is true for strong and electromagnetic interactions, it is actually not so in weak interactions. This has rather far-reaching consequences so let us discuss it in some detail.

We need to introduce two concepts: *helicity*, and *chirality*. The spin component along the direction of motion is called the *helicity* of that particle. In the case of quarks and leptons, whose spins are $s = \frac{1}{2}$, these components are either $m_s = \frac{1}{2}$ or $-\frac{1}{2}$. The former is known as a *right-handed helicity*, because when you stand behind the particle, the angular motion of its spin turns like a right-handed screw. The latter is called a *left-handed helicity*. (See Sec. 3 of Chapter III and Fig. III.6 for more discussions).

If the mass of this spin $\frac{1}{2}$ particle is zero, then *chirality* is defined as twice the helicity, so it takes on values $+1$ (right-handed chirality) or -1 (left-handed chirality). When the particle is massive, the concept of chirality is still mathematically well-defined, but it is hard to describe in a physical language. One may think of it simply as twice the helicity to an observer who sees the particle moving close to the speed of light, for in that case the particle appears to the observer as virtually massless. Then the usual theory of special relativity would inform us of its helicity content as measured by any other observer. Though it can be defined this way, it is much more cumbersome than it has to be when one is allowed to use mathematics. So for our purpose, let us just think of chirality as twice the helicity for a massless particle; and for a massive one, it contains a known amount of admixture of the other helicity, an admixture that goes up with increasing mass of the particle.

A right-handed screw looks like a left-handed screw in a mirror. Thus the parity transformation reverses helicity and chirality. Since the late 1950's, the weak interaction has been known to be *left-handed.* This means that only quarks and leptons of left-handed chirality, and only anti-quarks and anti-leptons of right-handed chirality, participate in the interactions. This preference of one hand over another clearly violates parity ($\mathcal{P}$) symmetry. It also violates *charge conjugation symmetry* ($\mathcal{C}$) as we will explain, though it preserves the combined $\mathcal{CP}$ symmetry.

Charge conjugation symmetry means that if we change all particles to their anti-particles (γ, Z^0, H^0 are their own anti-particles, and W^+ and W^- are anti-particles of each other), without altering any other quantum number, then the new process is still an allowed one occurring with the same probability. The weak interaction violates charge conjugation invariance because interactions occur for left-handed quarks and leptons but not for left-handed anti-quarks and anti-leptons. It is the right-handed anti-quarks and anti-leptons that participate in the interactions. So $\mathcal{C}$ symmetry is violated. Nevertheless, the weak interaction is still invariant under a combined $\mathcal{CP}$ operation, for this changes the left-handed quarks and leptons to right-handed anti-quarks and anti-leptons.

Strictly speaking, there is one decay process observed in the neutral *kaons* K^0 and $\bar{K}^0$ (kaons are mesons of the following kind: $K^0 = \bar{d}s$ and $\bar{K}^0 = \bar{s}d$) which violates $\mathcal{CP}$ at the level of 0.1% of the other weak processes. Whether this should be counted as a part of the weak interaction, or a new superweak interaction, we do not know for sure at the moment. Experiments have been planned to look for other $\mathcal{CP}$ violating processes in the hope of shedding some more light on its mechanism. One of the most promising places to look is the system involving b and $\bar{b}$ quarks.

Strong and electromagnetic interactions have both $\mathcal{P}$ and $\mathcal{C}$ symmetries.

There is a third symmetry, called *time reversal invariance* ($\mathcal{T}$). Under this symmetry, dynamical laws remain the same if the direction of time is reversed. There is a famous theorem in quantum field theory, known as the

CPT *theorem*, which states that the combined operation CPT always leaves every (relativistic local) quantum field theory invariant. Consequently if CP is conserved, so will T be.

This is a little off the topic, but while we are at it, there is another important theorem of quantum field theory. This is known as the *spin and statistics theorem*. Particles are of two types: *fermions* and *bosons*. What distinguishes them is the following. If we put a number of identical fermions in a box, then they will tend to avoid each other. If we put a number of identical bosons in a box, then they will tend to stick together. In other words, they have different statistical properties. A more exact statement is as follows. The wave function of a number of identical fermions is antisymmetric in all its variables, whereas the wave function of a number of identical bosons is symmetric in all its variables. A consequence of the former is that no two fermions with identical quantum numbers can occupy the same space-time point. In this latter form, it is known as the *Pauli exclusion principle*. (See Appendices B and C for more discussions).

The spin and statistics theorem states that particles of integer spins ($s = 0, 1, 2, \cdots$) are bosons, and particles of half-integer spins ($s = \frac{1}{2}, \frac{3}{2}, \cdots$) are fermions (actually it simply states that the opposite is not possible). Thus quarks and leptons are fermions, and photons, gravitons, pions, K, $W^{\pm}$, and Z^0 are all bosons.

This is even more off the topic, but it concerns fermions and bosons. There is a *fermionic number conservation law* which states that the fermionic number, which is defined to be the total number of fermions minus the total number of anti-fermions, is conserved. This corresponds to a $U(1)$ symmetry. So although matter is not strictly indestructible, because a fermion can annihilate with an anti-fermion, it remains indestructible if there are no anti-fermions around. This fermionic number conservation law is therefore the modern counterpart of the indestructibility of atoms in Dalton's atomic theory. It is therefore reasonable to identify fermions in our minds with 'matter'.

Actually, as far as we know, the *quark number* (number of quarks minus the number of anti-quarks), and the *generation leptonic number* (number of leptons of one generation minus the number of anti-leptons in the same generation) are separately conserved as well.

However, bosons can be created and absorbed. As a matter of fact, that is why they can transmit forces. For that reason, there is no conservation law for the number of bosons. This difference then can be viewed as another fundamental difference between fermions and bosons.

Summary

A dynamical symmetry leads to a conservation law according to the theorem of Noether. Conservation laws may be exact or approximate. There is also a class which is exact at high temperatures and approximate at the present temperature.

Conserved quantum numbers may be additive, or vectorial. The former add arithmetically, but the latter also contains non-additive quantum numbers which do 'add', but in a peculiar way. The corresponding symmetry groups are abelian for the former, and non-abelian for the latter.

Possible internal symmetries are discussed, including those given by the groups $SO(n)$, $SU(n)$, and $U(1)$. The number of parameters parametrizing each of these groups (the dimension d of the group) is given, and so is the number of additive quantum numbers (the rank r of the group). If $r < d$, the group is non-abelian, and we need r more non-additive quantum numbers (the Casimir invariants) to specify a multiplet.

Also discussed in the last subsection are the discrete C, P, T symmetries and their violations, the concepts of chirality and helicity, fermions and bosons, the TCP theorem, the spin and statistics theorem, and the various fermion number conservation laws.

---

5. GAUGE THEORIES

Through years of experimentation, a long list of conserved and approximately conserved quantum numbers has been assembled. Some of these, like energy, momentum, and charge, are exactly conserved. Others, like strong isospin, are conserved only in strong interactions. Still others, like parity and charge conjugation, are conserved in strong and electromagnetic interactions, but not in weak processes. There is quite a rich variety of them. If we are forced to gulp down all these delicacies at once, we will probably get a rather severe case of indigestion. So let us skip the details, and pretend now that we know them all.

The central problem is to find out the dynamics of the elementary particle interactions. Conservation laws are useful in that regard because, using Noether's theorem, they can be translated into dynamical symmetries. Unfortunately, there are many dynamical theories possessing the same set of symmetries, so more input is needed before we can pin down the details. Throughout the years, various guesses have been made, and many tests were performed. Finally, a *Standard Model* of strong, electromagnetic, and weak interactions emerged, which is so successful that it has passed all the tests to which it has been subjected with flying colors. As for gravity, we know that Einstein's theory of general relativity works for macroscopic bodies, but owing to the weakness of its strength, quantum gravity cannot be tested.

The three fundamental forces in the Standard Model are all described by *gauge theories*. The fourth, Einstein's theory of gravity, is also a kind of gauge theory though technically sometimes not regarded as such.

A gauge theory is a very special theory of force. It is probably the most elegant theory around. We do not know why our fundamental forces are all of the gauge type, except to say that this confirms once again the apparent attachment of Nature to beauty and elegance.

The simplest and the oldest of these gauge theories is the Maxwell theory of electromagnetism. It will serve as a template for the other gauge theories, so let us study it in some detail.

5.1 Electromagnetic Theory

Classical electricity and magnetism were unified and completed by James Clerk Maxwell in the late 19th century. Its quantum version, QED, debuted in the late 1920's. From then on it was just a matter of obtaining numbers from it to compare with experiments. Calculations of this sort actually ran into a snag which did not get resolved until the late 1940's. The solution goes by the name *renormalization*, a topic which we shall come back to in the next section.

In order to expose the elegance of the theory, and to motivate how similar considerations can be applied to other gauge theories, let us now pretend not to know the Maxwell theory, and see how far a few simple facts can lead us.

From the long range nature of the inverse-square law, we deduce that the exchange particle is massless. It must be electrically neutral, for otherwise the charge of the electron will be altered when a photon is emitted. It must also be a boson, for otherwise the electron would change into a boson after the (fermionic) photon emission on account of conservation of fermionic numbers; being a boson, the electron would get absorbed and lost. Now that we have established photon to be a boson, its spin has to have integer values. The fact that it transmits both an electric and a magnetic field shows that it must carry a non-zero spin. In fact, its spin is known to be 1.

We will now argue that this spin 1 nature is extremely important in keeping its mass zero *, and hence the range of the Coulomb force infinite.

A particle at rest with mass m has an amount of energy equal to $E = mc^2$. The value of energy levels, according to quantum mechanics, shifts with interactions, and this leads to a change in the value of the mass m. This at first might seem contradictory because the mass is an intrinsic and unchangeable quantity for a particle. How can it change? It does not, but we are talking about the hypothetical changes if we could turn on and off forces (which we cannot except mathematically).

Now we can either believe that the forces of the universe are so finely tuned so as to make the mass of the photon come out to be exactly zero, or that this is no accident and there is a fundamental mechanism behind it. Let us pursue the latter.

We will now explain a mechanism by which the mass of a particle with non-zero spin could remain zero and unaffected by the interactions.

A particle *at rest* with spin s has $2s + 1$ possible spin orientations. All of these $2s + 1$ states can be realized because we can get to each one of them by a suitable rotation. A *massive* particle in motion also has all of the $2s + 1$ helicities, because this particle would appear to be at rest to an observer travelling with the same velocity. According to the theory of special relativity, if this is true for one observer it must be true for all observers.

This argument of having all $2s + 1$ helicities, however, is no longer valid if the particle is *massless*. In that case it travels with the speed of light, and no (massive) observer can ever catch up with it. As a matter of fact, a massless particle is allowed to have fewer than $2s + 1$ realizable helicities.

Conversely, if somehow we can keep the number of physically realizable

---

* By invoking the Nambu-Goldstone mechanism explained at the very end of Sec. 5.4, it is also possible to make a spin 0 particle massless. However, the force so generated will not be of the electric and magnetic types.

helicities of a particle fewer than $2s + 1$, then it must be massless. This is
the mechanism we will use to keep the photon massless. For this to work for
a boson the spin s clearly must be at least 1.

We will now discuss the mechanism whereby we can force the photon to
have fewer than $2s + 1$ helicities.

The photon has spin 1 and a priori $2s+1 = 3$ spin orientations. We must
now find a way to restrict it to only 2 helicities. The best way to accomplish
that is not to suppress the third helicity, but rather to write down the theory
in such a way that no matter how much we add or subtract to this third
component of helicity, the theory remains invariant. If we succeed in doing
that, it must mean that the third component is physically decoupled and
irrelevant, for how else can it have an arbitrary value?

A theory with this invariance can actually be written down. It is called
a *gauge theory*, the exchange particle is called a *gauge particle*, or a *gauge
boson*, and the invariance is called a *gauge invariance*. See Sec. 4 of Chapter
III for a more formal discussion of this concept.

Gauge invariance is a *local invariance*, or a *local symmetry*, because the
invariance of the third spin component must occur at all time wherever the
photon is in space, *viz.*, at every point of space-time. This is to be contrasted
with *global invariance*, which is the kind of symmetry we obtain from conservation laws using Noether's theorem. That is global because we require
the *total* quantum number to be conserved, and 'total' means the sum of
that quantum number everywhere. As mentioned at the beginning of this
section, global invariance is not sufficient to tie down the dynamics. This is
because dynamics means equations of motion, which is usually described by a
differential equation to be valid at every point in space-time. In other words,
dynamics requires local information which cannot be supplied by global symmetries alone. On the other hand, a gauge symmetry is a local symmetry,
so it is conceivable that it would provide the complete dynamics. Indeed
it does, and it is unique if we impose on it the additional requirement of
renormalizability (see Sec. 6).

The gauge theory so formulated has one additional important feature.
The photon couples only to the conserved (electric) charge (actually current) in the theory. Moreover, the interaction strength is proportional to the
charge, so there will be no interactions for neutral particles. The fact that

interaction recognizes only the charge, and not anything else, means that the electric force (at distances large compared to the proton size) from an electron is exactly opposite to that of a proton, in spite of the fact that the proton has strong interactions, has a finite size, can emit and absorb pions, and the electron does not and cannot. Without this property, there would have been a net electric force exerted by a neutral atom at large distances, and the whole world would have behaved very differently. For one thing, the gravitational force would probably be completely overshadowed by the stronger electric forces. We might therefore say that we owe our existence in the present form to the fact that the electromagnetic theory is a gauge theory.

This property of universal coupling, that it recognizes only the charge of the particle but not its other attributes of it, be it 'race, color, or religion', is a great democratic and important principle that is respected in all gauge theories. It is called the *gauge principle*.

The gauge theory of electromagnetism, obtained in the way outlined above, turns out to be nothing but the Maxwell theory. The two spin orientations of the photon turn out to correspond to the left-handed and the right-handed circular polarizations of an electromagnetic wave. We have therefore 'derived' the Maxwell theory from the desires to have a long range force of the electromagnetic type. Please note however that such a 'derivation' contains some implicit assumptions and a large dose of hindsight and should be taken with caution. Nevertheless, this exposé does bring forth the beauty and the elegance of the theory.

The symmetry corresponding to charge conservation is a $U(1)$ symmetry. It is an abelian group, so the gauge theory considered above is also known as an *abelian gauge theory*, or a $U(1)$ gauge theory.

5.2 Yang-Mills Theory

Chen Ning Yang and Robert Mills were able to generalize, in 1954, the abelian gauge theory to a *non-abelian gauge theory*, nowadays also known as the *Yang-Mills theory*. In this generalization, the conserved quantum numbers come from a non-abelian symmetry group.

The gauge bosons in the Yang-Mills theory again have spin 1 and carry only two allowed spin orientations. The coupling of the gauge boson to particles is also universal and democratic, in that it recognizes only the quantum

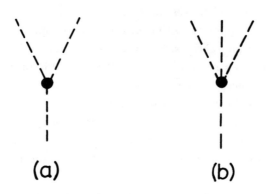

FIG. VI.5 *Vertices for three (non-abelian) gauge bosons (a) and four gauge bosons (b).*

number of that *non-abelian group* carried by the particles, and nothing else. The coupling is proportional to the internal quantum number of the gauge group, and is zero when the particle belongs to a singlet multiplet of that group. The constant of proportionality is essentially the coupling constant g. In other words, all the gauge interactions associated with a given gauge group is parametrized by a single coupling constant g. If the gauge group is the direct product of several groups, then the number of coupling constants will be equal to the number of members in the direct product.

All these are properties similar to those of the abelian gauge theory.

There is however some difference between the two. If the non-abelian group has dimension d and rank r, there will be d gauge bosons, and not just one. Moreover, the additive $U(1)$ quantum number (electric charge) is zero for an *abelian* gauge boson, but the r additive quantum numbers for the *Yang-Mills* gauge bosons are not trivial. As a result, because of the universality of gauge interactions, non-abelian gauge bosons can be emitted and absorbed from the gauge bosons themselves, whereas the same cannot happen in an abelian gauge theory. This leads to the Feynman diagram vertices shown in Fig. VI.5 which do not exist in the abelian gauge theories.

We shall see in Sec. 6.3 that these additional vertices produce an important qualitative difference between the abelian and the non-abelian theories.

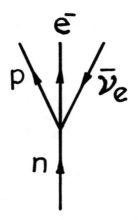

FIG. VI.6 *Neutron decay Feynman diagram. If we change the neutron (n) and the proton (p) in the diagram to the down (d) and the up (u) quarks respectively, then we obtain Fig. VI.3e which may be thought to be the fundamental process inducing the neutron decay.*

The number of gauge particles in an $SU(n)$ theory is $d = n^2 - 1$, which is equal to 3 for $SU(2)$, and 8 for $SU(3)$.

5.3 Electroweak Theory

• *Fermi theory*

The weak interaction was discovered at the turn of the century in beta decays (see Sec. 2). The first theory to describe it was written down by Enrico Fermi in 1934 shortly after Pauli postulated the existence of the neutrino. Indeed it was Fermi who christened this elusive particle emerging from the beta decay of the neutron: $n \rightarrow p + e^- + \bar{\nu}_e$. This theory has a vertex given by Fig. VI.6.

Fermi's theory was very successful in predicting the energy distribution of the electron, but when parity violation was discovered, it had to be modified to take into account parity and charge-conjugation violations. This was carried out successfully in the late 1950's.

The resulting theory resembles a gauge theory, except that the mass of the would-be gauge particle has to be very very large to explain the extremely short range of the weak forces. In fact, it would have been nice if this were indeed a gauge theory, for both the original and the modified theories are *non-renormalizable* (see Sec. 6), a trait which prevents the process from being calculated very accurately. In contrast, gauge theories are *renormalizable* and do not suffer from this drawback.

We motivated gauge theories in Secs. 5.1 and 5.2 to be theories designed to keep the mass of the gauge particles zero. One would therefore have thought that there should be no such thing as a gauge theory with a large gauge boson mass. Happily, there is a way around this seemingly insurmountable obstacle for the modified Fermi theory to become a full-fledged gauge theory. The magic allowing this to be done goes under the name of a *Higgs mechanism*.

• *Higgs mechanism*

The Higgs mechanism enables the gauge particle to become massive.

To make this mechanism work we need to have some spin 0 particles present, whose fields are called the *Higgs fields*. We need to have as many Higgs fields as the number of gauge bosons we want to make massive, and with electric charges identical to those of the gauge bosons.

What makes the gauge boson massless in the first place is its lack of the third spin orientation. To make it massive we must somehow supply this missing degree of freedom, and this is exactly the role of the Higgs field if it is properly coupled to the gauge field.

But something else is needed in order to make this mechanism work.

Massless particles travel at the speed of light, but a massive particle always travels more slowly. Thus to make a massless particle massive, we must somehow slow it down. We know how to do so in ordinary life: objects slow down when they encounter friction or resistance. The motion of a car is slowed down by the friction of the brake; the speed of an airplane is slowed down by the drag of the atmosphere. Similarly, if our vacuum is filled with a high density of Higgs field, sometimes called a *vacuum condensate*, or more specifically a *Higgs condensate*, gauge bosons travelling through it would be slowed down through their mutual interactions. This would then make the gauge bosons massive.

It might seem to be a contradiction to have a *vacuum* filled with a *condensate*. Does a vacuum not mean a state with nothing in it? The answer is, classically this is so, but not quantum mechanically. A quantum mechanical vacuum is a very complicated object. Even if we put nothing there to start with, quantum fluctuations allow virtual particles to be continuously created and annihilated. If you like, there is always a constant mist in the quantum mechanical vacuum. Ordinarily, this mist of virtual particles comes and goes. However, if a strong attraction exists among them, then they might just stick together, in the process releasing enough of the borrowed energy to enable them a permanent existence. This is how a *vacuum condensate* is formed. By definition, the vacuum state is the state with the lowest possible energy. In fact, unless the Higgs field develops a mutual repulsion when its *density* gets too high, there is nothing to stop a runaway situation with the continuous creation of the Higgs field. Therefore, the condition for a Higgs condensate to be formed is to have enough attraction at low density, but a repulsion at high density.

How much attraction is enough to produce the condensate? That depends on a number of factors, including the ambient temperature of the surroundings: the thermal energy must be low enough not to tear the condensate apart. The temperature at which the condensate first appears is known as the *transition temperature*. We consider the vacuum with or without the condensate to be in two different thermodynamic *phases*, so a *phase transition* for the vacuum occurs at the transition temperature.

This transition is like the transition between water vapour and liquid water. The water in our oceans is formed from the condensation of water vapour, which occurs when the temperature is below 100° C. Condensation occurs because water molecules have a tendency to stick together; but a low temperature is also needed to reduce the thermal motions which tend to tear them apart.

If the Higgs fields contain a variety of electric charges, then only the electrically neutral ones may condense in the vacuum. Otherwise the vacuum would be so full of charges that it would cost nothing to give up one or to take in an extra one from a reaction process occurring on top of it, which means that charge would not be conserved in that process.

To be able to give the gauge bosons a mass, the vacuum condensate must interact with the gauge bosons. This means that the Higgs field must belong to a non-singlet multiplet of the gauge group in question.

The members of a Higgs multiplet may or may not all have the same electric charge. If they do not, then the original symmetry is broken, for it is only the electrically neutral component among them that is allowed to condense, thereby creating an asymmetry. This kind of symmetry breaking, induced only by the presence of the condensate, is known as *spontaneous symmetry breaking*. The surviving symmetry is given by those original symmetry transformations which leave the condensate unchanged.

• Glashow-Salam-Weinberg theory

With the Higgs mechanism, there is now hope to make the weak interaction theory into a gauge theory. It turns out that not only this can be done, the eventual theory also combines weak interaction and electromagnetism into a single theory, nowadays known as the *Glashow-Salam-Weinberg* (GSW) theory, or the *electroweak theory*.

The theory is a non-abelian gauge theory with a high temperature symmetry $SU(2) \times U(1)$. The quantum number of the $SU(2)$ symmetry is called the *weak isospin*, and the quantum number of $U(1)$ is called the *weak hypercharge*, and is usually denoted by Y. If I_3 denotes the value of the z-component of the weak isospin, then the weak hypercharge Y of a particle is determined by its I_3 and its electric charge Q through the formula $Q = I_3 + Y/2$.

The theory contains two coupling constants, one each for $SU(2)$ and $U(1)_Y$, and 3+1=4 gauge bosons. The gauge bosons for $SU(2)$ form a weak-isotriplet, with weak hypercharge 0. It is denoted by (W^+, W^0, W^-), with the z-components of their weak isospin to be $+1, 0, -1$ respectively. The gauge boson for $U(1)$ is a weak-isosinglet, with weak hypercharge 0, and will be denoted by X^0.

The Higgs fields form an $SU(2)$ doublet denoted by $\phi = (\phi^+, \phi^0)$, with $I_3 = \frac{1}{2}, -\frac{1}{2}$, and $Q = +1, 0$, respectively. Thus their hypercharge is $Y = +1$. The antiparticle of this pair is denoted by $\bar{\phi} = (\bar{\phi}^0, \phi^-)$; its weak hypercharge is -1. In total, there are four Higgs fields.

The matter particles consist of the quarks and the leptons. Among spin $\frac{1}{2}$ particles, only fermions with left-handed chirality and antifermions with right-handed chirality participate in the $SU(2)$ interactions. In other words, the $SU(2)$ interaction is *left-handed*. For that reason it is sometimes written as $SU(2)_L$.

To make sure that the interaction is left-handed, the right-handed fermions and the left-handed antifermions are made to be weak-isosinglets. With $I_3 = 0$, the hypercharge Y is $2Q$, so the weak hypercharges are $\frac{4}{3}$ for the right-handed t, c, u quarks, $-\frac{2}{3}$ for the right-handed b, s, d quarks, and -2 for the right-handed leptons. The hypercharges for the left-handed antifermions are just the opposite of these.

The left-handed fermions form a number of weak-isodoublets as follows: $\ell_1 = (\nu_e, e^-)$, $\ell_2 = (\nu_\mu, \mu^-)$, $\ell_3 = (\nu_\tau, \tau^-)$, $q_1 = (u, d')$, $q_2 = (c, s')$, and $q_3 = (t, b')$. The subscripts are the generation labels. The primes appearing with the charge $\frac{-1}{3}$ quarks indicate that they are not quite the same as the usual (unprimed) ones which have definite masses. Instead, (b', s', d') are mixtures of the usual (b, s, d), parametrized by four parameters. The matrix mixing them is called the *Cabibbo-Kobayashi-Maskawa matrix*, or CKM matrix for short. One of these four parameters, when it is not zero, gives rise to $\mathcal{CP}$ violation. All presently known $\mathcal{CP}$ violation data can be described in this manner.

The weak hypercharges of these doublets are $Y = \frac{1}{3}$ for the quarks and $Y = -1$ for the leptons. The additive quantum numbers of the right-handed antifermions are opposite to those of the left-handed fermions.

The additive quantum numbers for the Standard Model crowd is summarized in the Table of the following page, where subscripts L and R refer to Left-handed and right-handed chiralities respectively.

Electroweak Quantum Numbers

Particles	I_3	Y
$(u, d')_L$	$(\frac{1}{2}, -\frac{1}{2})$	$\frac{1}{3}$
$(c, s')_L$	$(\frac{1}{2}, -\frac{1}{2})$	$\frac{1}{3}$
$(t, b')_L$	$(\frac{1}{2}, -\frac{1}{2})$	$\frac{1}{3}$
u_R, c_R, t_R	0	$\frac{4}{3}$
d'_R, s'_R, b'_R	0	$-\frac{2}{3}$
$(\nu_e, e^-)_L$	$(\frac{1}{2}, -\frac{1}{2})$	-1
$(\nu_\mu, \mu^-)_L$	$(\frac{1}{2}, -\frac{1}{2})$	-1
$(\nu_\tau, \tau^-)_L$	$(\frac{1}{2}, -\frac{1}{2})$	-1
e_R, μ_R, τ_R	0	-2
(W^+, W^0, W^-)	$(1, 0, -1)$	0
X^0	0	0
(ϕ^+, ϕ^0)	$(\frac{1}{2}, -\frac{1}{2})$	$+1$

The fermions and the Higgs interact with the gauge bosons in the usual way, with a strength related to their $SU(2)$ and $U(1)_Y$ quantum numbers. The Higgs are supposed to interact also directly among themselves, giving rise to an attractive interaction at low densities and a repulsive interaction at high densities, so as to allow a condensate to form at low temperatures. In order to agree with experimental findings, the transition temperature for the Higgs condensate turns out to be a few hundred GeV.

Using group theoretical techniques, one can also show that the $SU(2)$ symmetry forbids the fermion to have masses when this high temperature symmetry is exact. However, the fermions are allowed to interact with the Higgs field.

At a temperature higher than the transition temperature, the Higgs condensate is absent. All fermions are massless, and all gauge bosons are massless. The high temperature symmetry $SU(2) \times U(1)_Y$ is exact, and the corresponding conservation laws are valid.

When the temperature goes below the transition temperature, the Higgs condensate forms. It is made out of a mixture of $\bar{\phi}^0$ and ϕ^0. The existence of this condensate breaks the $SU(2) \times U(1)_Y$ symmetry down to the $U(1)_{em}$ symmetry. This means that the weak isospin and the weak hypercharge are only (very) approximately conserved, but the combination of I_3 and Y which gives rise to the electric charge Q is still exactly conserved. Hence the subscript em (for electromagnetic) for the $U(1)$ group.

The gauge bosons $W^{\pm}$ acquire a mass by interacting with this condensate. Since the original symmetry is broken down to $U(1)_{em}$, there is nothing preventing the two neutral gauge bosons W^0 and X^0 to mix. In fact, one of the mixtures gains a mass, and this particle is called the Z^0. The other mixture remains massless, and it is nothing but the photon γ. In order for the gauge bosons $W^{\pm}$ and Z^0 to acquire masses, one extra degree of freedom must be supplied to each to make up for the third spin orientation. This is supplied by $\phi^{\pm}$, and a mixture of ϕ^0 and $\bar{\phi}^0$. Effectively, these three Higgs fields have been captured by the W and Z fields to be their longitudinal polarizations. The other mixture of the two neutral Higgs is not attached to anybody, so it forms a new particle called H^0, or the *Higgs boson*.

Fermions are massless above the transition temperature, but generally they also interact with the Higgs fields. Below the transition temperature, the interaction with the Higgs condensate allow the fermions to acquire masses in much the same way as the gauge particles. The amount of masses they acquire depends on the strength of the coupling between the fermion and the Higgs. The top quark t, which has not been found but is certainly by far the heaviest of the fermions, will have to have the strongest coupling with the Higgs.

Just like in the Fermi theory, neutrinos are assumed to be massless in the GSW theory. It can be shown that the Higgs mechanism giving masses to other fermions does not affect the massless nature of the neutrinos.

The interaction through γ obtained in this theory is identical to that in the Maxwell theory. The interaction through $W^{\pm}$ essentially leads to the modified Fermi theory at low energies. This is sometimes known as a '*charged current*' interaction. The interaction through Z^0 gives rise to a new weak force, and new reactions known as '*neutral current*' processes. If we bombard a neutron with a neutrino, the usual (charged current) process mediated through the charged W particles is the inverse β-decay process

$\nu_e + n \rightarrow e^- + p$. The neutral current process mediated by the exchange of Z^0 is $\nu_e + n \rightarrow \nu_e + n$. The subsequent experimental detection of the neutral current processes added greatly to the credibility of the theory. More recently, the $W^\pm$ and the Z^0 were detected at exactly the masses predicted by the GSW theory.

Whenever γ can be exchanged, so can Z^0. The latter however is a weak interaction process which violates parity. Thus its existence can already be detected if we look for parity violations in atomic physics transitions. This has been carried out and the result agrees with the theory. In fact, every prediction of the GSW theory has since turned out to be true.

The two coupling constants for $SU(2) \times U(1)_Y$ are of the same order as the electromagnetic coupling constant e. The reason why weak interaction is weak at low energies, as seen in beta decays, is solely because of the large mass of $W^\pm$. As a result of this mass, the range of the weak force is extremely short. Although the strength of the force within the range is not particularly weak compared to electromagnetic force, low energy particles have a long wavelength and only part of it can be within the range of the force. For that reason the average force is highly diluted and the effective interaction seems weak. If we carry out a high energy experiment where the wavelength is at least as short as the range of the weak force, then weak interaction will no longer appear to be weak compared to the electromagnetic interaction.

5.4 Strong Interactions

Nuclear forces between nucleons were first proposed by H. Yukawa to be due to the exchange of pions (see Sec. 3.1). Since then, many heavier *mesons* (mesons are hadrons made out of quark anti-quark pairs. They have integer spins and are bosons) have been discovered. They can be exchanged between the nucleons as well to produce additional forces of shorter ranges.

The proliferation of these hadrons led to a problem of classification and organization, not unlike that faced by Mendeleyev before he discovered the periodic table for chemical elements. Many unsuccessful attempts at their classification had been made, before the successful $SU(3)_F$ symmetry was by Murray Gell-Mann and Yuval Ne'eman.

• *Approximate flavor symmetry*

The $SU(3)_F$ symmetry is an approximate symmetry valid when the electroweak forces are turned off. It is believed that the u, d, s quarks would then have the same masses, so in the parameter space spanned by these three quarks, there is an $SU(3)_F$ symmetry. The subscript F here stands for *flavor*, a term which is invented to label different kinds of quarks. In the 1960's only three quarks were known, so there were only three flavors. Now we know of the presence of five quarks, and expect the existence of the sixth, so we might talk about an $SU(5)_F$ or an $SU(6)_F$ symmetry. The problem is, we really cannot turn off the electroweak interaction. In its presence, the masses of the quarks are different. The u and the d quarks have small and similar masses, about 0.3 GeV, so the $SU(2)_F$ symmetry from these two is a very good (though still approximate) symmetry. This is nothing but the strong isospin symmetry discussed in Sec. 4.3. Next in mass comes the s quark, at about 0.5 GeV, and the $SU(3)_F$ is still quite an acceptable approximate symmetry. This is the one discovered by Gell-Mann and Ne'eman which in turn led to the discovery of quarks. If we go to $SU(4)_F$, the c quark has a mass about 1.5 GeV. The mass difference between the c and the s, u, d quarks become so large and the symmetry is so badly broken that it becomes quite useless already. For that reason we do not usually talk about the $SU(n)_F$ symmetry for $n > 3$.

The $SU(3)_F$ symmetry is also known as the *eightfold way*, because under this symmetry, the *nucleons* belong to a multiplet with multiplicity eight, and the *pions* and *kaons* (kaons are mesons with quark contents $\bar{s}u, \bar{s}d, \bar{u}s, \bar{d}s$) belong to a separate multiplet also with multiplicity eight. We should not conclude from this that all hadronic multiplets have multiplicity eight. When a nucleon is bombarded by mesons like pions and kaons, certain unstable baryons are formed. They have spin $\frac{3}{2}$, and at the time when Gell-Mann studied them there were nine of them known. There are however no multiplets in $SU(3)$ with multiplicity nine, though there is one with multiplicity ten. Gell-Mann then predicted that there should indeed be a tenth one, called Ω^-, which was later found in 1964, thereby consolidating the success of the $SU(3)_F$ symmetry. We will come back to this *decuplet* of unstable baryons in the next subsection.

• $SU(3)_C$ *color symmetry*

Though the $SU(3)_F$ symmetry is very successful in many aspects, there is one very puzzling thing about it. Quarks have spin $\frac{1}{2}$ so they should be fermions.

This demands any wave function of several quarks to be antisymmetric in all their variables. On the other hand, if we consider a decuplet baryon, its wave function in the spin and spatial variables turns out to be symmetric, for reasons to be discussed below. Thus either the celebrated spin and statistics theorem (Sec. 4.5) is violated, or else there must be one hitherto unknown variable in the wave function so that the total wave function becomes antisymmetric when this new variable is taken into account. The latter turns out to be true and the new variable is *color*.

The decuplet baryons have spin $\frac{3}{2}$. Since a baryon has three quarks and each quark has spin $\frac{1}{2}$, this can be achieved only when the three spins all line up along some direction. The spin wave function is therefore symmetric in the three spin variables. The spatial wave function of a decuplet baryon is also known to be totally symmetric, a property which is quite common among states of low energies. Thus the combined wave function in the spin and the spatial coordinates are symmetric. If we now introduce a new variable *color*, and demand that the wave function in the color variable be *antisymmetric*, then the total wave function, being the product of the spatial, the spin, *and* the color wave functions, will be antisymmetric as demanded by the spin and statistics theorem.

To make an antisymmetrical color wave function from three quarks, we need at least three different colors. One of the simplest groups working on three variables is $SU(3)$, and to distinguish it from $SU(3)_F$, the former is usually called $SU(3)_C$.

The existence of color has since been confirmed experimentally in many different ways.

- ## Asymptotic freedom

In the late 1960's, an experiment performed by Jerome Friedman, Henry Kendall, Richard Taylor, and collaborators at the Stanford Linear Accelerator (SLAC) changed the outlook of the strong interaction theory completely.

In this experiment, high energy electrons thrown at a nucleon target were detected at *large angles* (a more precise technical description is *large 'momentum transfers'*). As a result of the collision energy is often taken from the electron to create additional particles, and they were interested in those electrons which suffered a great loss in energies. Collision of this type is known as *deep inelastic scattering*.

The probability of detecting an electron depends on two variables, the energy loss of the electron as well as the angle at which it is detected. James D. Bjorken predicted earlier that in the deep inelastic scattering region, the probability depends effectively only on one *independent* variable x, which is a known function of the electron angle and the energy loss. If two sets of angle and energy loss give the same x, then the probabilities of detecting an electron at these two sets of variables are related in a known way. This prediction of Bjorken is nowadays known as the *Bjorken scaling*.

When the experiment was analysed in the deep inelastic region, Bjorken scaling was seen to be obeyed. Richard Feynman then realized that this scaling pheonomenon could be understood *physically* if the nucleon was composed of many constituents, which he called *partons*, provided that the partons do not interact among themselves nor do they radiate other particles when they are knocked around in the deep inelastic scattering process. Since nucleons are made up of quarks bound together by gluons, these partons must be nothing but quarks, antiquarks, and gluons.

There are two puzzling features in the parton picture. First of all, we mentioned before that a nucleon is made up of three quarks, but the parton idea of Feynman requires the nucleon to contain many many partons. So how many quark partons are there? Secondly, quarks and gluons interact strongly, so how can partons fail to interact with one another and stop radiating?

To understand the first puzzle, let us return to the theory of force explained in Sec. 3. According to local quantum field theories, particles are constantly being created and absorbed. If an exchange particle created by particle A finds its way to particle B before the allotted time is up, a force will be transmitted from A to B. What happens if there are no particles in the vicinity of A to absorb the exchange particle? Then A must take it back. For this reason, we must find a steady stream of exchange particles being emitted and reabsorbed by A. As a result, there is always a halo of exchange particles around each particle A. This halo is called the *Yukawa cloud*. Its extent R is given by the Yukawa formula, and its density is determined by the coupling constant g.

Since the exchange particle itself can virtually decompose into a particle anti-particle pair, the Yukawa cloud contains these pairs as well.

Now a question arises as to what one means by a particle. Would it

be a particle with or without its Yukawa cloud? That must depend on the
context, and the resolution of the probe. If we try to probe the particle with
an object possessing a long wavelength, resolution is poor and we cannot help
but take in the whole Yukawa cloud as well. On the other hand, if we try
to probe the particle with an object possessing a wavelength shorter than
the extent of the cloud, then with this improved resolution it is possible to
see a part of the cloud. We may then pick up objects hidden in the cloud
without picking up the original particle or the whole cloud. So what is meant
by a particle in an experiment depends on the resolution of the probe: low
resolution probes pick up the particle and the whole cloud and call that a
single particle, whereas high resolution probes can pick up specific objects
hidden in the cloud, and call this part of the cloud a particle. See Sec. 6 for
a further discussion of this topic.

When we talk about a baryon being composed of three quarks, and a
meson being composed of a quark and an antiquark, we were referring to the
quarks with their full share of the Yukawa clouds. To emphasize this point
these are sometimes called the *constituent quarks*. They are the only ones a
long wavelength probe can see.

In electron scattering, the nucleon is being probed by a virtual photon
emitted by the incoming electron. This virtual photon has a *short* wavelength
in the *deep inelastic region*. Therefore it has good spatial resolution, and is
able to look into a small part of the constituent quark. As a result it may pick
out individual gluons, quarks, and antiquarks in the cloud. There could be
infinite number of these hidden deeply in the cloud, and they are the partons.
This solves the first puzzle.

The second puzzle is not so easy to understand. In fact, the resolution of this puzzle led to the discovery that the dynamical theory for strong
interaction is an $SU(3)_C$ Yang-Mills gauge theory.

Constituent quarks interact strongly among themselves. But is it possible that when one of them is bombarded by the energetic virtual photon
unleashed by an electron undergoing deep inelastic scattering, the partons
within the constituent quark are set free under this onslaught? This is indeed possible, and the technical jargon justifying it is called the *impulse
approximation*. What is hard to prevent though is the radiation of gluons by
the partons so bombarded. In fact, from ordinary electrodynamics, we know

that the quarks so bombarded will radiate photons, and the harder the bombardment the more photons they radiate. It is true that the electromagnetic coupling constant is small and photon radiation can be ignored. But why should gluon radiation not behave the same way? These are the radiation that cannot be ignored. If we take them into account, then we lose one of the essential ingredients of Feynman's parton picture, and we cannot understand the outcome of the SLAC experiment.

We are of course getting a bit ahead of ourselves. In those days gluons were not yet discovered; we are simply using the term 'gluon' to describe whatever the partons radiate when struck by a hard blow.

In light of this dilemma a theoretical effort was on to find a compromise; a cake that you can have and eat, so to speak. One must find a fundamental interaction of quarks that has this strange property, that the harder it is struck, the less it radiates. Nowadays this property is called *asymptotic freedom*; 'asymptotic' in the sense of being struck infinitely hard, 'freedom' because of the absence of interaction and radiation.

After many unsuccessful attempts, it was finally realized by David Gross and Frank Wilczek, and by David Politzer, that this is possible if quarks interact through a non-abelian gauge theory. Since we were forced by the Pauli exclusion principle to introduce this new quantum number 'color' for the quarks and not for the leptons, since the other quantum numbers 'flavor' and 'charge' have been used to describe weak and electromagnetic interactions, it is reasonable to assume that 'color' is the quantum number that governs the new Yang-Mills gauge theory for strong interactions. This $SU(3)_C$ non-abelian gauge theory is nowadays accepted as the standard theory of strong interaction between quarks.

This theory has since become known as *quantum chromodynamics*, or QCD for short, to indicate the involvement of color. Its gauge particles are called the *gluons* (g). Since the group is an $SU(3)$, there are eight gluons in total, as stated in Sec. 3.1. The gauge group $SU(3)_C$ is supposed to be independent of the electroweak gauge group $SU(2) \times U(1)_Y$, hence the gluon must not carry weak isospin nor weak hypercharge, and hence not any electric charge. This last aspect is actually seen directly in deep inelastic scattering experiments, where the *struck* partons are seen to carry only about 50% of the momentum of the nucleon. The other 50% is being carried by the

gluons, which because of its electric neutrality, do not interact directly with
the incoming virtual photon.

Leptons do not carry color so they do not experience nuclear forces.

We have now obtained the *Standard Model* of strong, electromagnetic,
and weak interaction. It is a gauge theory based on the group $SU(3)_C \times SU(2) \times U(1)_Y$.

The property of asymptotic freedom makes the interaction of quarks
and gluons relatively weak when they scatter from one another with a large
momentum transfer. It is this property that suppresses the gluon radiation
in the deep inelastic scattering region, and make it feasible to be computed
using only a few Feynman diagrams. However, even at the largest momentum
transfer available today, the coupling constant is not zero, so radiation from
the struck quarks cannot be avoided. As a result, the Feynman picture is
not strickly valid, and Bjorken scaling must be slightly violated. Improved
analysis and increased experimental accuracy show that this is indeed the
case. The amount of observed violation, however, is perfectly consistent with
what is computed from the Feynman diagrams using QCD.

It is interesting to note that in an abelian gauge theory like QED,
fermions radiate strongly when struck hard, whereas in a non-abelian gauge
theory like QCD, it is the other way around. The reason for this marked
contrast will be discussed in Sec. 6.3.

• *Confinement*

Quarks have never been seen in isolation, so QCD must carry with it the confinement property (see Sec. 3). Ordinary hadrons like nucleons and mesons
are not subject to confinement because the quarks and antiquarks within
them conspire to form color singlets. As such they no longer interact directly
with the gluon, which is why the dominant nuclear force between two nucleons is carried by the pion and not by the gluon. See Sec. 3.1 for more
discussions on this point.

The mechanism behind confinement is still not clear. This ignorance
lies in our inability to calculate reliably strong interaction field theories, except when the momentum transfer is large and the effective coupling small,
as is the case in the deep inelastic scattering region. There are however approximate numerical calculations (known as *lattice gauge theory* calculations)
which indicate that confinement indeed occurs in QCD.

We will now describe a qualitative model of confinement which has some intuitive appeal, but whose quantitative validity is not yet known. Nevertheless, it will serve to illustrate what kind of mechanisms might induce confinement.

According to this model, at temperatures below about 0.2 GeV, a *colorless gluon condensate* is formed in the vacuum. Like the Higgs condensate, this will happen only when the vacuum with the condensate is more stable than the one without. In other words, the condensate should have a negative energy density. Because of relativistic invariance, it turns out that any vacuum condensate with a negative energy density should exert a positive pressure on its surrounding. Let ϵ be the difference in energy density between the vacuum without the condensate and with the condensate. If a color electric and magnetic field were present within a volume V, then the condensate within that volume would first be polarized, then ionized and torn apart by these fields, and this would require the supply of an amount of energy ϵV. For energetic reasons, and because of the positive pressure mentioned earlier, the color electric and magnetic fields originating from a quark will be squeezed inside as small a volume as possible, so as to minimize the energy deficit ϵV. In this way the fields between two quarks would be squeezed into a cylinder, and confinement would be achieved (see Sec. 3.2).

If this scenario is correct, one might be able to achieve *deconfinement* by high energy heavy ion collisions, if a temperature of ~ 0.2 GeV can be created and trapped by these ions to melt away the condensate. Experiments of this kind are underway.

• Chiral symmetry

Imagine now turning off the electroweak interactions. Then the resulting masses of the quarks will be zero, unless there are other mechanisms present that we are not aware of for generating masses. In the absence of masses, a new symmetry called *chiral symmetry* emerges.

Recall from Sec. 4.5 that chirality is simply twice the helicity for a massless fermion. Chirality is conserved for a massless fermion theory under gauge interactions, but a massive fermion with a fixed helicity causes a steady oscillation of the two chiralities, so chiral symmetry is violated.

For massless quarks of n flavors, the flavor symmetry $SU(n)_F$ is now enlarged to a *chiral flavor symmetry* $SU(n)_L \times SU(n)_R$, because the flavor

symmetry is preserved independently for the left-handed quarks (L) and the
right-handed quarks (R).

Since in reality the electroweak forces cannot be turned off, this chiral
symmetry is at best an approximate one. However, if we look at the u and
the d quarks, their masses are both small, so the chiral symmetry will work
best when $n = 2$. We shall concentrate on this one from now on.

In the presence of chiral symmetry, there should be two sets of (one for L
and one for R) (approximate) conserved quantum numbers rather than one.
Since the charged current weak interaction is left-handed, it can be used to
probe whether the left-handed flavor quantum numbers are approximately
conserved. They are. This confirms the existence of chiral symmetry, at
least for two flavors.

There is however one problem. There do not seem to be enough hadrons
around with the same mass and spin as would be required for the existence
of a chiral $SU(2)$ symmetry. Take the three pions, for example. Their quark
contents are given by $\pi^+ = \bar{d}_\uparrow u_\downarrow$, $\pi^0 = \bar{d}_\uparrow d_\downarrow - \bar{u}_\uparrow u_\downarrow$, $\pi^- = \bar{u}_\uparrow d_\downarrow$. The novel
feature in the notation here is the arrow subscripts attached to the quark
symbols. Imagine the pion and its quarks travelling along the z-direction,
then the two arrows indicate the two possible helicity states the quark could
be in.

The pion is a spin 0 particle, so its spin projection along the z-direction
must be 0. This is why the two arrows appearing with the quarks always point
in opposite directions. Actually, to do a proper job in quantum mechanics,
we need to add other terms as well but we will not bother to do so because
these extra terms do not change the argument. Now we notice that another
spin 0 particle can be defined: $\sigma = \bar{d}_\uparrow d_\downarrow + \bar{u}_\uparrow u_\downarrow$. The difference is that the
three π's form a strong-isospin triplet, but the σ is a strong-isospin singlet.
In fact, it is to keep this isospin separation that we must take the particular
linear combinations of $\bar{d}_\uparrow d_\downarrow$ and $\bar{u}_\uparrow u_\downarrow$ that appear before.

An $SU(2)_F$ strong-isospin transformation mixes the three π's but leaves
the σ invariant. An $SU(2)_R$ transformation, however, mixes σ and the π's.
This is because $SU(2)_R$ affects only the down-arrow quarks (say) and not
the up-arrow ones. If the transformation is to interchange $u_\downarrow$ and $d_\downarrow$, for
example, we would get $\sigma \rightarrow \bar{d}_\uparrow u_\downarrow + \bar{u}_\uparrow d_\downarrow$, which clearly becomes a mixture
of π^+ and π^-. Particles that can be transformed into one another by an

internal symmetry transformation must have the same mass, for otherwise they could never be symmetric. If we only have an $SU(2)_F$ symmetry, then the three π's must have the same mass, but σ may have another mass. Under chiral symmetry, all four of them must have the same mass.

So to summarize, if chiral $SU(2)_L \times SU(2)_R$ symmetry is valid, then there should also be a strong-isospin singlet particle σ with the same mass as the pions.

The problem is, such a particle does not exist; not near the pion mass anyway.

There is one way out of this dilemma if a condensate is present to cause a spontaneous symmetry breaking of the chiral symmetry. It turns out that this symmetry breaking does not affect drastically the observable feature of a left-handed charged current weak interaction, but it affects the σ mass.

Since strong-isospin symmetry $SU(2)_F$ is known to be valid, its transformations must leave the condensate invariant, so this condensate must be a strong isospin singlet. the field for σ then seems to be a perfect candidate for this *chiral condensate*. Now that the symmetry is reduced from the chiral symmetry to $SU(2)_F$, we no longer need to have a σ particle at the mass of the pions. Moreover, the quarks will become massive by interacting with the chiral condensate, in much the same way that gauge bosons and fermions acquire masses in the presence of the Higgs condensate.

The surprising and the interesting thing is, the pion mass now becomes zero.

This is the famous *Nambu-Goldstone theorem*, which states that massless and spinless particles (the *Nambu-Goldstone bosons*) are always present when a continuous symmetry is spontaneously broken. If the original symmetry group G has dimension d_G, and the residue symmetry group H (after the symmetry breaking) has dimension d_H, then there are $d_G - d_H$ Nambu-Goldstone bosons.

In the present case, $G = SU(2)_L \times SU(2)_R$, and $H = SU(2)_F$. Thus $d_G = 3 + 3 = 6$, $d_H = 3$, and there are three Nambu-Goldstone bosons. These are the pions in the theory.

By the mechanism of spontaneous symmetry breaking, we have killed two birds with one stone. We have gotten rid of the necessity of having too many hadrons of the same mass and spin, and we have solved a longstanding puzzle of why the pions have such low masses compared to other hadrons.

For a comparison of the hadronic masses, consider the nucleon, whose mass is about 1 GeV/c^2. This makes the mass of each constituent quark about 0.3 GeV/c^2. A meson is made up of a quark and an antiquark, so the pion would be expected to have a mass somewhere around 0.6 GeV/c^2— but it is only 0.14 GeV/c^2. The reason as we know it now is because the pion is a Nambu-Goldstone particle, the reason why this mass is not exactly zero is because chiral symmetry is not exact.

The actual physical mechanism to make the pion massless will be discussed towards the end of this subsection.

The transition temperature at which this chiral condensate occurs has been estimated to be again about 0.2 GeV.

It is interesting to see how the quarks steadily increase their masses as the surrounding temperature is lowered. The quarks start out to be massless at high temperatures, then an electroweak mass is developed after the condensation of the Higgs, at a few hundered GeV. When the temperature goes below 0.2 GeV where the chiral condensate materializes, the quarks would gain an even larger mass. Conversely, if we heat up the quark surroundings, say by a heavy ion collision, then the quark would get progressively lighter each time it undergoes a phase transition to evaporate a condensate. This lightening of the quarks may have observable consequences which will help reveal experimentally the presence of these condensates and the phase transitions.

The reader might have noticed that the transition temperatures for deconfinement and chiral restoration are both quoted to be about 0.2 GeV. At present, it is not known for certain whether these two are exactly the same or not.

So far these various condensates have not been directly detected experimentally. As mentioned above heavy ion collisions might help to do that. This is an important task because it would help to nail down one corner of a serious problem, as to why our universe is so large. This is formally known as the *cosmological constant problem* and will be discussed in Chapter VII on Cosmology.

Let us now discuss in more detail why the pion has a zero mass in the absence of the electroweak interaction. In other words, the *Nambu-Goldstone mechanism.*

Instead of lowering the temperature to facilitate the formation of the chiral σ condensate, we may simulate the same effect by lowering the σ mass m. Both have the effect of making it easier for the would-be chiral condensate to form. In the high temperature phase, which is sometimes known as the *Wigner phase*, $m^2 > 0$ and a σ condensate cannot be created spontaneously from the vacuum. As a result, chiral symmetry is exact and the mass m is common to the pions and the σ. The pions and the σ also interact similarly because of the chiral symmetry.

However, as we lower m^2, a phase transition may occur and a σ condensate may be spontaneously created when $m^2 < 0$. This new phase is known as the *Nambu-Goldstone phase*.

Particles of negative m^2 are sometimes called tachyons. They travel with a speed faster than the speed of light, and as far as we know they do not exist in Nature. What happens when $m^2 < 0$ is the following. The σ field requires no effort to be created from the vacuum so lots of it will be produced. If there is nothing to stop this production we will have a runaway situation and the vacuum will collapse altogether. On the other hand, if we now introduce an interaction between the σ's which abhors overcrowding, then a point will be reached where no more σ's can be created, and an equilibrium value of σ will be established. This is the σ condensate. The mechanism for its creation is very similar to that of the Higgs condensate discussed before.

Small fluctuations about this equilibrium value of the σ field in the condensate produce the σ particle, with a mass depending on the details of the interactions. At the same time, the interaction with this condensate gives rise to the massless Nambu-Goldstone bosons which are the pions.

We might think of this promotion of masses, from a negative m^2 to a positive one for the σ, and to zero for the pions, again through the same intuitive mechanism as discussed before in connection with the Higgs condensates. At negative m^2 the particles are tachyons and travel faster than light. Their interactions with the condensate, however, slows them down. The amount of this slowing down depends on the interactions with the condensate. For the pions, which are isospin triplets, mathematics show that they would be slowed down just to the speed of light, meaning that the resulting pions become massless. The σ particle, which is an isospin singlet, would be slowed down further to gain a positive mass.

Since the same condensate generating mechanism is at work in the electroweak theory, we can also use the language of the Nambu-Goldstone particles to analyse that situation. In that case, the original symmetry group is $G = SU(2) \times U(1)_Y$ and the broken symmetry group is $H = U(1)_{em}$, thus $d_G = 3 + 1 = 4$ and $d_H = 1$, and there are three Nambu-Goldstone bosons. The only additional kink here is that these three Nambu-Goldstone Higgs bosons would then be eaten up by three gauge particles to fatten themselves into three massive particles, the $W^{\pm}$ and the Z^0. A single fasting gauge particle remains massless and becomes the photon. The fluctuations of the Higgs field along the direction of its condensate then become a separate particle like the σ particle, but this time it is the Higgs particle H^0.

Summary

Gauge theories are theories where forces are transmitted by massless spin 1 particles. To keep them massless a gauge mechanism is introduced, in which one of the three spin degrees of freedom is decoupled. Such decouplings can be achieved only when the gauge boson is coupled to a conserved quantum number. The symmetry group for that conserved quantum number is then the gauge group for the theory. Gauge invariance is a local invariance, and as such it can, and does, give rise to a complete set of equations of motion.

Abelian gauge theory is just the Maxwell theory. Non-abelian, or Yang-Mills, gauge theories are more complicated. They differ from abelian theories in having direct interactions among the gauge bosons.

The electroweak interaction is given by the gauge theory $SU(2) \times U(1)_Y$. To make it work, a Higgs condensate is required.

The strong interaction is given by the gauge theory $SU(3)_C$, also known as QCD. Asymptotic freedom exists in QCD; confinement may be present there as well.

The Standard Model of strong, electromagnetic, and weak interaction is given by a gauge theory based on the gauge group $SU(3)_C \times SU(2) \times U(1)_Y$.

An approximate flavor and an approximate chiral symmetry exist for the light quarks. The chiral symmetry is spontaneously broken to the flavor symmetry. As a result, Nambu-Goldstone particles appear in the form of pions.

6. RENORMALIZATION

Has it ever occured to you that it is a miracle that we can discover any physical laws at all?

Newton managed to discover his three famous laws of mechanics and many others without ever knowing any atomic or molecular physics, because he did not have to. *Microscopic details* are really not important to macroscopic physics. This is not to say that there is absolutely no relationship between the microscopic and the macroscopic worlds. The mass of an object, its thermal conductivity, its tensile strength, etc., are controlled by the details of atomic and nuclear physics. Macroscopic theories, however, treat these quantities as experimentally measured parameters. If we want to *compute* these parameters, then a detailed knowledge of atomic and/or nuclear physics will be needed. This is the normal world.

Imagine, if you will, a completely different world, in which the *details* of atomic and nuclear physics are required to describe *macroscopic* physics — the *details*, and not just a *finite* number of parameters. What would happen in such an imaginary world? Everything down to the tiniest details would be interlinked; the equations governing the physical laws would be so horribly complicated that they could never have been discovered. Physics as we know it would not exist any more.

Such a strange world is luckily not our world, but it is a conceivable world. It is a frightful world, and the theories describing it have a frightfully technical name: *non-renormalizable*. In contrast, theories for the kind of normal world we know are said to be *renormalizable*.

A *renormalizable* theory is one in which the details of a deeper scale (smaller distance or larger energy) are not needed to describe the physics at the present scale, save for a few experimentally measurable parameters. In one problem, the 'present' scale could be macroscopic physics and the deeper scale could be atomic physics or, the present scale could be atomic physics and the deeper scale could be nuclear physics in another problem. Within the world of elementary particles which we will concentrate on in the rest of this section, the 'present' scale would be some relatively low energy (or large distance) scale, and the deeper scale would be a much higher energy (or shorter distance) scale. It turns out that all *gauge theories are renormalizable*, but the *Fermi theory* of weak interaction, as well as the present theory of *quantum gravity*, are *not renormalizable.*

There is a question of how easy it is to *renormalize* a renormalizable theory. This means, how do we transform the deeper physics of a deeper scale into the physics at the present scale. Alas, this in general is not an easy task, as we can see in the following example. The motion of water in the macroscopic scale is described by the 'Navier-Stokes equation', but in the atomic and molecular scale, the motion of the water molecules would be given by the quantum-mechanical Schrödinger equation. It would be foolhardy to treat the waves in the ocean with 10^{26} coupled Schrödinger equations for the water molecules; we will never get anywhere with such a complexity. So we resort to the simpler Navier-Stokes equation, which is derived classically and contains a viscosity parameter. This parameter is determined classically by measurements, though a knowledge of atomic physics would allow it to be computed. This then is the hallmark of a renormalizable theory — the connection with the underlying deep-level physics resides only in a finite number of measurable parameters.

Now the Navier-Stokes equation bears very little resemblance in appearance and *form* to the Schrödinger equation, though the former must somehow follow from the latter. The difficulty of deriving the Navier-Stokes equation from the Schrödinger equation illustrates the difficulty of carrying out the renormalization procedure in detail, a difficulty which is present in many problems. However, in relativistic theory of elementary particles, we are lucky! Relativity is highly restrictive to the *form* that the equations can take, whatever the energy scale, as long as it is relativistic. For that reason it is actually not difficult to carry out the renormalization procedure in a relativistic theory, at least in the perturbation theory framework.

Renormalization is sometimes stated as a way of getting rid of the unwanted infinities arising in local quantum field theories. Historically that is how it came about, in QED, but it is probably not a good way to look at it as such. We should regard renormalization in the way explained above. If we do it right, then these unwanted infinities will automatically disappear.

We will start the discussion with a brief recounting of the history.

Soon after the inception of QED back in the late 1920's, a serious difficulty was encountered, whose solution did not come until the late 1940's and the early 1950's. The method which made it possible to overcome this difficulty is *renormalization*. As we have explained, the central idea of that is simply to summarize uncalculable (infinite) quantities into a few measurable parameters. The prominent names associated with this effort are Richard Feynman, Julian Schwinger, and Sin-itiro Tomonaga, although very important contributions have been made by many others as well.

The coupling constant in QED is the electronic charge e. The corresponding *fine structure constant* $\alpha \equiv e^2/\hbar c$ which appears in perturbation calculations (see Sec. 3.4) is approximately equal to $1/137$, a small number. Perturbation calculation with a few Feynman diagrams should therefore be valid.

However, when Feynman diagrams were actually calculated, it was found that diagrams with loops sometimes gave rise to *divergences* (mathematical infinities), although there was never any divergence in tree diagrams. Moreover, numbers computed from the tree diagrams alone agree quite well with the experimental data. From this empirical fact rose Heitler's famous rule: just forget about the loop diagrams.

This is fine as far as it goes, but there are two problems. First, logically the rule should not have worked, because the terms neglected are infinitely large. Secondly, with the neglect of these higher order diagrams it is impossible to get highly accurate results needed to compare with precision experiments.

The challenge to Heitler's rule finally came in the late 1940's. After the Second World War, radar technology became available, and it was put to good use in microwave spectroscopy. Using it, Willis Lamb and Robert Retherford found an unexpected energy shift in an atomic hydrogen line, a shift that is very small but nevertheless cannot be accounted for by computations using Heitler's rule. In response to the challenge to explain this *Lamb shift*, an intense theoretical effort was undertaken to find a way to calculate beyond the tree diagrams. This effort finally led to the formulation of the method of renormalization, and a precise calculation of the Lamb shift that agreed with the experiment. Subsequently, many other high precision experiments have been performed and each time the result agrees with the calculation using the renormalization technique.

6.1 Removal of Infinities

It was mentioned in Sec. 3.4 that physically measurable quantities can be computed from T-matrix elements, which in turn can be computed from Feynman diagrams. As mentioned above, divergences appear when certain loop diagrams are so computed.

If we examine the origin of these infinities, we will find that they can be traced back to the singular nature of the inverse-square law $1/r^2$ at $r \to 0$. Since the inverse-square law is the universal force law (see Sec. 3.2) at short distances r, these divergences are pretty much unavoidable in local quantum field theories.

However, there is no *direct* physical evidence why the divergence should occur, because *experimentally* we do not know the inverse-square law to be valid below a distance of some 10^{-18} m or so. What if the force is not so singular at ultra-short distances? Then divergences would be avoided. If we assume the force not to be so singular, then what should we assume it to be? That depends on whether QED is renormalizable or not. If it is, then we know that the outcome should not depend on the details at the deep scale, so it should not matter what detailed form we take it to be. Different ways of making the force law less singular at $r = 0$ are known as different *regularization schemes*. The final physical outcome should be independent of any particular scheme, so for the sake of definiteness let us choose a very simple one in the following. Let us replace the inverse-square law at any distance $r \leq r_0$ by its value at $r = r_0$. It does not matter what we choose r_0 to be, as long as it is much less than the minimal distance that present-day experiments can probe at.

It turns out that QED is indeed renormalizable, and that renormalization can be carried out successfully for it. To understand how that works, let us first consider the following.

• Degree of divergence

If a T-matrix element diverges like $r_0^{-D} \ln^n(r_0)$ as $r_0 \to 0$, for some $D \geq 0$, then its *degree of divergence* is said to be D. It turns out that the divergences of the T-matrix element are all contained in $D + 1$ (divergent) parameters, and it is essentially this fact that allows the theory to be renormalizable and the Lamb shift to be computed.

A simple recipe is available to determine the degree of divergence D for any Feynman diagram. We shall not go into the derivation of this recipe, but the final formula is: $D = 4 - \frac{3}{2}F - B - kV$. In this formula, F and

B are respectively the number of external fermion and boson lines, V is the number of vertices in the Feynman diagram, and k is the energy dimension of the coupling constant.

The coupling constants g for any gauge theory is dimensionless, so $k = 0$. The coupling constant in the Fermi theory of weak interaction (Sec. 5.3) is $G_F = 1.17 \times 10^{-5} (\text{GeV})^{-2}$, which has the unit $(\text{energy})^{-2}$, so its energy dimension is $k = -2$.

Depending on whether $k = 0$, $k > 0$, or $k < 0$, the corresponding theories are *renormalizable*, *'super'-renormalizable*, or *non-renormalizable*.

• *Renormalizable theories*

All gauge theories have $k = 0$ and are renormalizable. According to the formula for D, only a small variety of graphs in a renormalizable theory are divergent. In order for $D \geq 0$, the only values allowed for the pair of numbers (F, B) are: (a). $(0,0)[D = 4]$ (*vacuum diagram*); (b). $(0,2)[D = 2]$ (*photon self-energy diagram*); (c). $(2,0)[D = 1]$ (*electron self-energy diagram*); (d). $(2,1)[D = 0]$ (*vertex diagram*); and (e). $(0,4)[D = 0]$ (*light-light scattering diagram*). For definiteness, we have named these diagrams using the terminology in QED, but the result can be generalized to any gauge theory. The simplest of these divergent diagrams are drawn in Figs. VI.7a–VI.7e, respectively. None of the other Feynman diagrams with different number of external lines have any (primitive) divergence.

The word 'primitive' in the parenthesis is a caveat. What it means is that there is no divergence for the diagram as a whole, but there may be divergences hidden in a smaller part of the diagram. For example, the electron-electron scattering diagram in Fig. VI.7f has no primitive divergence, as $(F, B) = (4, 0)$ in this case. However, there is a subdiagram in its upper corner that looks like diagram (d), which is known to have $D = 0$. In other words, divergences in subdiagrams are not counted as primitive divergences for the whole diagram.

A gauge theory has a high degree of symmetry, which allows some primitive divergences to be cancelled. This is the case for the photon self-energy diagrams, whose degree of divergence D are actually 0 and not 2, and for the light-light scattering diagrams, which are actually primitively convergent. So all in all, only the following diagrams in QED are primitively divergent: the vacuum diagrams with $D = 4$, the vertex diagrams with $D = 0$, the electron

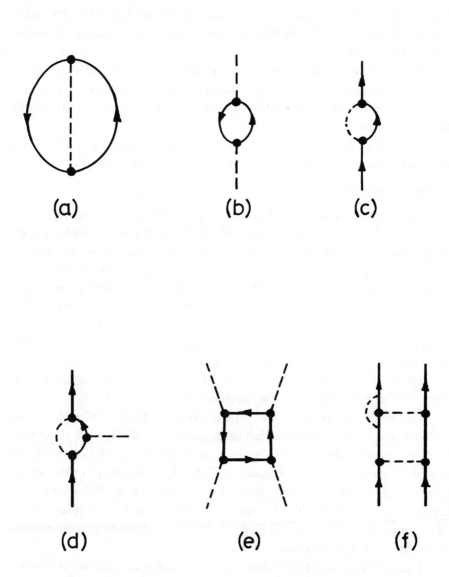

FIG. VI.7 *Feynman diagrams for electrodynamics. (a) vacuum diagram; (b) photon self-energy diagram; (c) electron self-energy diagram; (d) vertex diagram; (e) light-light scattering diagram; (f) electron-electron scattering diagram.*

self-energy diagrams with $D = 1$, and the photon self-energy diagrams with $D = 0$.

The vacuum diagrams and their divergences are usually ignored, on the grounds that they are not tied in with any external particles and cannot be measured in any experiment involving the collision of these particles. We shall not discuss them any further in this chapter. See, however, Sec. 7.3 on Supersymmetry and Chapter VII on Cosmology for their significance.

Recall that the divergences of a diagram are summarized in $D + 1$ parameters. Hence there is one divergent parameter in the vertex diagrams, usually denoted by Z_1^{-1}, two divergent parameters in the electron self-energy diagrams, usually denoted by m and Z_2^{-1}, and one divergent quantity in the photon self-energy diagrams, usually denoted by Z_3^{-1}. Each of these quantities, as well as everything else computed using Feynman diagrams, depends on the two parameters defining the theory: the *bare mass* m_0 of the electron, and its *bare charge* e_0.

The word *bare* is used to describe the situation when the electromagnetic interaction (more generally, any other interaction) is turned off. The word *dressed* is sometimes used to describe the situation when the interactions are on. Thus, the bare mass and the bare charge are the mass and the charge of the electron in the absence of the electromagnetic interaction. They differ from the physical mass (dressed mass) $m = 0.51 \text{MeV}/c^2$ and the physical charge (dressed charge) $e = \sqrt{\alpha \hbar c}$ for the following reason. Interaction causes energy to change. Since mass is nothing but the energy divided by c^2 when the particle is at rest, the mass is shifted as well when the electromagnetic interaction is turned on. Hence $m \neq m_0$. Moreover, interaction creates a Yukawa cloud around the electron (see Sec. 5.4). The electron-positron pairs in the cloud shield off part of the original charge e_0 of the electron, just like what the molecules in a piece of solid do in response to a free charge put amongst them, by polarizing themselves to reduce the electric fields thus placed in the solid. The effective charge of the electron measured at a distance from the bare electron is the shielded charge e, which is smaller than the bare charge e_0.

Let us now return to these divergent quantities, $Z_1^{-1}, Z_2^{-1}, Z_3^{-1}$ and m. Calculation shows that the parameter m should actually be identified with the observed mass of the electron, but how can this be when the calculated value is divergent and the observed value is finite? Similarly, the fully shielded electron charge e should be calculated from the formula $e = \sqrt{Z_3}e_0$, which would be zero (because Z_3^{-1} is divergent) and not the observed electron charge. Is there something we can do to correct these disastrous discrepancies?

There is, if we realize that all along we have made the tacit assumption that the original parameters e_0 and m_0^{-1} are both finite. Then m is indeed divergent and e is indeed zero. However, if we allow e_0 and m_0^{-1} to become divergent as $r_0 \rightarrow 0$, we can adjust things so that m and e take on the observed values. In other words, we can pass the infinity 'buck' from m and e^{-1} onto m_0^{-1} and e_0.

Are we ahead in any way by this buck passing game? We are, because m_0 and e_0 are not measurable at the moment and we actually do not know what their values are, although we know very well what the values of m and e should be. We can never turn off the electromagnetic interaction, so the only way of measuring m_0 and e_0 is to probe the electron with a very high energy device, an energy far greater than what is available now. When this is done, we will probably find out that the inverse-square law is not valid down to $r = 0$, and that e_0 is not infinite. What m_0 and e_0 really are need not concern us now, for they would be answered by physics of the future.

Since m and e are the measurable parameters, we should express all measurable quantities in terms of them, rather than m_0 and e_0. The nice thing is, once we do so, all the T-matrix elements turn out to be finite and independent of r_0, *i.e.*, independent of the regularization scheme! This is the essence of renormalization; this is how the infinities disappeared, or more accurately made irrelevant; and this is how the small corrections in the Lamb shift can be calculated.

• *Super-renormalizable and non-renormalizable theories*

For a super-renormalizable theory, $k > 0$, so there are even fewer divergent diagrams than the renormalizable theories. Moreover, using the formula for D, we see that all diagrams are primitively convergent if V is large enough. The way to renormalize a super-renormalizable theory is the same as the way to renormalize a renormalizable theory, so we will not discuss this any further.

A non-renormalizable theory has $k < 0$, so D grows indefinitely with V. Since we need $D + 1$ divergent parameters in each case, a non-renormalizable theory contains an infinite number of divergent parameters. Even if we had an infinite number of bare parameters to pass the infinity buck on and to turn the infinite number of originally divergent parameters finite, like what we did for e and m in QED, we would still need an infinite number of parameters to express the T-matrix in. With two parameters in QED, they can be measured. With an an infinite number of parameters, it would take a long time to measure all of them, and even if we could do so, the theory with so many parameters would not have much of a predictive power anyway. For that reason it is impossible to make any reliable loop calculations for a non-renormalizable theory, so more or less we are compelled to stick to the tree diagrams and Heitler's rule.

6.2 Renormalizability

Is there no hope for non-renormalizable theories? Are these theories condemned to be incomprehensible forever?

The answer to the second question, fortunately, is probably *no*. The fact that the coupling constant of a non-renormalizable theories have a (negative) energy dimension is the saving grace.

If all goes well, meaning that things will eventually fit into this historical trend of renormalizability, the non-renormalizable theory presumably must turn renormalizable at a higher energy scale. The fact that it is non-renormalizable at the present simply reveals our temporary ignorance of all the necessary information to build a renormalizable theory on. So how much more information do we require and when does it turn renormalizable? Here the energy scale of the coupling constant gives the clue: presumably it is at this scale that the transition will take place. This guess is confirmed at least for the Fermi theory of weak interaction.

The coupling constant for the Fermi theory (see Sec. 5.3) is $G_F = 1.17 \times 10^{-5}$ $(\text{GeV})^{-2}$, and the theory is non-renormalizable. The energy scale defined by this constant is $G_F^{-\frac{1}{2}} = 292$ GeV. Indeed, as we now know, the Fermi theory is a low energy approximation to the GSW electroweak theory, which is renormalizable. The Fermi theory is no longer a good approximation to the GSW theory when an energy scale comparable to the gauge boson masses (~ 100 GeV) or the Higgs condensate (~ 250 GeV) is reached. Qualitatively this agrees quite well with the energy scale of 292 GeV obtained from the coupling constant.

Quantum gravity, the quantized version of Einstein's theory of general relativity, is also non-renormalizable, because the Newtonian gravitational constant, $G_N = 6.7 \times 10^{-39}$ (GeV)$^{-2}$, also has a negative energy dimension. The corresponding energy scale $G_N^{-\frac{1}{2}} = 1.2 \times 10^{19}$ GeV is usually called the *Planck mass*, or more accurately, the *Planck energy*, and is denoted by $M_P c^2$. According to the previous arguments, presumably some new physics will happen at this huge energy scale to turn quantum gravity into a renormalizable, or better still, a finite theory. This, for example, is what the string theory suggests (see Sec. 7.5). Incidentally, the distance scale corresponding to M_P is $R_P = \hbar/M_P c = 1.6 \times 10^{-35} m$.

6.3 Running Parameters

Let us now confine ourselves to elementary particle physics, and ask what would we get when we subject a particle to probes of a different energy scale, or equivalently, a different spatial resolution.

A (dressed) particle is made up of its bare self, and a Yukawa cloud of virtual particles around it (see Sec. 5.4). We see the whole dressed particle when we probe it with intruments of low spatial resolution, but we see only part of the dressed particle with intruments of high spatial resolution. The result is that the parameters for this particle so measured would appear to change with the spatial resolution of the probe. For that reason, these parameters are known as *running parameters*.

Take, for example, QED. A dressed electron has around it a Yukawa cloud of photons and e^+e^- pairs. Its mass ($m = 0.51$ MeV/c^2) and charge ($e = \sqrt{137\hbar c}$) are traditionally measured with a large probe, which registers the whole electron as a whole. Imagine now probing the electron with a short wavelength virtual photon, as can be done in a deep inelastic scattering experiment. Then the probe sees only part of the Yukawa cloud, and the corresponding mass and charge it registers would be different. One expects the mass to decrease, but the charge to increase from the values quoted above. The first is easy to understand, because we are weighing only part of cloud, so we might expect it to weigh less. The second is slightly harder to understand, but it is so because the e^+e^- pairs in the cloud shield off the original bare charge of the electron. The phyical charge e one measures at large distances is the completely shielded charge, which is smaller than the bare charge e_0, as mentioned before in Sec. 6.1. As we penetrate deeper into the cloud with a smaller probe, we begin to leave behind some of the shielding and see a larger effective charge.

Thus when we continue to reduce the size of the probe from a large one, the effective charge in QED increases from the physical value e to the bare value e_0, and the effective mass decreases from the physical mass m to the bare mass m_0.

QED is an abelian gauge theory. In a non-abelian gauge theory the running fermion mass m still behaves the same. However, the running coupling constant g in QCD varies in the opposite way than the coupling constant e in QED: g actually decreases when the size of the probe is reduced. Eventually g goes to zero at very short wavelengths, which is what is meant by 'asymptotic freedom' (see Sec. 5.4). The reason for the opposite behavior between the coupling constants of QED and QCD will be explained below.

The Yukawa cloud around a quark contains gluons and quark antiquark pairs, just as the Yukawa cloud around an electron contains photons and e^-e^+ pairs. The difference is, photons do not interact with themselves, because they do not carry any electric charge, whereas gluons do because they carry color spins. This basically marks the difference between an abelian and a non-abelian gauge theory. If we look at the vertices describing gluon interactions (Fig. VI.5), we would see the possibility for a gluon to divide itself into two or three gluons. This reproduction of the gluons increases their population and tends to spread them around, but must be done in such a way to conserve the total color.

When we penetrate deeper into this cloud with a smaller and smaller probe, two competing effects occur. The probe sees less color shieding from the quark anti-quark pairs in the cloud, just as the photon sees less charge shielding from the e^-e^+ pairs. This effect tends to increase the effective color registered. However, this penetration also leaves behind more and more of the gluons in the cloud, hence the color they carry, this tends to decrease the effective color the probe registers. For an abelian theory, the second effect is absent, so the net result is the increase of the charge. For a non-abelian theory, the second effect is present, so it comes down to a competition between the two opposing effects. The outcome depends on the details of the non-abelian gauge group, and the number of flavors (variety) of quark pairs present. A larger group tends to increase the spreading of the gluon population, and more flavors tend to increase the shielding. For QCD, the group is $SU(3)$, the net result is that if there are less than 17 flavors present, which is certainly the case in reality, the gluon spreading effect wins out, and the effective running coupling constant g decreases when the probe is reduced. This then leads eventually to asymptotic freedom.

What has all this got to do with renormalization? Plenty. Renormalization is a question of relating the physics at different scales. In particular, it relates the measurable parameters such as mass and coupling constants at different scales. That is just what the running parameters are.

Summary

Physics as we know it can be divided into different levels: classical physics, atomic physics, nuclear and elementary particle physics, etc. These levels are separated by energy scales. Physics at one level depends on the physics at a deeper level only through a number of measurable parameters. In this sense physics is renormalizable. The parameters are often not reliably calculable until some of the details of the physics at the next level is known.

If we force a computation of these parameters in local quantum field theories using Feynman diagram techniques, these parameters will become divergent. However, if we leave these parameters at their experimentally measured values, and express all other physical quantities in terms of them, then infinities will no longer appear. This is the essence of the renormalization program, which enables infinities to be removed and accurate physical quantities to be calculated.

We also argued that non-renormalizable theories will become renormalizable when the fuller details of their physics are known.

A dressed particle consists of its bare self and a Yukawa cloud around it. Depending on the spatial resolution of the measuring probe, it may see the whole dressed particle or only a part of it, so the effective parameters like the charge and the mass of the particle registered by the probe would vary with the size or energy of the probe. Such varying parameters are called running parameters. The behavior of running masses and running coupling constants are discussed both for the abelian and for the non-abelian gauge theories.

7. OUTLOOK

We have reviewed in the last six sections some of the exciting developments in particle physics. The importance of this field can be gauged from the number of Nobel Prizes awarded: more than thirty people mentioned in the previous sections are Nobel Laureates. Moreover, in order to keep this chapter to a reasonable length, we have been forced to leave out the important work along with the names of some other Nobel Laureates in the field.

As interesting as it is to understand the academic importance of the field, and what ultimately makes us tick at the most fundamental level, some might also want to know whether this deep knowledge could have any practical applications. This is an unanswerable question exactly because particle physics is such a frontier field. Like any other human endeavors, the efforts involved in the beginning are mostly in the ground-breaking. What structures one can build on it, and what application one can make of it, are up to the future generations to decide. This is the nature of pure science. An example might be useful to illustrate the point. We all know that Lasers (see Chapter I) are immensely useful nowadays, and with the advent of high temperature superconductors (see Chapter II), their applications will be increasingly important in the days to come. Yet the working of both of these depends critically on the coherence of a single quantum state. Without the discovery of quantum mechanics in the 1920's, it is doubtful that one could ever invent the Laser. One might be able to find the superconductors but one would never be able to understand enough of their properties to make efficient use of them. So in the field of high tech, something very practical often depends on something very fundamental, in this case the quantum theory. And, the practical application often comes years after the fundamental principles are discovered (see Chapters I and II for the history of these two topics).

Particle physics is presently at a cross road. What it faces is not unlike what physics went through at the end of the 19th century. Most of the major theories of classical physics had been developed by that time, every calculation one ever made was found to agree with the experimental data, and this led to statements by famous people to the effect that there was nothing more to do in physics except to measure a few physical parameters more accurately. How wrong they were! To be sure, there was already a small number of ominous signs which very few people paid much attention to. This includes the nature of black-body radiation, the photoelectric effect, the discrete atomic spectral lines, and radioactive decays, etc. It was precisely these phenomena that bursted open the gate to the modern era. To paraphrase Mark Twain, the demise of physics predicted at the end of the 19th century was greatly exaggerated.

Nowadays everything in particle physics that can be calculated using the Standard Model agrees also with the experimental findings. Again there are parameters in the Standard Model that require more accurate determination, but does that mean that we are at the end of the road? Judging from the experience at the end of the 19th century, one can almost certainly say no. What shape the future development will take nobody knows, but we shall spend this section discussing some of the speculations.

The final judge for these speculations is of course experiment. We will therefore discuss also some of the high energy accelerators which will become available in the next ten years or so that can be used to unravel the mystery.

As we know, the Standard Model of strong and electroweak interactions has been enormously successful. The t quark and the Higgs boson, however, are still missing, so the first item of business is to find them.

The top quark is estimated to have a mass somewhere between 100 to 200 GeV/c^2. It is being looked for at the proton-proton collider at Fermilab near Chicago. If it is not found there, we might have to go to a higher energy machine. The machine at Fermilab has a center-of-mass (c.m.) energy of 2 TeV (1 TeV=1000 GeV). The Superconducting Supercollider (SSC) being constructed in Texas has a c.m. energy of 40 TeV; it is scheduled to be completed towards the end of this century or at the beginning of the next. There are also a Soviet machine (UNK) at Serphukov, with 6 TeV c.m. energy, probably to be completed towards the end of this century, and the Large Hadron Collider (LHC) at CERN in Geneva, with 17 TeV c.m. energy, to be completed by 1997.

Then there is the Higgs boson H^0. According to the standard electroweak theory, we need at least one such particle, but more could exist. In the Standard Model this particle is assumed to be elementary, but there is nothing preventing it from being made up of a fermion antifermion pair. We know even less about its mass than the top mass, though the guess is that it is heavier than the top. However, since it is a boson, and as such can be produced singly rather than in pairs, as the tops must be, there is a chance for other machines to see it as well.

There are many kinds of high energy accelerators. Some problems are better solved by observations in proton-proton machines, because we can produce there a higher energy and a more intense beam of protons than anywhere else. However, these machines have their drawbacks in that the proton is a hadron, so most of the reactions will proceed through strong interactions. This is not always advantageous, for the great strength of the strong interaction produces a large abundance of hadrons which can easily overshadow whatever we are looking for. Depending on the physics, some experiments are more suitably done in an electron-proton machine, and others in an electron-positron machine. A big machine of the electron-proton type will hopefully begin taking experimental data in early 1992. It has a 30 GeV electron beam colliding with a 820 GeV proton beam, giving a c.m. energy of 314 GeV. It is located at Hamburg, and has the acronym HERA, standing for Hadron-Elektron Ring Anlage. As for electron-positron machines, there are currently several. The highest energy ones have a 50 GeV electron beam colliding with a 50 GeV positron beam, making a c.m. energy of 100 GeV. One of them is located at Stanford, California. It is called the Stanford Linear Collider (SLC). The other one, the Large Electron-Positron Collider (LEP), is located at CERN in Geneva and has much more intense beams. The c.m. energy of LEP is scheduled to be upgraded to 200 GeV sometimes in the mid 1990's. Besides these two, there are also other lower energy machines, including one with a c.m. energy of 60 GeV (TRISTAN) located at Tsukuba, Japan, and one with a c.m. energy of 5 GeV (BEPC) located at Beijing, China.

Besides these high energy leaders there is also a number of lower energy machines of various types. Then there are heavy ion machines, of which the largest will be built at the Brookhaven National Laboratory in Long Island, New York. This Relativistic Heavy Ion Collider (RHIC) can accelerate two protons up to 250 GeV each, and two gold nuclei up to 100 GeV each *per proton*. This machine is particularly useful in generating high temperatures in the colliding region needed to study questions of condensates, confinement, and chiral restoration.

All of these machines are engaged to find new particles, new phenomena, and new physics. They are also used to determine the mechanisms and the parameters used in the Standard Model. The latter may seem mundane but these two problems are actually related, in that one is able to pick out what is new only when one understands thoroughly what is old. New discoveries are made in two ways. The best way is to see the new phenomenon or the new particle directly. This often requires a high energy machine with sufficient energy to produce these particles. An indirect method would be to do a low energy experiment with great precision, and look for deviations from the predictions of the Standard Model. But to be able to do that we must know the Standard Model very well, and this includes knowledge of the precise values of its parameters, as well as being able to calculate with it accurately.

This leads to the problem of calculations. The only general and analytical technique available for calculations is the perturbation theory, using Feynman diagrams. This unfortunately is not suitable for strong interactions because of the strength of the coupling constant. Numerical methods, called *lattice gauge theory* calculations, are available and have given interesting results on the static properties of hadrons. However, computers will not be powerful enough in the near future to deal with high energy collisions in which many particles are produced. This drawback is not too serious when the momentum transfers are high, for then we can use asymptotic freedom and perturbation theory. It is when the momentum transfer is low that it is difficult.

Even when we are dealing with electroweak processes, our inability to calculate strong interactions also comes back to haunt us whenever hadrons are involved. This makes the developments of calculational techniques for strong interactions one of the most important theoretical problems.

The lack of calculational tools also affects our understanding of the condensates and phase transitions. Modern particle theory relies heavily on the idea of spontaneous symmetry breaking, but the formation of the condensates leading to it is often a non-perturbative problem which we cannot deal with from first principle.

Going beyond these immediate problems, there are other questions to be settled. For example, a large number of parameters are used in the Standard Model. These include the masses of every lepton and every quark, the four mixing parameters in the CKM matrix, the magnitude of the condensates, and the strengths of the various coupling constants. Can they somehow be calculated? From the discussions in Sec. 6, we do not expect so until a more fundamental theory at a higher energy scale becomes available.

Then there is the question of divergences. We have dealt with it successfully through renormalization, but renormalization is only a buck-passing device. It blames the inability to calculate certain parameters on our ignorance of the physics at ultra-high energies. Surely in a *fundamental* theory we cannot make that excuse, and the ultimate theory of Nature had better be finite. This theory may be deeply hidden at an impossibly high energy. The fact that quantum gravity points already to the huge scale of 10^{19} GeV also does not inspire any optimism that this ultimate theory will appear in the near future. One such candidate is the Superstring theory, which we will discuss in Sec. 7.5.

We will now discuss some of the other theoretical ideas and speculations. There is a large amount of these, so we can only pick out a few to give the readers a taste of the menu.

7.1 The Strong CP Problem and Axions

The vacuum is the lowest energy state. Classically it represents empty space. Quantum mechanically, there are vacuum fluctuations in which virtual particles can be spontaneously created and annihilated, like those shown in Fig. VI.7a, so the vacuum is by no means empty and devoid of structure. We shall now see that in a gauge theory the vacuum is even more bizarre and complicated.

Being the lowest energy state of a system, one might think of the vacuum as analogous to the ground state of an electron in an isolated atom. In this analogy, particle states would correspond to excited states of the atom.

If we use this analogy for the vacuum state in an ordinary quantum field theory, then the vacuum for QCD is like an electron in the ground state of a *crystal* made out of these atoms. This comes about because of the special mathematical structure of a gauge theory. Classically, the ground state energy of an electron in a crystal might differ somewhat from the ground state energy of an isolated atom, because of the force exerted by the nearby atoms on the electron. Quantum mechanically, the difference goes far beyond that. With quantum mechanical fluctuations, the electron attached to one atom would occasionally find itself with enough energy to jump over to another atom. This phenomenon is called *quantum mechanical tunnelling*. It allows an electron to wander around the crystal and interact with all of its atoms. As a result, the electrons from different atoms are inextricably mixed up, and their ground energies spread out into a contiuous band. In the case of the QCD vacuum, the states in the band are labelled by a single parameter called θ_{QCD}. The difference is, all the levels in the band now has the same lowest energy, so we do not really know which of them is preferred and which is the true vacuum we live in. In other words, there is an unknown parameter θ_{QCD} that labels our vacuum.

We know, however, that CP symmetry is violated if $\theta_{QCD} \neq 0$. Since all strong interaction processes conserve C and P, and the only known CP violation process occurring in nature is at a level of 0.1% of the strength of the weak processes, we conclude on phenomenological grounds that θ_{QCD} is either zero, or exceedingly small. But why should that be so?

Roberto Peccei and Helen Quinn came up with a mechanism in which θ_{QCD} can be *naturally* tuned to 0. In this mechanism, another weak isodoublet of Higgs field is introduced to enable the symmetry of the electroweak interaction to be enlarged to $SU(2) \times U(1)_Y \times U(1)_{PQ}$, in which the extra $U(1)_{PQ}$ gives the new symmetry. The symmetry transformation for this new symmetry can be exploited to rotate the parameter θ_{QCD} to zero. So by introducing this new Higgs doublet one has eliminated θ_{QCD} from being a meaningful physical parameter so that it can be set equal to zero. However, since this extra PQ symmetry has not been observed, it must be spontaneously broken. When this happens, a Nambu-Goldstone particle of zero mass and zero spin must be present, according to Nambu-Goldstone's theorem. This particle is called an *axion*. Actually, this multiple-vacuua structure in QCD gives a small mass to the Nambu-Goldstone particle, so it is not exactly massless. Moreover, the coupling of axions to matter is directly related to its mass. The axion has been looked for but not found. What this possibly means is that the axion has a very tiny mass, so that it has too feeble an interaction with matter to be detected. For this reason it has been dubbed an *invisible axion*; how invisible it is depends on its exact mass value. In any case, even if it is not detectable in the laboratory, there may be astrophysical implications which enable its existence to be inferred.

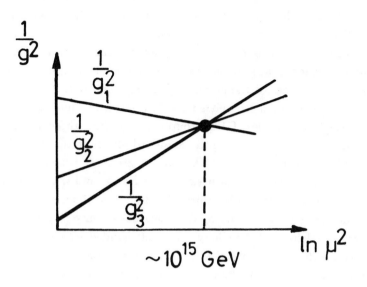

FIG. VI.8 *Running coupling constants g_1 (for $U(1)$), g_2 (for $SU(2)$), and g_3 (for $SU(3)$) as a function of probe resolution μ.*

7.2 Grand Unified Theories

There are three coupling constants in the Standard Model, one each for the three gauge groups $SU(3)_C$, $SU(2)$, and $U(1)_Y$. Let us denote these coupling constants by g_3, g_2, and g_1 respectively. We know that they all run with the energy μ of the probe (see Sec 6:3). A small probe with a short wavelength is a high energy probe. We shall write $g_i(\mu)$ $(i = 1, 2, 3)$ to indicate this running dependence.

For the abelian group $U(1)_Y$, $g_1(\mu)$ increases with increasing μ; for the non-abelian groups $SU(2)$ and $SU(3)_C$, it is the other way around, $g_2(\mu)$ and $g_3(\mu)$ decrease with increasing μ (*asymptotic freedom*). Moreover, $SU(3)$ is a larger group than $SU(2)$, so $g_3(\mu)$ changes faster than $g_2(\mu)$. At low energies and small μ, we know from experiments that $g_3 > g_2 > g_1$. So these three running couplings behave like what is sketched in Fig. VI.8. Note that it is g_i^{-1} and not g_i that appears in the ordinate.

We see from this graph that the three coupling constants are moving towards one another at large μ. It is therefore possible that they all come together at some μ. Indeed that happens at an energy of the order of $\mu = 10^{15}$ GeV, resulting in a common coupling constant $g = g_1 = g_2 = g_3$ at that μ*. The existence of this common coupling constant naturally led to the speculation that forces at such short distances are unified into one. Theories advocating that are called the *grand unified theories*, or GUT for short.

The Standard Model group has rank 2+1+1=4, *i.e.*, four additive quantum numbers. The simplest GUT group to unify these three symmetries is then a rank 4 group G containing the Standard Model group $H = SU(3)_C \times SU(2) \times U(1)_Y$ as a subgroup. The rank 4 group $G = SU(5)$ is such a candidate. See Sec. 4.4 for a discussion of the $SU(n)$ and the $SO(n)$ groups.

If we count only left-handed chiralities, there are 15 fermions per generation. In doing so we are assuming the neutrinos to have zero mass, so that they have only left-handed helicity. The left-handed fermions of the first generation are the left-handed u and d quarks, and the left-handed $\bar{u}$ and $\bar{d}$ antiquarks, each coming in three colors, making a total of 12. Then there are 2 left-handed leptons, and one left-handed antielectron, making a grand total of 15. The antineutrino has no left-handed helicity so it is not counted here.

In the standard model group H, the left-handed quarks belong to an $SU(3)_C$ triplet (quarks have three colors) and a doublet of $SU(2)$ (quarks are weak isodoublets); the left-handed antiquarks belong to a triplet of $SU(3)_C$ and two singlets of $SU(2)$ (see Sec. 5.3). The left-handed leptons belong to a singlet of $SU(3)_C$ (leptons have no color) and a doublet of $SU(2)$ (they are weak isodoublets). Finally, the left-handed antilepton is a singlet of both.

Several of these multiplets under H will combine into a single multiplet under the larger group G. In fact, these 15 particles are contained in two $SU(5)$ multiplets, one with multiplicity 5, and the other with multiplicity 10.

---

* The slopes of the lines in Fig. VI.8 depend on the masses of the existing particles. Recent high precision measurements of the coupling constants show that these three lines actually do not quite meet at the same point, as previous data indicated, if we take into account only the known particles. This is sometimes interpreted as an evidence for the existence of additional particles, yet undiscovered. There is no shortage of such candidates, supersymmetric particles (see Sec. 7.3) being one of the favorite one.

Both multiplets contain quarks as well as leptons.

$SU(5)$ has 24 gauge bosons, whereas the Standard Model group H has only 8+3+1=12. A spontaneous symmetry must occur in which the Higgs mechanism makes $24 - 12 = 12$ of them too heavy to have been seen. To do that we need new Higgs bosons but we will not discuss the details.

This symmetry breaking can occur anywhere equal or below this unification scale of 10^{15} GeV. When this happens the heavy gauge bosons will be expected to acquire a mass of the order of the symmetry breakdown energy. The heavier this mass, the shorter the range of the force mediated by this particle, and the less the effect it might have on low energy physics.

Every particle in a gauge multiplet is coupled to every other inside the same multiplet through a gauge particle. The fact that both quarks and leptons belong to the same multiplets of $SU(5)$ implies that they are coupled together. Since they are not coupled together by the Standard Model group H, they must be coupled through the heavy gauge bosons in $SU(5)$. Let Y be such a gauge boson, then $u \to e^+ + Y$ and $u + Y \to \bar{d}$ can both occur. By exchanging a Y between the two u's in a proton $p = uud$, it is possible for the following reaction to proceed: $p = uud \to e^+ \bar{d}d \to e^+ + \pi^0$. Through this mechanism a proton can decay into a π^0 and a positron.

The proton is known to be very stable, with a life time more than 10^{32} years. The decay rate computed from the mechanism above is also very small, owing to the heavy mass of Y. But is the decay rate small enough to allow the proton to live that long? Unfortunately, the answer is no, *if* the simplest-minded $SU(5)$ theory is used. This does not mean that $SU(5)$ is dead, because this problem can be bypassed if more complicated versions of the theory are adopted. However, the resulting theory will have no prediction that can be checked with present-day experiments. This example underlines the general difficulty of testing any theory at an energy scale much higher than what we can presently achieve. It is very hard to find observable consequences of these theories. When one is found, there are usually sufficient uncertainties in the theory to prevent us from making any firm conclusions.

Another popular GUT relies on the group $G = SO(10)$, which actually contains $SU(5)$, and therefore also the Standard Model group H, as subgroups. It has rank 5, so there is one more high temperature additive quantum number. However, it allows the incorporation of a massive neutrino. It has 45 gauge bosons; all left-handed leptons of one generation are now contained in a single multiplet of multiplicity 16. This includes the 15 in $SU(5)$, and a left-handed antineutrino which is allowed if the neutrino is massive.

Many experiments have been undertaken to search for non-zero masses for the neutrinos. There had been indications that they might be present but none of them have been definitely established. So it is still an open question whether the neutrinos have masses or not.

The GUT models unify and enlarge the Standard Model symmetry based on the group H. There are other models which enlarge the symmetry group H but do not unify it. As such the extra symmetry is allowed to surface at an energy scale much lower than the GUT scale of 10^{15} GeV, and will therefore hopefully be observable in the near future. We will not discuss them in any detail. A typical feature of an enlarged symmetry of this kind is that more fermions and other particles are often required to fill out the enlarged multiplets. Since these extra particles have not been observed, they must be heavy, but they should be observable at approximately the energy scale where the extra symmetry surfaces. A common risk of an enlarged symmetry is that with the extra particles, new forces will be present which may induce processes that are experimentally known not to occur, or to happen very rarely.

7.3 The Hierarchy Problem

If new physics is opened up at some high energy scale Λ, it is expected to have an effect even on low energy quantities. The size of the effect depends on the details of the problem. For proton decay, the effect is very small. For fermionic mass shifts, the effect goes like $g'^2 \ln(\Lambda/\mu)$, where g' measures the coupling to the new physics, and μ is the energy scale at which effects are to be calculated. The Higgs boson has spin zero, it cannot be 'protected' by any spin-related symmetry such as gauge symmetry and chiral symmetry, its mass(-squared) shift is large and is of the order of $g'^2(\Lambda/\mu)^2$.

For large Λ, such as those encountered in GUTs, this could turn out to be a very large number — much larger than the observed Higgs mass. Since renormalization is always necessary, one might simply tune the bare mass parameter m_0^2 to yield whatever mass one observes for the Higgs boson. But that will appear very unnatural if the observed mass is much much smaller than the shift, for we must then arrange to have the subtraction of two big numbers to get the small observed mass. This calls for a lot of fine tuning, and appears to be very unnatural. This problem goes under the name of a *fine tuning problem*, or a *hierarchy problem*.

There are two ways to avoid this unnaturalness. The first is to assume the Higgs to be a composite rather than an elementary particle. In that case, the quadratic mass-shift formula mentioned above will no longer be valid, and the problem disappears.

This seems a natural solution, for compositeness is just what happens to the Higgs field in superconductivity. Before the microscopic theory was devised by Bardeen, Cooper, and Schrieffer (BCS), a Higgs version for superconductivity was suggested by Landau and Ginzburg. What that Higgs turns out to be, in the BCS theory, is a pair of electrons (*Cooper pair*) weakly bound together by the phonons (lattice vibrations). See Chapter II for more details. Following this lead, one might think that the same could happen to the Higgs in the electroweak theory. There are a number of suggestions as to what fermion one should use to substitute the Cooper pair of electrons in this scenario, but as yet there are no consensus.

The second mechanism to cure the hierarchy problem relies on *supersymmetry*, or SUSY for short. It postulates the existence of a new symmetry between bosons and fermions. If the symmetry were exact, bosons and fermions of *equal masses* must be present in pairs. An electron would have a spin 0 partner called a *selectron*, and a quark would have a spin 0 partner called a *squark*. Similarly, the photon would have a spin $\frac{1}{2}$ partner called the *photino*, etc. Since these SUSY partners, if they exist, must have heavier masses than the existing particles to avoid detection, SUSY must be broken. The energy at which it is broken would give an estimate of the masses of the SUSY partners. This energy scale is not known, but if it is too small, it will not lead to a heavy enough mass for the SUSY partners to have avoided detection; if it is too large, then it would not have helped to solve the hierarchy problem which SUSY is designed for. Something of the order of 10^3 GeV has been suggested for this scale. If SUSY partners do exist, most people believe that they should be found at the SSC, if not sooner.

The reason why SUSY helps the hierarchy problem comes from a simple property of the Feynman diagrams: bosonic loops and fermionic loops come in with opposite signs. If exact SUSY is present, a perfect cancellation exists between fermion loops and boson loops, leading to the absence of any mass shifts. In particular, the Higgs mass shift will be zero. Even in the presence of a broken SUSY, this shift could still be small and the hierarchy problem would thereby be avoided.

This sign difference between a boson loop and a fermion loop may sound very mathematical. The additional -1 sign in a fermion loops comes from the statistics of fermions, because the wave function of identical fermions changes a sign when two fermions are interchanged.

In the case of a one-loop vacuum diagram, this minus sign can be seen directly from the Dirac theory of electrons. We have not discussed this theory in the previous sections, but it represents one of the greatest theoretical triumphs in the twentieth century. Let us now sketch what it is.

By the late 1920's, non-relativistic quantum mechanics had been worked out. Paul Dirac then went about trying to combine that with the theory of special relativity. He succeeded in writing down an equation describing the relativistic motion of an electron, an equation which came to be known as the *Dirac equation*. This equation however has a very unexpected feature which presented a serious difficulty. Dirac himself solved the difficulty a year later (1929), and in the process, he predicted successfully the existence of the positron (antielectron) several years before it was experimentally found by Carl Anderson in 1932. The same argument also predicts the existence of an antifermion for every fermion of spin $\frac{1}{2}$.

The solutions of the Dirac equation for a mass m electron allow a range of energies from $+mc^2$ to ∞. This is just as it should be. When the electron is at rest it has an energy mc^2. When it moves close to the speed of light its energy will approach infinity.

What was surprising and not understandable, was that the electrons are also allowed energies from $-mc^2$ to $-\infty$. How can a free electron have negative energies, and possibly $-\infty$ at that? Physical systems always seek to occupy a lower energy state; if $-\infty$ is available, every electron in the world would tumble down to this deep hole and be lost to us forever. The whole world would collapse, and this is certainly not right.

Dirac himself gave a way out of this difficulty a year later. He said, suppose for some reason all the negative energy states were occupied. Since the electrons obey Fermi statistics and the *Pauli exclusion principle* (see Appendix B), it is not possible for two electrons of the same spin to have the same energy and momentum. For that reason if every negative-energy state is already occupied, no positive-energy electron can find its way down there, and the world would remain stable.

This sea of negative-energy electrons is now known as the *Dirac sea*. In this picture, the vacuum state, or the ground state of the world, is the state with the Dirac sea completely filled.

If there are so many (negative energy) electrons in the great empty space in front of us, how come we do not keep on bumping into them when we walk? We do not, because for all intents and purposes these electrons are transparent to us on account of Pauli's exclusion principle. We can meaningfully collide with something only when energy and momentum can be exchanged between us and the object we collide with, but because of the full occupancy of the Dirac sea, there is no way for the electrons in the vacuum to lose or gain energy, unless the energy gained by one of them is at least as large as $2mc^2$. We can certainly not provide that kind of energy as we walk through space.

Now suppose an amount of energy larger than $2mc^2$ can be imparted to an electron down in the Dirac sea (whose energy is less than $-mc^2$) to elevate it to a positive- energy state (whose energy has to be larger than $+mc^2$). What is left behind in the sea is a hole. If we think of the vacuum with the negative-energy Dirac sea filled as having all additive quantum numbers 0, and that q is any one of the original additive quantum numbers of the negative-energy electron that got promoted upstairs, then the hole, or the *absence* of this electron, must have a quantum number $-q$. Thus a hole behaves like a real particle with positive energy and a charge opposite to that of an electron. In other words, it is the antiparticle of the electron, or a positron.

The second consequence of this picture of the vacuum, in which all negative-energy states are fully occupied, is that the ground state energy is negatively infinite. This is the same conclusion that one would come to if we calculate the one-fermion loop vacuum diagram Fig. VI.9b.

The simplest one loop vacuum Feynman diagrams are those shown in Fig. VI.9, which do not have any vertices. The one in diagram (a), with

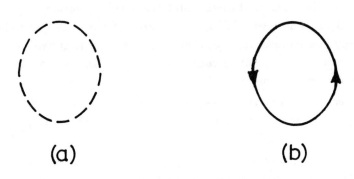

FIG. VI.9 *Simplest vacuum diagrams involving bosons (a) and fermions (b).*

a broken line, represents the contributions from bosons to the vacuum energy; the one in diagram (*b*), with a solid line, represents the contributions from fermions to the vacuum energy. Both of these loops are divergent (see Sec. 6.1); the boson loop has a positive sign, and gives $+\infty$; the fermion loop has a negative sign, and gives $-\infty$. The latter agrees with the result obtained in the last paragraph from the Dirac sea argument. The origin of the $+\infty$ for the bosonic vacuum loop comes from the so-called *zero point energies*. We will not go into it here.

If SUSY is present, the net result of these two loops in Fig. VI.9 is zero. Moreover, even when higher loops with vertices and interactions are included, the net contribution would still be zero. This illustrates the SUSY cancellation mechanism and how SUSY abhors changes. It is the same mechanism that renders the mass shift of the Higgs boson zero in the presence of an exact SUSY.

7.4 Kaluza-Klein Theories

A large number of conserved internal quantum numbers are encountered in particle physics. They are related to dynamical symmetries (Sec. 4) operating in an internal space. What is this internal space? Is it simply a mathematical construct to help us to visualize the symmetry operations, as we asserted in Sec. 4, or could it be something more physical?

One might argue that the internal space could not be physical, for everybody knows that we live in a space of three spatial and one time dimensions. To say that there are real internal spaces beyond that would imply that there are extra dimensions beyond the usual four. We have never seen that except in science fictions. So how could an internal space be real?

It could, if the size of the internal space is much smaller than what we have been able to measure, for then we would never have detected it. From the current limit of measurement, anything smaller than 10^{-20} m will certainly do.

The first and probably the most important internal quantum number is the electric charge. It was suggested by Kaluza and Klein, in the 1920's, that charge is indeed associated with a small internal space. By writing down a general relativistic theory in a five dimensional space, with the usual four-dimensional spacetime plus a fifth internal dimension, they were able to unify four-dimensional electromagnetism and gravity, after a fashion. The charge degree of freedom represents motion in the internal space; the extra gravitons that would have been present in a five-dimensional space become the photon and an extra spin-zero particle in the four-dimensional spacetime.

This idea was later generalized to internal spaces of higher dimensions to allow for more varieties of internal quantum numbers, and unification of gravity with theories beyond electromagnetism. Such higher-dimensional theories are nowadays known as *Kaluza-Klein theories*.

We have mentioned in Sec. 4.5 the far reaching consequences of the discovery of parity violation. One of them is to allow the formulation of the current electroweak theory, which is a *chiral theory*. This term means that the left-handed particles and the right-handed particles in the theory do not always have the same internal quantum numbers. It turns out that a Kaluza-Klein higher-dimensional gravitational theory will generally not be chiral. If we start with a higher-dimensional theory, and expect it to yield a four-dimensional chiral theory, then we may have to make it not purely gravitational. Some internal quantum numbers and Yang-Mills interactions would have to be there already in the higher-dimension. A theory of this type is the *superstring theory*.

7.5 Superstring Theories

A superstring theory is a string theory with supersymmetry. A string theory is a theory in which the elementary constituents are not point objects like an electron, but are one-dimensional objects like a string.

The most promising string theory is called the *heterotic string* theory, in which the string is a little closed loop. The size of the loop is very small, the corresponding energy scale is probably of the order of the Planck mass. The theory is originally formulated as a ten-dimensional theory, with a rank 16 symmetry group on top. The internal consistency requirements of the theory dictates the dimension ten, as well as the possible rank 16 symmetry groups. Two such groups are allowed, one is related to $SO(32)$, and the other is related to an *exceptional group* $E_8 \times E_8$ which we did not discuss in Sec. 4.4. This extra symmetry group makes it possible to bypass the difficulties encountered by the non-chiral nature of the classical Kaluza-Klein theories. The string theory does not specify the size of the ten dimensions, so we can take four of them to be our spacetime, and six others to be small and be the dimension of the internal space.

An important difference between a string and a point particle is that the string can admit all sorts of excitations, in the form of rotations and vibrations, each of which corresponds to a different particle as we know them. The higher the excitation energy, the larger the mass of that particle. Although a string accommodates a huge number of particles, almost all of their masses are so heavy that they will never be produced in our accelerators. Only the few that have small masses can ever be directly observed. Nevertheless, because of the enormous number of these heavy particles, their combined (virtual) contribution to low energy phenomena is not negligible. The interesting thing is, this contribution also conspires to make the string loops finite! If one traces through the origin of this conspiracy, one finds that it is due to the enormous amount of local symmetry that the string enjoys.

Among the massless particles in a string, there is a graviton and a large number of gauge bosons. Thus this theory possesses some of the attributes of an ultimate theory, in that it contains and is capable of unifying the four fundamental forces, and more importantly that it is also a finite quantum field theory! Among other things, even quantum gravity is finite here.

The problem is, there are many versions of the superstring theory, depending largely on the characteristics of the six-dimensional internal space which we are so ignorant of. Often the theory contains too many gauge bosons, as well as many other unwanted particles. It is conceivable that a heterotic string without these extraneous particles can be found, but nobody has produced one yet.

It is possible that the heterotic string theory is a correct theory when the energy density of the universe is at the Planck mass scale, and as the universe expands and cools various phase transitions and accompanying symmetry breakings take place. If we knew how to do these extremely difficult dynamical calculations, we might end up with a unique string theory at the present temperature that agrees with the Standard Model. This calculation, however, is so difficult that it is unlikely that it can be done in the forseeable future. This is like, but immensely more difficult than, the following problem. Can you predict superconductivity when you are given the forces between electrons and/or nuclei to be Coulomb forces, but otherwise knowing not even the theory of normal metals or solids?

The string theory, if it is correct, may be an important theory for the discussion of the early universe, when the energy density was high and many string modes could be excited. We have not mentioned the formulation of the string theory, but it is not a theory formulated in our usual spacetime, like a local quantum field theory. For that reason a thorough understanding of the string theory might even help to answer the question of the origin of our spacetime. See Sec. 7.4 of Chapter VII on Cosmology for a discussion of that.

Summary

We have outlined in this section some of the immediate tasks, some of the long term problems, as well as some of the theoretical problems and speculations in high energy physics.

CHAPTER VII

COSMOLOGY

CONTENTS

1. HUBBLE'S LAW

Modern cosmology is founded on the 1929 discovery of Edwin Hubble that all galaxies are flying away from us.

This *Hubble's law* states that a galaxy at a distance s from us recedes with a velocity $v = Hs$. The proportionality constant H is called *Hubble's constant*. Its value, $H = h_0 \times 100$ km/s per Mpc, is not very well measured, with h_0 known only within the range $0.4 < h_0 < 1$. Nevertheless, for the sake of illustration, we shall set $h_0 = 0.7$ when we need to quote a number that depends on h_0. The unit 'Mpc' stands for *megaparsec*; it is a distance unit which we will discuss below.

Hubble's law means that a galaxy $s = 1$ Mpc away recedes from us with a velocity $v = 70$ km/s, and a galaxy $s = 2$ Mpc away recedes from us with a vecolcity $v = 140$ km/s, etc. The more distant the galaxy, the faster it runs away. Since this is the most important discovery in modern astronomy, it is worthwhile to spend some time finding out how this law was arrived at.

To establish the law, both the distance to a galaxy and its recession velocity have to be measured. We shall discuss separately how these are done.

1.1 Astronomical Distances

Distances between cities are measured in kilometers (km), but this is too small a unit for astronomy. Astronomical distances are so large that even light, which has a speed of $c = 3 \times 10^5$ km/s, is the fastest object in the universe, and can go around the earth more than seven times in a second, must travel 500 seconds from earth to reach the sun, and some four years to reach the nearest star. The *distance* covered by light in one second is called one light second, that covered by light in one year is called one *light year*. While the nearest star is four light years from us, the center of our Milky Way galaxy is some twenty eight thousand light years away. In ordinary units, a light year is equal to 9.46×10^{12} km, or in words, almost ten million million kilometers.

Distances of nearby stars are measured by triangulation. As the earth moves in its orbit, the position of a *nearby star* against the background of distant stars appears to shift, for much the same reason that nearby scenary appears to move against distant mountains when you look out of the window of your car. If the star is at a distance such that the shift from the average position is 1 second of arc each way, then the star is said to have a parallax angle of one second, and the distance of the star is defined to be 1 *parsec*. This distance turns out to be 3.26 light years. A thousand parsecs is called a kiloparsec (kpc), and a million parsecs is called a megaparsec (Mpc). It is this last unit that has been used in the value of the Hubble constant above.

The more distant the star, the smaller the parallax angle, so after some 30 pc or so, the parallax angle will become too small for this method of measuring distances to be useful. Other methods are required.

One such method was discovered by Henrietta Leavitt and Harlow Shapley at the beginning of the twentieth century.

During his exploration in 1520, Magellan found two luminous 'clouds' in the sky of the southern hemisphere, two 'clouds' that turned out to be two nearby irregular galaxies, called the *Large Magellanic Cloud* (LMC) and the *Small Magellanic Cloud* (SMC). In 1912, Leavitt found in the SMC twenty five variable stars of the type called *Cepheid variables*. A variable star is a star whose brightness changes with time. A Cepheid variable is a kind of variable star which has a fixed period of brightness variation, and this period usually falls within a range of a few days to a month or more. They are called Cepheid variables because a typical variable star of this type is found in the constellation Cepheus.

By studying these Cepheid variables in the LMC, Leavitt found that the period of variation of these twenty five Cepheid variables follows a simple law. The period of a star is related in a definite way to its average brightness observed on earth: the longer the period, the brighter the star.

The brightness of a star as observed a distance R away is not its intrinsic brightness. The light emitted by a star is spread over a spherical shell whose area grows with R^2, thus making the (apparent) brightness per unit area seen a distance R away fall like $1/R^2$. If we know the distance of the star from earth and its apparent brightness as seen here, then we can use this relation to compute its *intrinsic brightness* (technically known as the *absolute luminosity*). This is important because it must be the intrinsic brightness that is controlled by physics, and not the *apparent brightness*.

At the time of Leavitt's discovery, the distance to the SMC was not known. In 1917, Harlow Shapley was able to use a statistical method of stellar motions to estimate this distance, as well as the distances to some globular clusters containing Cepheids. The distance to each individual star in a galaxy or a globular cluster is different for different stars, but since a galaxy or a globular cluster is very far away from us, the difference in distances between individual stars within a galaxy or a globular cluster is negligible in compasrison, so we may treat them all to be at some average distance to the galaxy or globular cluster.

This distance measurement allowed Shapley to convert the apparent brightness of these stars to their intrinsic brightness. Like Leavitt, he plotted the period of these Cepheids against their brightness, but this time he used the *intrinsic* brightness. He found in this way that the period of variation of a Cepheid variable bears a *definite* relationship with its *intrinsic brightness*, whether the star is in one galaxy or another globular cluster. In other words, he found this to be a physical law independent of the location of the star. This celebrated law is called the *period-luminosity relation* for the Cepheid variables.

Once established, this period-luminosity relation can be used as a yardstick to measure the distance to a new Cepheid variable. What one has to do is to measure its period as well as its apparent brightness on earth. From the period one gets its absolute brightness through this relation. Knowing now both the absolute and the apparent brightness, one can calculate its distance R.

In 1923, Hubble found twelve Cepheid variables in a nearby nebula in the Andromeda constellation. This nebula is actually a galaxy and is now called the *Andromeda galaxy*, or by its Messier catalog number M31. He also found twenty two other Cepheid variables in another nebula, M33. In this way, he determined both of their distances to be several hundreds of kiloparsecs away, placing them well outside of the confines of our Milky Way galaxy, and settling the debate once and for all in favor of these objects being distant galaxies in their own rights, rather than some luminous clouds within our own galaxy.

The distances to nearby galaxies can be determined by the brightness of their Cepheid variables. For remote galaxies whose individual stars cannot be clearly identified, this method is useless. Instead, one has to rely on the assumption that galaxies (or globular clusters) of the same type have approximately the same intrinsic brightness. This assumption has been checked for nearby galaxies and is found to be true. Using it, one can determine the distance to a remote galaxy by its apparent brightness and its intrinsic (universal) brightness, as explained before.

1.2 Velocity Measurements

The velocity of an object along the line of sight can be measured using a physical phenomenon known as the *Doppler effect*. This is what highway police use with a radar to catch speeding cars. It is also why the whistle of an approaching train has a higher pitch, while the whistle of a departing train has a lower pitch.

When a train moves away from us, it stretches the sound wave between the train and our ear, thereby elongating its wavelength. An approaching train does the opposite and compresses the wavelength. Since a longer wavelength gives a lower pitch sound, and a shorter wavelength gives a higher pitch sound, that is why the pitch of the whistle behaves the way described above. The faster the velocity of the train, the more the stretch or the compression of the wave, and the more the change of the pitch occurs. This change of wavelength, or frequency, with the speed of the train is the Doppler effect.

A similar effect also occurs on light waves, except that this time what we experience is not the musical sensation of pitch, but the visual sensation of color. A shift towards a longer wavelength in the visible spectrum moves light towards the red color, and is thus called a *red shift*. A shift towards a shorter wavelength moves light towards the blue color and is therefore called a *blue shift*. Again, the amount of the shift indicates the velocity of the object; red shift for recession, and blue shift for approach. It is by measuring such shifts in the light reaching us that the velocities of stars and galaxies are determined.

To make the method useful, we must know the original color of the light in order to figure out how much it has shifted. Since we are unable to go to the star to find out the original wavelengths of its light, we must resort to other means. For this we make use of a remarkable property of the atoms:

the existence of *spectral lines.*

Just as finger-prints and dental records can vary from individual to individual, atoms and molecules can also be distinguished by their 'finger-prints' — now in the form of their characteristic spectral lines. An object, when it is heated, emits two kinds of electromagnetic radiation. One has a broad spectrum of wavelengths; it reflects the ambient temperature of the surrounding but not the identity of the constituent atom (see App. C for discussions of radiations of this type). The other kind of radiation appears only at discrete and fixed wavelengths characteristic of the atoms present. When analysed through a spectrograph, this kind of light appears as a series of (spectral) lines on the photographic plates. This radiation is caused by the electrons inside an atom jumping from one discrete orbit to another. The wavelengths characterize the energy differences of the orbits, which characterize the atom. It is for this reason that spectral lines can be used as a tool to identify atoms.

By looking at the light coming from a star or a galaxy, two interesting points were discovered.

First, except for possible Doppler shifts, the spectral lines of light from distant stars are identical to the spectral lines of the appropriate atoms on earth. This must mean that physics on a distant star is identical to physics on earth! This is an extremely important discovery, for it allows us to apply whatever knowledge of physics we learn here to other corners of the universe.

Secondly, the spectral lines from many galaxies are red-shifted (but seldom blue-shifted). This means that these galaxies are flying away from us. The velocity of this flight can be determined by measuring the amount of the red shift.

Between 1912 and 1925, V.M. Slipher measured the aforementioned red shifts for a number of galaxies. Using Slipher's compilation of redshifts of galaxies, the receding velocities can be computed. With his own measurements of distances, Hubble was then able to establish his famous law that the receding velocity increases linearly with the distance of the galaxy.

This law revolutionizes our thinking of the universe, as will be discussed in the rest of this chapter.

Summary

The distance to a star can be measured by triangulation if the star is nearby, and by using the period-luminosity relation if it is farther away, pro-

vided the star is a Cepheid variable. Other methods of measuring distances are available but not discussed in the text. The distance to a nearby galaxy can be measured by the distance to its Cepheid variables, but the distance to a remote galaxy can be determined only by assuming all galaxies to have the same intrinsic brightness. The velocity of an object approaching us or receding from us can be measured by the amount of blue shift or red shift of its spectral lines. Using these two kinds of measurements, Edwin Hubble established the most important law in modern cosmology, that galaxies are receding from us with velocities proportional to their distances.

---

2. THE BIG BANG

Since the galaxies are flying away from us, they must have been closer to us a million years ago, and closer still a billion years ago. Running this 'movie' *backwards*, at some point all these galaxies must come together to a point. That must have been the beginning of the universe!

In fact, Hubble's law $s = vH^{-1}$ is just like the elementary physics formula $s = vt$, which gives the distance s covered in time t by an object travelling with the speed v. Comparing this with Hubble's law, we see that we can interpret all the galaxies to have originated from a single point at a time $t = H^{-1}$ ago. At that time, a big explosion (the *big bang*) occurred, throwing everything radially outwards. The galaxy with a velocity v will then end up at the distance s it has today.

Actually, this formula overestimates the *age of the universe*. This is because the velocity v of a galaxy has not always been the same. The gravitational pull between the galaxies holds them back to slow down the expansion of the universe. This means that the expansion rate was faster in the past, and so it does not take quite as long as $t = H^{-1}$ after the explosion to reach the present stage. We will come back to a more detailed discussion of this slowdown in Sec. 3.

Given that 1 Mpc=3.08×10^{19} km, H^{-1} works out to be 14×10^{9} yrs if $H = 70$ km/s/Mpc. This gives a ballpark estimate for the age of the universe if we discount the slowdown.

From Hubble's law every galaxy is flying away along its line of sight. This suggests that we are at the center of the universe! Are we? Well, yes, if it helps our ego to think so, but we had better realize that everybody else in the other galaxies can claim the same honor for themselves as well.

There are two standard analogies to help us visualize how everybody can claim to be at the center of the universe. The first is to imagine the galaxies to be raisins in a raisin bread. The dough rising in the oven models the expanding universe. Pick a raisin to be our Milky Way galaxy. As the dough rises, all the other raisins fly away from us just as Hubble said that they would. However, if we put ourselves at any other raisin, the same would still appear to be true. If we are at the center of the universe, then so is everybody else.

The second analogy is to imagine our universe to be on the surface of a balloon. The expansion is achieved by inflating the balloon. If we are now a speck of dust on the surface and other galaxies are other specks of dust, then these other specks will all fly away from us as the balloon is inflated. This analogy allows us to see the curvature of the universe and the possibility of having a finite universe without boundaries. Its disadvantage is that this analog universe is only two dimensional.

The fact that we are not the only ones at the center of the universe may come as a blow, but then the blow had really been struck long time ago. In ancient times it was really easy to be convinced that we were at the center of the universe. After all, the sun and the moon, the stars and the planets, all appear to revolve around us in the sky. Copernicus insisted that this was only an illusion, and he turned out to be right although he suffered greatly for this claim at the time. Now times are different. This Copernican principle, that nobody in the universe is more privileged than the rest, and that the universe looks the same in all directions to an observer anywhere*, is now universally accepted as the *cosmological principle*. This principle is supported by the isotropy of the expansion of the universe as seen on earth, by the appearance of the same physics everywhere as revealed by the spectral lines from distant stars, and by the existence and the isotropy of the cosmic background radiation which we shall discuss in Sec. 4. The principle is also consistent with the picture of the big bang discussed at the beginning of this

---

* This is true only when we average our observations over a large enough scale. Locally, there are stars and galaxies, and the universe will not look homogeneous and isotropic.

section.

The idea that the universe started from a big bang is certainly a very revolutionary one. Many interesting questions immediately come to mind. What caused the explosion? What happened before the explosion? How was the universe created? Where do you put it when it is created? Why is the universe created in a three dimensional space? How did it evolve after creation? And how is it going to end up? We do not have all the answers, but we shall discuss some of them in the rest of this chapter.

The big bang beginning of the universe solves an old paradox put forth by Heinrich Wilhelm Olber in 1823. He realized that the night sky should not have been dark if the universe was infinite in extent, which he assumed. In that case, no matter in which direction you look, you will encounter a star sooner or later. If the star is far away its light may be very faint, but then there should be more stars at that distance within the same solid angle to compensate exactly for this individual faintness. As a result, the night sky should appear to be very bright rather than very dark.

A suggestion had been put forth to resolve the paradox, which relies on the interstellar dust to absorb the distant starlight. However, it turns out that the amount of dust present is not enough to give rise to the necessary absorption required to have a dark sky.

The modern resolution of *Olber's paradox* relies on the big bang beginning of the universe. The universe has been around for only some ten to twenty billion years or so, and not forever. Therefore there is no starlight coming in from more than ten to twenty billion lightyears away, and this is probably enough to make the sky dark. The red shift might also help the argument a bit, but probably not too much.

Summary

The universe started out from an explosion, known as the big bang, some ten to twenty billion years ago. On a large scale, the universe is homogeneous and isotropic, and is believed to obey the cosmological principle. This picture of the universe creates many interesting philosophical and physical problems, which will be discussed in the rest of this chapter. Olber's paradox for a dark night sky can probably be resolved by the big bang.

3. THE FATE OF THE UNIVERSE

3.1 Open, Critical, and Closed Universes

How is the universe going to end up?

There are three possible scenarios. The slowdown by the gravitational pull may not be sufficient to stop the expansion; it may be just enough to stop the expansion after an infinitely long time; or, it may be so large to be able to stop the expansion after a finite amount of time, after which the universe will reverse itself and start contracting under the continuing gravitational pull. A universe of the first kind is called an *open universe*, a universe of the second kind is called a *critical universe*, and a universe of the third kind is called a *closed universe*. Which is ours?

There are two ways to answer this question by observations. We can either measure directly how fast the universe is slowing down, or we can determine the strength of the gravitational pull and calculate the slowdown from that. The strength of the pull in turn is determined by the average density of the universe.

Let us consider the first method. We cannot determine how fast the universe is slowing down by making an observation today and another one ten years from now. The time span of ten years is simply too short for the expansion rate to have any measurable change. What we need is a cosmological time scale between measurements, say in the hundreds and thousands of millions of years. We definitely do not live that long to repeat these measurements, but we do not have to, thanks to the cosmological principle. The point is that the light from a distant galaxy reaching us today actually left there a long time ago. The further the galaxy is, the further back in time we see. So galaxies at different distances act like time machines, in that their observed recessions give the expansion rates of the universe at various times. In this way we can in principle compare the rates at different times to decide how fast the expansion is slowing down. Unfortunately, the present accuracy of these measurements are far too poor to tell us the fate of the universe by this method.

Let us now turn to the other approach, namely measuring the average energy density ρ of the universe. Let us call a density *critical* if it produces a critical universe. Densities larger than the critical density would produce a

closed universe, and densities smaller than the critical density would produce
an open universe.

The critical mass density is $\rho_c/c^2 = 0.92 \times 10^{-35}$ g/m^3, assuming $h_0 = 0.7$. This is about ten million-million-million-million-millionth the density of
water. This is equivalent to having 0.14 suns in a volume of a cubic mega-
parsec, or about five hydrogen atoms per cubic meter.

What is the actual density of the universe? Unfortunately this is a very
hard question to answer because of the local density fluctuations and because
of the presence of dark matter. The average density of the earth is 5.5 g/cm^3,
the average density of the sun is 1.4 g/cm^3, and the density of water is 1
g/cm^3. If we ask about the density of the solar system, then it is way down.
The mass of the solar system is concentrated almost completely in the sun,
but the volume of the solar system is large. For the sake of illustration, let us
suppose the solar system includes all space out to Pluto. Then the density of
the solar system works out to be 0.5×10^{-17} g/m^3, or still some 10^{17} times
the critical density. Now if we go to the scale of our galaxy, then there are
even more empty spaces between stars, and the density will decrease to 10^{11}
times the critical density, assuming we take the galactic mass to be 10^{12} that
of the sun and the radius to be 15 kpc. This incidentally is the density of 14
billion solar masses per cubic mega-parsec. From the sun to the solar system
to our galaxy, the ratio of the present density to the critical density decreases
from 10^{29} to 10^{17} to 10^{11}. If we now go to the scale of clusters of galaxies,
then the density of the *luminous* matter becomes roughly 3×10^{-37} g/m^3, or
only a few percent of the critical density.

The universe is also known to have non-luminous or *dark matter* which
exerts a gravitational pull on stars and galaxies just like the luminous matter.
There are two kinds of dark matter: *hadronic*, and *non-hadronic*. Nucleons
(neutrons and protons) and other particles subject to nuclear forces are called
hadrons. Hadronic matter refers to matter consisting of hadrons, and non-
hadronic matter refers to matter that cannot be affected by the nuclear forces
(see Sec. 1 of Chap. VI for a more detailed discussion). The total amount of
hadronic matter in the universe, luminous and dark, can be estimated from
data on primordial nucleosynthesis (see Sec. 6). This corresponds to having
0.4 to 0.04 hydrogen atoms per cubic meter, a number which is still quite a
bit below the critical density value of five hydrogen atoms per cubic meter.

If this were all, the universe would be open. There are however reasons (see Sec. 8) to suggest that the universe is critical. If that is the case, then there must be plenty of non-hadronic dark matter out there waiting to be discovered. There is no shortage of suggestions as to what they might be, but unfortunately there is as yet no experimental data to confirm any of these.

3.2 Gravitational Physics

We shall study in this subsection in greater detail how gravity affects the expansion of the universe. The discussion is quite technical and somewhat involved, and it is really not needed for the rest of the chapter. It is put here to provide more quantitative information for those who want to know, and to provide additional knowledge to support some of the statements made is Sec. 5. The readers who do not wish to go into such details can skip directly to Sec. 4.

Combining the cosmological principle with general relativity, a model of the universe (called the Friedmann-Robertson-Walker universe) can be constructed. It goes as follows.

The matter and energy in a homogeneous and isotropic universe of size R can be thought of as a 'fluid' confined to a sphere of radius R, a fluid which is characterized by its energy density ρ and its pressure P. The pressure P is generally regarded as a known function of the density ρ (this relation is called the *equation of state*), determined separately from the detailed structure of the fluid. With this, the universe is described by two functions of time t: its radius $R(t)$, and its energy density $\rho(t)$.

These two functions are determined by two equations. For want of a better name, we shall refer to these two equations as the *'force equation'* and the *'energy equation'*. These two equations are derived from general relativity. The force equation is the equation given by the 'time-time component' of Einstein's equation, and the energy equation can be derived from the 'space-space component'. However, one can obtain a rather good intuitive understanding of them by using elementary physics arguments.

To do so, single out a small chunk of matter on the surface of the sphere with radius R. (see Fig. VII.1).

This chunk is attracted to the rest of the universe by gravity, so it possesses a negative potential energy. This potential energy is given by

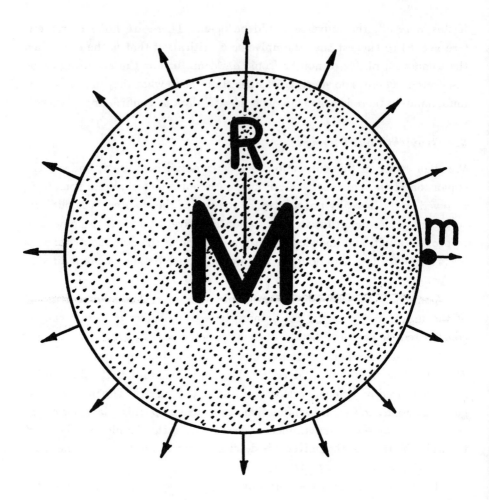

FIG. VII.1 *A ball of radius R and mass M depicting the universe, at the edge of which a small mass m is placed.*

$-G_N M m / R$, where $G_N = 6.67 \times 10^{-8}$ cm^3/g/s^2 is the *Newtonian gravitational constant*, M is the total mass of the universe and m is the mass of the small chunk on the surface of the sphere. The minus sign in the potential energy comes about because the gravitational force is attractive, so that energy has to be supplied to pull the small chunk of matter away from the big mass M.

The volume of the sphere is given by $\frac{4}{3}\pi R^3$, so the mass of the universe is given by $M = \frac{4}{3}\pi R^3 \rho / c^2$, where ρ is the *energy* density of the universe. This means that the potential energy can also be written as $-\frac{4}{3}\pi G_N \rho R^2 m / c^2$.

As the universe expands, the small chunk of matter on its surface moves outward and this motion acquires for it a positive kinetic energy. This kinetic energy is $\frac{1}{2}m\dot{R}^2$, for the radial velocity of the chunk is $\dot{R} \equiv dR(t)/dt$.

The *energy equation* is simply the equation expressing the conservation of energy of this mass chunk: the sum of its kinetic and its potential energies must be the same at all times. Denote the total, time-independent, energy of the chunk by $-\frac{1}{2}mc^2 k$, where k is a parameter. Then the energy conservation equation is $\frac{1}{2}m\dot{R}^2 - \frac{4}{3}G_N \rho R^2 m / c^2 = -\frac{1}{2}mc^2 k$. Cancelling the common factor m, this equation can be written as

$$(\dot{R}/R)^2 = \frac{8}{3}\pi G_N \rho / c^2 - kc^2 / R^2.$$

This is the *energy equation*. The left hand side of this equation is sometimes written as $H(t)^2$, for it is nothing but the square of the Hubble constant at time t.

Energy units can be chosen so that k is either $+1$, 0, or -1, depending on whether the total energy of the chunk is positive, zero, or negative. These three cases correspond respectively to a closed, a critical, and an open universe. Note that in order to get the right answer agreeing with general relativity, we must use non-relativistic kinematics in our 'derivation'. Note also that in the case $k = 0$, the density is by definition the critical density $\rho_c = 3H^2c^2/8\pi G_N$. This is the equation used before to compute the numerical value for ρ_c.

The *force equation* is just Newton's second law $F = ma$ familiar in elementary mechanics: a is the acceleration of the mass m, and is equal to the acceleration of the radius R; F is the gravitational force the universe is exerting on the mass chunk m.

In Newtonian gravity, the gravitational force between two objects depends on their masses. Since mass and energy are related in special relativity, one might expect the gravitational force in a relativistic theory to depend on energies rather than masses. Using that, Einstein predicted that the starlight grazing the surface of the sun should be deflected by the gravitational attraction of the sun. This would not have happened if gravity depended only on mass, for the photons making up the light beam are massless. This bending of light leads to a shift of the known position of a star when it appears behind the edge of the solar disc. Because of the glare of the sun, such a star cannot be seen except during a total solar eclipse. This prediction was indeed verified at the solar eclipse of 1919 by a team led by A.S. Eddington.

It turns out that pressure P also makes a contribution to F, and the correct prescription is to replace the mass density of the universe by $\rho + 3P$. The Newtonian gravitational force on the mass chunk is $-GMm/R^2$, and its acceleration is $m(d^2R/dt^2)$. Newton's second law therefore demands $\frac{4}{3}G_N\pi\rho = -(d^2R/dt^2)/R$. With the replacement of ρ by $\rho + 3P$, we obtain in this way the *force equation*

$$-(d^2R/dt^2)/R = \frac{4}{3}G_N\pi(\rho + 3P).$$

By differentiating the energy equation and combining it with the force equation, one can arrive at the equation

$$-d[(\frac{4}{3}\pi R^3)\rho c^2] = [(4\pi R^2)P]dR,$$

which we shall refer to as the *work equation* for the following reason. The left hand side of this equation is minus the change of energy of the universe, and the right hand side gives the work done by the pressure to expand the universe. So this equation simply states that energy has to be taken from the universe to provide the work done to blow up the universe against the pull of gravity.

We may reverse the calculation, start from the energy and the work equations, and derive the force equation. In particular, the somewhat odd-looking combination $\rho + 3P$ in the force equation will appear automatically.

Two years after his introduction of the general theory of relativity, Einstein introduced an extra *'cosmological term'* into his equation. This term requires the introduction of a parameter Λ, called the *cosmological constant*. What this cosmological constant term does is effectively to change the density ρ and pressure P to new values $\tilde{\rho}$ and $\tilde{P}$ respectively, with $\tilde{\rho} = \rho + (c^2/8\pi\hbar^2 G_N)\Lambda$, and $\tilde{P} = P - (c^2/8\pi\hbar^2 G_N)\Lambda$. Note that P decreases by the same amount that ρ increases, if $\Lambda > 0$.

Before the discovery of Hubble the universe was thought to be *static*: what is true today is true a hundred billion years from now, or a hundred billion years ago. This picture of the universe is certainly consistent with the cosmological principle discussed in Sec. 2. A static universe, however, is impossible to be maintained without the cosmological term because of the constant pull of gravity to collapse it. With this cosmological term, a static universe can be achieved. We merely have to choose Λ so that $\tilde{\rho} + 3\tilde{P} = 0$. In that case there is no net gravitational force on the chunk and there is no deceleration. If we also choose the size R of the universe appropriately, namely $R = [4\pi G_N(\rho + P)/c^4]^{-1/2}$, then there will neither be a kinetic energy nor a potential energy and we will satisfy the energy equation as well.

Now that the universe is known to be expanding and not static, this solution is no longer relevant.

We may however still take the approach of introducing the cosmological constant term, and ask observation to determine how big it is. The result is (see Sec. 7.1) that the cosmological constant Λ, if it is not zero, is actually very very small. It must not be bigger than $10^{-121}M_P^2$, where $M_P \equiv \sqrt{\hbar c^5/G_N} = 1.22 \times 10^{19}$ GeV $= 1.22 \times 10^{28}$ eV is called the *Planck mass*. With such a small bound it is hard to believe that it is not exactly zero. On the other hand, it is extremely difficult for the modern theory of particle physics (see Sec. 7.1) to get a small enough value for the cosmological constant. This underlies one of the most fundamental difficulties of modern times.

Summary

Because of gravitational attraction the expansion of the universe is slowing down. Whether it will eventually stop and reverse itself, or whether it is going to expand forever, is a question that has not yet been settled by observations. The amount of luminous matter and hadronic matter in the universe alone is not enough to stop the expansion, but then there is dark matter around.

The detailed dynamics of how gravity affects the expansion of a homo-

geneous and isotropic universe is discussed in the second subsection. The important quantities are the radius and the energy density of the universe at any given time. Two equations can be derived to solve for these two quantities, as described in the text. These equations are dubbed the 'force equation' and the 'energy equation', respectively. Alternatively, one can replace the force equation by the 'work equation'.

---

4. COSMIC BACKGROUND RADIATION

The age of communication satellites brought to us an unexpected gift from heaven. This happened in 1965, when Arno Penzias and Robert Wilson, of the Bell Telephone Laboratories in New Jersey, discovered the *cosmic background radiation.*

They used a radio horn, originally designed for linking up with communication satellites, to make astronomical observations at a microwave wavelength of 7.35 cm. After they eliminated all known sources of background and interference, an annoying noise remained which could not be removed even after they dismantled the horn and cleaned out the bird droppings on it. Moreover, this noise remained the same no matter in which direction the horn pointed. It turns out that this unwanted noise became one of the most important discoveries of astronomy, and the second window for our view into the universe.

It was predicted by George Gamow in the late 1940's that a reverberating radiation from the big bang should remain even today. He estimated the temperature of this radiation to be about 5 K*. By looking at the distribution of the wavelength of the radiation, one can determine the temperature of the radiation by fitting it to the so-called *Planck's black body radiation law* (see App. C for a discussion of the law). The resulting temperature measured this way turned out to be 2.7 K.

This radiation also turned out to be highly isotropic, to the precision of several parts in 10^5. What this means is that at the time (several hundred

---

* Add 273 to degrees in centigrade to convert it to degrees in the Kelvin scale.

thousand years after the big bang) this radiation was emitted, the universe was highly uniform. See Sec. 5 for a more detailed discussion of this point.

The photon energy density at $T = 2.7$ K can be computed from the *Stefan-Boltzmann law* (see App. C). It turns out to be $\rho_\gamma = 4 \times 10^{-13}$ erg/cm^3=0.25 eV/cm^3; the number density of photons turns out to be 400 per cubic centimeter.

Summary

The universe is filled with an electromagnetic radiation at 2.7 K, left over from the original big bang. This radiation has an energy density of $\rho_\gamma = 4 \times 10^{-13}$ erg/cm^3=0.25 eV/cm^3, and contains 400 photons per cubic centimeter.

---

5. THE THERMAL EVOLUTION OF THE UNIVERSE

We shall try to trace through the history of the universe in this section. For that we need some knowledge of thermal physics and statistical mechanics discussed in App. C. It would also be useful to have some idea about gravitational physics discussed in Sec. 3.2. The discussions in App. C and Sec. 3.2 are useful mainly to provide some quantitative details for Sec. 5.2 below. For those who are willing to accept the numerical claims, very little of App. C and practically none of Sec. 3.2 are needed, and they can skip Sec. 5.2.

5.1   A Brief History of the Universe

Right after the big bang some ten to twenty billion years ago, the universe was tiny, but in its infancy it already contained all the energy. A small volume containing a large amount of energy means a huge energy *density* ρ and correspondingly a very high temperature T. Like the expanding fireball from a big explosion, this high energy density and high pressure drove the universe to expand outwards. As the universe expanded, the energy density was diluted by the increased volume, and this brought down the temperature. In the mean time, the gravitational pull was working relentlessly to try to hold back the expansion of the universe. As a result, the rate of the expansion gradually decreases.

In the beginning, when the universe is hot, the particles moving in it acquire a large amount of thermal energy. They move fast, with a speed close to the speed c of light. We call such fast particles *relativistic*. Photon, for example, is always relativistic, but in this early universe, all other particles we know are relativistic as well, like light. The universe in this era is said to be *radiation dominated*.

As the universe expands and cools further, the particles in it lose energy and all except the massless ones slow down. This occurs first to the heavy particles, but gradually even the light ones are affected. At some point most of the energy in the universe becomes concentrated in the *non-relativistic* particles, and the universe is said to be *matter dominated*. The present universe is matter dominated, and has been so since some hundred thousand years after the big bang.

In the radiation dominated era, the radius R of the universe grows with the time t, measured from the time of the big bang, like $\sqrt{t}$. The temperature T of the universe decreases with time like $1/\sqrt{t}$. A simple and easy-to-remember relation exists: if T is measured in MeV (1 MeV$=10^6$ eV$=1.2\times10^{10}$ K) and t in seconds, then $T \simeq 1/\sqrt{t}$.

In the matter dominated era, the radius R of the universe grows with time t like $t^{2/3}$.

To understand how this vast amount of energy in the universe is shared, we appeal to a fundamental principle of *statistical mechanics*: the *equipartition principle*. This says roughly that the thermal energy at a given temperature T is equally shared by all the particles, so if you like, it is a great democratic principle. The amount of energy per *degree of freedom* is of order T. As we shall see, because of this principle, the particle content of the universe changes as the universe cools down. This in turn affects the expansion rate of the universe.

A particle with mass m requires an amount of energy mc^2 to be created. This can be supplied by the thermal energy if T is of the order of mc^2 or bigger. If the particle is a *boson*, then it can be created singly; if it is a *fermion*, then it must be created in pairs of fermions and antifermions. Like chemical reactions, it is important to note that whatever mechanism that is used to create particles, the same reaction can be run backward to annihilate them. This tug-of-war between creation and annihilation continues until a *thermal equilibrium* is established, from which point on there will be as many creations and annihilations. If there are too many particles of a particular kind, then there are more to be annihilated, the annihilation rate would surpass the creation rate, and the number of such particles will decline towards the equilibrium number. If there are too few such particles, then it will be the other way around.

As the temperature of the universe declines, the massive particles whose rest energy exceeds T by a large amount become extinct. This happens because once they are annihilated, there will no longer be a sufficient amount of thermal energy to create them again. The temperature of the present universe is 2.7 K in the deep interstellar space, though it is considerably higher near a star. In any case, this amount of thermal energy is not sufficent to create any particle that is not massless. This creates an interesting dilemma as we shall see.

Given that there is an equal number of fermions and antifermions at the beginning of the universe, as dictated by the equipartition principle, all of the massive fermions should have been annihilated with the antifermions by now. That is manifestly untrue for there are still plenty of protons, neutrons, and electrons around, with no evidence for the presence of an equal number of antifermions. In spite of this large amount of matter in the universe, if we compare the ratio η of the number of nucleons to the number of photons in the universe, this ratio is very *small*, being of the order of $10^{-9} \sim 10^{-10}$. So relatively speaking it is true that the nucleons have *almost* completely disappeared.

Nevertheless in absolute number there are still many nucleons around. Where do they come from and how do they escape annihilation? There is not yet a definite answer to this puzzle but we will discuss some of the speculations in Sec. 7.3.

Let us now return to the early universe at about 1 second after the big bang, when the temperature was about 1 MeV. As we shall see, that was the time when some complex nuclei (meaning anything heavier than hydrogen) were first manufactured. To distinguish it from nuclei created at the center of a star at later epochs, the manufacture of nuclei at the beginning of the universe is known as *primordial nucleosynthesis*.

1 MeV is roughly the energy released in a radioactive beta decay (see

Sec. 2 of Chap. VI for a discussion of beta decay). It is also roughly the energy scale needed to ionize the nucleons in a nucleus. For the latter reason complex nuclei cannot exist at a temperature much higher than 1 MeV. For the former reason, when the temperature is finally lowered to 1 MeV to enable complex nuclei to be formed, many of the neutrons (whose number equals to the number of protons at the beginning of the universe) will have decayed away and cannot be recreated. This is the reason why some 80% of the elements in the universe remains to be hydrogen. There are simply not enough neutrons anymore to do anything else. A more thorough discussion of the primordial nucleosynthesis of nuclei will be given in Sec. 6.

Another important landmark for the history of the universe occurs at a temperature 1 million times lower. Above the temperature of 1 eV or so, there is no way for neutral atoms to form because the excessive thermal energy will tear away the electrons. At such temperatures the universe consists of a plasma of negatively charged electrons and positively charged nuclei. At roughly the temperature of 1 eV and below, electrons will combine with nuclei to form neutral atoms. Since photons interact only with charged objects, their interaction with matter is greatly weakened when neutral atoms appear. For that reason the time when neutral atoms occur is called the *decoupling time*. Old photons born at that era are still roaming around the universe unretarded!

Using the formula between temperature and time quoted before, decoupling occurs at about 10^{12} s after the big bang, which translates to tens or hundreds of thousands of years after the big bang.

Since that time, the electromagnetic radiation is basically decoupled from the rest of the matter in the universe. As the universe continues to grow, the temperature of the radiation continues to decline, eventually reaching the present value of 2.7 K. Because of the decoupling, the photons we catch from the cosmic background radiation today were already born at the decoupling time; they have just gotten older and weaker when we meet them. These photons are definitely older than the dinosaurs, the solar system, and all the galaxies! Nevertheless, they can still relate a few interesting stories of the good old days. For example, the universe at the decoupling time must have been very uniform, or else we would not be able to see the isotropy of the background radiation, known to be accurate to several parts in 10^5!

Let us now return to the expansion of the universe. As it continues to cool beyond the decoupling time, it will reach a point where the thermal motion is so reduced that gravity is able to clump matter together into galaxies, stars, planets, and satellites. The detailed mechanism for *galaxy formation* is a problem of great current interest. Observationally, very inhomogeneous structures are being discovered all the time, larger and larger in size and farther and farther out in space. Since it takes time for light from distant galaxies to reach us, this means to have inhomogeneous structures very far back in time. This, when coupled to the great uniformity of matter at the decoupling time, poses a serious difficulty which has not yet been solved satisfactorily. The main problem is how to find a clumping mechanism so that inhomogeneity of the right amount and the right shape as observed in galactic distributions can be built up in such a 'short' time after the decoupling.

Just as raindrops are condensed on seeds of dust particles, galaxies must start their built-up from something. That something could be derived just from the statistical fluctuation of matter, with the denser part in the fluctuation acting like a seed. Or else, the seed could be more exotic objects like a cosmic string (see Sec. 7.2 for a discussion of cosmic strings). Various computations and computer simulations have been carried out to test the viability of the various assumptions, but so far none of them seem to be entirely satisfactory.

With this temporary difficulty aside, it does seem rather incredible that 'astro-archaeology' can be so successful. Sitting in our armchairs, to be sure with a lot of observational data, we can figure out what happened to the universe billions of years back! This is then some of the great contributions of physics. It provides us with all the detective tools for remote sensing, to enable us look back to the decoupling era through the cosmic background radiation, and to look all the way back to the universe one second after the explosion through the abundance of the various elements created in primordial nucleosynthesis (see Sec. 6). The mere presence of nucleons and electrons at the present time poses a dilemma, whose eventual resolution almost certainly takes us back much further in time still.

Emboldened by the success in such archaeological findings, we might even dare to ask about the creation of the universe and the origin of our four-dimensional space-time. Some discussions along those lines will be given in Secs. 7.4 and 8.

This may be a good place to discuss a philosophical point. In its infancy, everything in the universe happened so quickly that, for example, all the vast amount of primordial elements were built in a few minutes. Yet what happened then actually sealed the fate of the whole universe. It seems odd and unsatisfactory then that the destiny of the whole universe can be decided even before the fastest computer has time to analyse it. Why should that happen?

When we think about this more carefully we just may decide that actually it is not so odd after all. Somehow this may simply be a question of what is the appropriate unit of time to compare things with. A *second* is a short time to us, but is it a short time way back then? We measure our time by an atomic clock, say, which relies on the frequency of a spectral line of cesium. In its infancy, everything in the universe was so incredibly hot that cesium can no longer exist, and the only sensible and relevant frequencies then were high frequencies. If a clock existed then we would never have based it on the cesium frequency standard; we would have used some frequency commensurate with its high temperature surroundings. If human beings existed in those high temperature surroundings, the 'chemical' and the 'biological' processes governing their lives would have proceeded at such an alarming rate that they could probably live only a small fraction of a second rather than a hundred years or so. If a computer existed then it would have operated at an incredible speed again commensurate with the quick tempo of all other events. Thus although in the present units of seconds, things seem to have happened too quickly, in the appropriate time units of the era, this may not have been so at all.

5.2 Quantitative Estimates

We shall supplement the discussions of the previous subsection with some quantitative information based on a knowledge of App. C and Sec. 3.2. Readers not interested in quantitative results can skip this subsection.

The expansion of the universe under gravity was described quantitatively in Sec. 3.2. The main thing to note is that the expansion is governed by two equations, which we can take to be the 'energy equation' and the 'work equation'. Three functions of time are involved in these two equations. They are the radius $R(t)$, the energy density $\rho(t)$, and the pressure $P(t)$ of the universe.

Given a composition of matter and energy, in principle one can calculate and express P and ρ as functions of the temperature T and the volume, $\frac{4}{3}\pi R^3$. These functions are discussed in App. C for relativistic free bosons and fermions. Using these functional relations, P and ρ can be eliminated so that we are left with only two unknown functions $R(t)$ and $T(t)$, which can be determined by the energy and the work equations.

In the radiation dominated era, both ρ and P are proportional to T^4 (the *Stefan-Boltzmann law*). The proportionality constant depends on the composition of matter, but the relation $P = \rho/3$ is always maintained. The solution of the energy and the work equations gives the result that $R(t)$ is proportional to $T^{-1}(t)$, which in turn is proportional to $\sqrt{t}$. The proportionality constant can be worked out as well, and the complete relation is $T \approx 2.4/(g^{1/4}\sqrt{t})$, if t is measured in seconds and T is measured in MeV. The number g is the *statistical degeneracy factor* which depends on the composition of matter (see App. C).

In the *matter dominated* era, the kinetic energy of a particle is quite negligible compared with its rest energy mc^2, so that P is negligible compared to ρ, and the total amount of energy in the universe is equal to the sum of the rest energies of the particles. Thus ρ decreases like $1/R^3$. Solving the gravitational equations, one obtains $R(t) \propto t^{2/3}$.

We recall from Sec. 4 that the cosmic background radiation has a temperature of $T = 2.7°$ K. Using the *Stefan-Boltzmann law*, this leads to a photon energy density $\rho_\gamma = 0.25$ eV/cm^3, and a photon number density $n_\gamma = 400/$cm^3. In contrast, the number of nucleons in the universe is smaller by a factor of $\eta \simeq 10^{-9} \sim 10^{-10}$ or so as can be estimated in the following way.

If $\Omega = \rho_M/\rho_c$ is the ratio of the actual density of matter in the universe to the critical density, then the energy density of matter is $\rho_M = \Omega\rho_c = \Omega \times 2.7 \times 10^3$ eV/cm^3, which is quite a bit larger than ρ_γ at any reasonable value of Ω. If we take $\Omega = 0.2$ then $\rho_M = 540$ eV/cm^3, which is about two thousand times larger than the total energy carried by the $2.7°K$ photons. Hence the present universe is matter dominated. Since the rest energy of a nucleon is about 10^9 eV, the number of nucleons per cubic centimeter is 5.4×10^{-7}. Compared to the photon number density of 400 per cubic centimeter, the number of nucleons in the universe is a factor $\eta \sim 10^{-9}$ down from the number of photons.

The total energy carried by the non-relativistic particles is independent of R, but the total energy carried by the relativistic particles such as photons varies like $1/R$ for small R. Clearly at a small enough R, the total photon energy will overtake the total nucleon energy in the universe, and the early universe therefore becomes radiation dominated. The crossover from one era to another occurs roughly when R is $1/z$ of the current size, and T is z times the current radiation temperature, where $z \simeq 2000$ is the current ratio ρ_M/ρ_γ. That puts the temperature at that time to be roughly 5400 K, which is roughly the same as the temperature at decoupling. Since $T \propto t^{2/3}$ in the matter dominated universe, that transition occurs at a time which is $z^{-3/2} \approx 1/90,000$ of the present age. Taking the present age to be 14 billion years, this works out to be about 150,000 years after the big bang when this happens. Before then, the universe was radiation dominated. Please note that more accurate calculations exist to improve on the very rough estimates given above.

Summary

The thermal history of the universe is briefly reviewed. In the first few hundred thousand years, the universe is radiation dominated. During this era, the radius R of the universe increases with time t like $\sqrt{t}$, and the temperature T of universe is given approximately by $T \simeq 1/\sqrt{t}$, where T is measured in MeV and t in seconds. Primordial nucleosynthesis occurs at a temperature about 1 MeV and a time about 1 second after the big bang. The decoupling of electromagnetic radiation occurs at a temperature of the order of 1 eV, or about 10^4 K, and at a time some 10^5 years after the big bang. This is also estimated to be close to the time when the radiation era turns into the matter era, where the present universe belongs. The energy carried by the nucleons at the present is estimated to be some 10^3 the energy carried by the photons, but the number of photons exceeds the number of nucleons by a factor $\eta^{-1} \simeq 10^9 \sim 10^{10}$.

6. PRIMORDIAL NUCLEOSYNTHESIS

The chemical element *helium* is so named because it was first discovered in the sun's atmosphere by J.N. Lockyer in 1868 (Helios is Greek for sun). It was one of the late comers as far as elements are concerned, but it is the second most abundant element in the universe!

We are so used to the richness of elements around us that sometimes we do not realize that most of what we see on earth is made up of extremely rare materials indeed. The universe is predominantly made up of hydrogen, then helium, with about one helium atom for every ten hydrogen atoms. The rest of the elements occur only in trace amounts.

In the beginning, right after the big bang, the universe is very hot. The intense heat prevents any compound nucleus from being formed, so the primordial soup consists of single protons and neutrons, as well as other particles. The only chemical element (rather, the nucleus of a chemical element) present is hydrogen, which is just the proton. As the universe expands and cools, at some point neutrons and protons can come together to form other nuclei, such as helium and a (very) few others. This formation is called *primordial nucleosynthesis*. Much later when stars appear, hydrogen and other fuels are burned in their interior to produce thermonuclear energy. This also produces heavier elements as a by-product. The larger the star is, the heavier the elements that can be produced. But no matter how large the star is, elements beyond iron can only be produced in supernova explosions (see Chap. V for more details).

Primordial nucleosynthesis occurs at the approximate time interval 10^{-2} s$< t < 10^3$ s (10 milli-seconds to 15 minutes) after the big bang. The range of temperature of the universe was then 10^{11} K$> T > 4 \times 10^8$ K. The primordial soup at that time consists of protons (p), neutrons (n), electrons (e^-), positrons (e^+), photons (γ), and the various kinds of neutrinos and antineutrinos $(\nu_e, \nu_\mu, \nu_\tau, \bar{\nu}_e, \bar{\nu}_\mu, \bar{\nu}_\tau)$ (see Sec. 1 of Chap. VI for a description of these particles). We know from the last section that the ratio η of the number of nucleons to the number of photons is very small, then and now, being of the order of 10^{-9} to 10^{-10}. Let us also remember that a temperature of 1 MeV corresponds to about 10^{10} K.

Since the neutron is 1.3 MeV/c^2 heavier than the proton, it will disappear via various guises of the beta-decay process $(n \rightarrow p + e^- + \bar{\nu}_e, n + e^+ \rightarrow p + \bar{\nu}_e, n + \nu_e \rightarrow p + e^-)$ at a temperature significantly below 1 MeV. The inverse processes, with $\rightarrow$ changed to $\leftarrow$, become rarer and rarer as the temperature goes down because of the lack of energy. If nothing else happened the neutrons would have disappeared completely. Fortunately, one in six or seven of them is captured by the protons to form complex nuclei (mostly helium) before they disappear. Once bound in a nucleus, it is no longer energetically favorable for neutrons to decay, so they remain there until today.

It takes two protons and two neutrons to form a helium nucleus, also known as an *alpha particle* (α). A ratio of two neutrons to twelve protons left in the universe means a ratio of one helium nucleus to ten hydrogen nuclei. This is approximately the ratio observed today.

The fact that one in six or so neutrons survive is to a large extent an 'accident' of the strength of the nuclear force. We saw in Sec. 2 of Chapter VI how our world would have looked completely alien if the strengths of the forces were altered. Here is another example of our dependence on just how things are. The nuclear force binds a proton and a neutron together to form a heavy hydrogen nucleus, called a *deuteron* (D); the corresponding atom is called *deuterium*. The binding energy of deuteron is 2.225 MeV, which is the energy needed to tear the proton away from the neutron in the deuteron. This means that the energy credit of 1.3 MeV obtained by neutron decay is not sufficient to pay for the deficit of 2.225 MeV to break up the deuteron, hence neutrons bound in deuterons become stable.

The 'accident' happens in the following way. When the universe reaches a *freeze-out* temperature of about 1 MeV, thermal energy is no longer sufficient to produce neutrons efficiently via inverse beta-decay. From that point on, it is a race against time for the protons to capture the neutrons to form complex nuclei, before the neutrons decay away and are lost forever. At that temperature, the density of nucleons is too small to allow three or more nucleons to collide all at once. The only reaction capable of neutron capture is therefore $p + n \rightarrow \gamma + D$, sometimes written $p(n, \gamma)D$ for short. The inverse reaction $D + \gamma \rightarrow n + p$, written also as $D(\gamma, n)p$, is possible as long as the temperature is not too low compared to deuteron's binding energy. How low is not too low? In this connection it is important to remember that the ratio of nucleons to photons, a ratio which we denote by η, is an extremely small number, being of the order of 10^{-9} to 10^{-10}. The fact that there are immensely more photons than nucleons favors the breakup of deuteron by photon. It is therefore a competition between the suppression of this breakup at low temperature and the enhancement of this process by the great abundance of photons. It turns out that as a result of this competition the concentration of D will increase only at a temperature lower than 0.1 MeV or so.

The deuterons so formed also collide with other particles in the environment to initiate further reactions: $D(n, \gamma)T$, $D(D,p)T$, $D(p, \gamma)^3He$, and $D(D,n)^3He$. Here T stands for *triton*, which is another isotope of hydrogen, with one proton and two neutrons in its nucleus. ^3He is an isotope of helium whose total atomic number is 3, meaning that it has in its nucleus a total of three nucleons, two of which must be protons because it is helium. This notation of having the number of nucleons written at the upper left hand corner of the chemical symbol is rather standard. In this notation, D is ^2H and T is ^3H.

Now other reactions also follow: $T(p, \gamma)^4He$, $T(D,n)^4He$, $^3He(n, \gamma)^4He$, $^3He(n, p)T$, $^4He(^3He, \gamma)^7Be$, and $^4He(T, \gamma)^7Li$. Of these, the normal isotope of helium, $^4He = \alpha$, is the most stable, with the largest binding energy of them all. All the others are either radioactive (*e.g.*, T, ^7Be), or are more fragile and have a smaller binding energy (*e.g.*, D). This is why almost all the nuclei formed this way end up to be ^4He. The absence of stable nuclei of atomic numbers 5 and 8 also prevents the destruction of helium once it is formed.

Practically no nuclei heavier than ^7Li are formed at this time. These heavier nuclei contain more protons, which repel one another electrostatically. This means that a high temperature is required to provide sufficient energy to overcome such electrostatic repulsion. On the other hand, we must first go through the bottleneck of forming D before having a chance to produce heavier elements, but deuteron is so fragile that it cannot be formed until the temperature is relatively low. For that reason, by the time the turn of the formation of heavier elements comes up, the temperature of the universe would no longer be high enough to produce them. Even the formation of helium will no longer be possible when the temperature gets below 10^8 K or so.

The final abundance of different nuclei depends on a number of factors. The factor η mentioned before is important, for it tells us the relative probability of formation (which is proportional to the density of nucleons) and photodisintegration (which is proportional to the density of photons). The

expansion rate of the universe is important too, for that tells us how fast
the temperature is going down and how much time we have to do the job.
From the energy equation in Sec. 3.2, the expansion rate increases with the
energy density in the radiation era, which in turn increases with the number of species of relativistic particles present (photons, and the number of
species $\mathcal{N}_\nu$ of neutrinos, for example). It turns out that ^{4}He production is
sensitive to $\mathcal{N}_\nu$: a larger expansion rate freezes the number of neutrons at a
higher temperature and increases the abundance of ^{4}He. The known value of
$\mathcal{N}_\nu = 3$ from particle physics experiments is consistent with the abundance
of primordial ^{4}He.

The abundance of the other nuclei is not sensitive to $\mathcal{N}_\nu$, but is sensitive
to η. A larger η gives less D because the increased abundance of nucleons
turns more of the fragile D into heavier nuclei. The abundance of other nuclei
is more complicated to describe but they can all be calculated. By comparing
these calculations with observations, one should be able to settle the various
parameters, including η. It is in this way that the number η is determined
to be somewhere between 10^{-9} to 10^{-10}.

The main difficulty in this program is to determine the primordial abundances of D, ^{3}He, ^{4}He, and ^{7}Li (^{7}Be is radioactive and decays into ^{7}Li) by
direct observation. In the solar system, abundances can be measured directly from the solar wind, lunar soil, and carbonaceous chondrites (a kind
of meteorite which had not ungone changes since the beginning of the solar system). Outside of the solar system the abundance can be measured
by spectroscopic means. The problem has always been to determine how
much of the measured quantities was produced primordially, and how much
was produced later in the stars. If produced primordially there should be
a rather uniform concentration in the universe and the element should not
congregate around the stars. Most of the helium seen in the universe obeys
this criterion which is why we are confident that most of it was produced
primordially. Beyond that, one might use stellar model calculations to estimate the amount that should be produced in stars and subtract them off to
get the primordial abundance. Various methods have been advanced in this
way but the different measurements do not always yield the same numbers.
However, the range of η values quoted above does seem to be in the right
ballpark for most of the measurements.

Summary

The most abundant element in the universe is hydrogen. Next comes helium, with about one helium atom to every ten hydrogen atoms. Most of the helium observed was formed primordially, at the beginning of the unvierse when it was only 10 milli-seconds to 15 minutes old. At that time other primordial elements, including D, ^{3}He, and ^{7}Li, were also formed. The abundance of these elements in the universe gives a method of measuring the ratio of nucleons to photons at that time, as well as other parameters.

7. COSMOLOGY AND PARTICLE PHYSICS

We have seen how gravitational physics and thermal physics find their way into questions of cosmology. We saw in the last section how important nuclear physics is to the study of nucleosynthesis. In this section, we will discuss some of the roles played by particle physics in this arena.

It might seem strange that the tiniest of particles can do much to influence the huge universe. Certainly if there are enough of them, they will exert a large gravitational force, but that is the trivial aspect which has been discussed before. The main point of the present section is that the universe was very hot in the beginning. Elementary particle physics (see Chap. VI for a review) is also called high energy physics, because many of these particles require a large amount of energy to create. Such an energy is available in the early universe right after the big bang, which can be used to create particles, including many of the exotic varieties not yet seen by mankind. To study the very early universe, knowledge of high energy physics is therefore necessary. If the universe were static and cold, then we might never have had to use our knowledge of high energy particle physics at all.

7.1 Cosmological Constant

The cosmological constant has been discussed in Sec. 3.2. It measures the amount of energy density and pressure in the vacuum. It is an incredibly small number as we shall discuss in the next paragraph. Those who are willing to accept this can skip to the paragraph after that. The main point is that any vacuum energy of a scale much larger than a few milli-Volts will give rise to too large a cosmological constant.

If Λ is the cosmological constant, then an amount $(c^2/8\pi\hbar^2 G_N)\Lambda$ is added to the energy density ρ and subtracted from the pressure P of the universe (see Sec. 3.2). The critical density of the universe is $\rho_c \simeq 2.7 \times 10^3$ eV/cm^3. The cosmological constant, if present, surely must not be allowed to add a density to the universe as large as that. This leads to the bound for Λ quoted in Sec. 3.2: $\Lambda < 10^{-121} M_P^2$, where the Planck mass is given by $M_P = 1.22 \times 10^{19}$ GeV (1 GeV$=10^9$ eV). This is an incredibly small number. For comparison with theory it is more useful to express the cosmological constant in terms of equivalent energy density, expressed in units of eV^4. The critical density ρ_c is expressed above in the unit of eV/cm^3. To convert it into unit of eV^4, we multiply it by $(\hbar c)^3$. Then $(\hbar c)^3 \rho_c = (3.8 \times 10^{-3}\text{eV})^4$. So the appropriate energy scale corresponding to the critical density is in the milli-Volt region.

Thus we would be in trouble if the scale of energy density of the vacuum were larger than a few milli-Volts. This unfortunately is what modern particle physics seems to find.

One might ask why the vacuum, which by definition is the lowest energy state with 'nothing' in it, can have a non-zero energy density. There are two reasons why it can. The first is quantum mechanics, whose uncertainty principle (see Appendix B) generally does not allow a ground state to have zero energy. The vacuum state is the state with nothing *permanently* present, but during small time intervals, there is no way to prohibit the quantum fluctuation of particles. There is no such thing in quantum mechanics as a state with nothing in it at all times. The second source of vacuum energy is the *condensates*. Modern particle physics holds the view that the universe underwent a series of *phase transitions* (see Sec. 5.3 of Chapter VI for details) as it cooled down to reach the present stage. Each time a phase transition occured, a new condensate permeating the vacuum was formed. The condensates as a whole act like a modern *aether*, except that relativity is not violated in this modern version. One might wonder again how come we have never noticed the presence of this aether in the vacuum when we are walking through it all the time. This is because we have always lived in it and do not know any differently. A fish is probably not consciously aware of the water around it until you drain the water. Then there is no longer any bouyancy and the fish can no longer swim. It will probably die as well. What happens if we could drain away the condensates making up the modern aether? All particles would probably become massless and quarks would be able to go freely all over the place. It would then in no way resemble the world we know.

These condenstates carry energy densities so they contribute to the cosmological constant. The energy density is usually given by an energy scale of the order of the *transition temperature* (the temperature at which the phase transition occurs). These temperatures are invariably far greater than the milli-Volt scale permitted by the critical density. In other words, these condensates invariably give rise to a large cosmological constant. This is a serious problem in elementary particle physics which has not been resolved. To be exact, there is a very distasteful solution. If the universe originally started out with a finite negative cosmological constant, which when combined with the contributions of the various condensates would give rise to a total value below the critical density of the universe, then things would have been alright. This is very distasteful both because it is contrived and because, even contrived, it is not easy to do it. There is a 'Higgs condensate' in the GSW electroweak theory (see Sec. 5.3 of Chap. VI) which has an energy scale of the order of 100 GeV. To cancel that so that the result is less than a milli-Volt, we must design the original cosmological constant energy scale accurate to one part in 10^{14}!

There is another way of putting this dilemma. If this vacuum energy density of energy scale $100 GeV$ were present, then the size of the universe, instead of being around 10^{28} cm as we know it, would be only a centimeter across *.

7.2 Topological Objects

As the universe cools and undergoes the phase transitions discussed in the last subsection, topological objects may be produced.

---

* A quick and dirty way to see this is to assume a static universe supported by a cosmological constant. The size of such a universe is given by the formula $R = [4\pi G_N \rho/c^4]^{-1/2}$ (see Sec. 3.2; P has been put equal to zero for our matter dominated universe). Putting the critical density for ρ we get $R = 10^{28}$ cm, which agrees with the present size of the universe. Now if we subsitute for ρ something which is a factor $(10^{14})^4$ bigger, the size R would become 10^{-28} smaller, hence $R = 1$ cm.

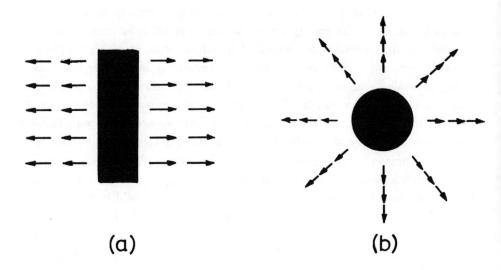

FIG. VII.2 *Topological objects (black) as transition regions of mismatched fields. (a) Domain wall; (b) cosmic string or magnetic monopole.*

Phase transitions are accompanied by the creation of condensates. These condensates provide for the lower energy of the vacuum in the new phase. They usually carry some internal quantum numbers, depicted by an arrow in Fig. VII.2. The length of the arrow indicates the magnitude of the condensate, which is usually fixed. However, the direction of the arrows is quite arbitrary.

Imagine now a new phase in which the arrows have a configuration indicated in Fig. VII.2a. Since the arrows on either side of the shaded object are mismatched, a buffer zone (of necessarily higher energy density) is created to bridge them. This is the shaded two-dimensional surface in the middle known as a *domain wall*. This wall is a topological object that cannot disappear because it is always needed to bridge the arrows on both sides. It contains a large energy density, again usually of the order of the energy scale of the transition temperature. If such an object existed at the usual phase transition energy scales, it would give rise to such a huge gravitational perturbation that it should have been observed by now. It is therefore likely that for some reason domain walls of a large energy scale are not present in the present universe.

The formation of domain walls can be illustrated with a practical example. When molten iron condenses into a solid, the magnetization directions of neighboring iron atoms tend to line up to save energy. However, since solidification takes place independently and simultaneously at different points of the solid, the magnetization directions in different regions are generally misaligned, thus forming different magnetic domains separated by domain walls.

Topological objects can take the form of a line as well, as depicted in Fig. VII.2b (the circle in that picture should be thought of as the section of a long thin cylinder sticking out of the plane of the paper). This is called a *cosmic string*. It might have helped in seeding the formation of galaxies.

If we read the shaded object in Fig. VII.2b as a three-dimensional sphere but not a cylinder, then we would have produced another topological object. This is how *magnetic monopoles* are formed. A lot of effort has been spent to find them, but so far none of them have turned up.

The formation of these topological objects are predicted in some *grand unified theories* (see Sec. 7.2 of Chapter VI). If however they were produced before the inflationary era (see Sec. 8), then they could be so diluted during inflation so as not to be observable.

7.3 The Asymmetry between Nucleons and Antinucleons

The number η, which is the ratio of the number of nucleons to the number of photons, is of the order of 10^{-9} to 10^{-10}. What does it mean to have it so small? Why isn't it zero anyway?

In the beginning of the universe, when the temperature was very high and all particles present were relativistic, the Stefan-Boltzmann law was at work. It requires the universe to have equal numbers of protons, neutrons, antiprotons, antineutrons, electrons, and positrons. Except for an extra factor of $\frac{4}{3}$ because it is a boson, the number of photons present must be the same as well (see App. C). Nucleons and antinucleons have masses of 1 GeV/c^2; they will combine to annihilate each other and be lost forever when the temperature drops significantly lower than 1 GeV. If nothing else happened before or since then to produce more nucleons than antinucleons, we would have ended up today with no nucleons nor antinucleons, with $\eta = 0$, and with no matter whatsoever in the universe! Then we would not be here to worry about this problem either.

The fact that η is $10^{-10} > 0$ means that a mechanism must have existed in the early universe to generate more nucleons than antinucleons. To do so the universe must not be in thermal equilibrium, for otherwise every reaction runs in both directions with equal likeliness, and whatever excess nucleons so generated will disappear in the next instant when the reaction runs backwards. Since the universe is expanding, reactions can actually be thrown off thermal equilibrium if the reaction rates are small compared to the expansion rate. One must also identify a reaction whereby such an excess of nucleons can be produced, and a mechanism of that kind might exist in grand unified theories (Sec. 7.2 of Chap. VI). Since the grand unification scale is very high, being of the order of $10^{15} \sim 10^{16}$ GeV, if this is the scenario then the extra nucleons are produced very early in the history of the universe. Such an excess can also be wiped out by later events at the electroweak scale. It is also possible that the nucleon excess occurs at a much lower energy, that of the electroweak scale, of about 100 GeV or so, through the intervention of a topological object known as the *sphaleron*. No definite conclusion has been reached yet at this point.

7.4 Superstring Theory

The theory of superstrings (see Sec. 7.5 of Chapter VI), if correct, is expected to be important at the beginning of the universe. At that time the temperature is so high that quantum gravity is expected to be important, and string theory is the only theory we know giving a sensible and finite quantum gravity theory. Moreover, there are many many very massive states in string theory, whose masses are of the order of the Planck mass and more. These affect the equipartition of energy and the subsequent thermal development of the universe. Unfortunately, the superstring theory as we know it today is incomplete because it is understood only in the perturbation theory framework. This difficulty prevents the superstring theory from making definitive predictions. Nevertheless, it continues to be a very interesting theory, and the only one which can potentially unify all the known forces into a single *finite* quantum field theory.

We shall now present a cosmological and somewhat philosophical argument to show, in a different way, why the string theory may be very important to the study of cosmology.

Our universe was created some ten to twenty billion years ago in a big bang. One might ask: what happened before the big bang, and where would the universe be put once it was created?

These are actually age-old questions, though in the old days they were asked in a somewhat different form: What happened before God created the world, and, where did God put the world once he created it?

Perhaps the best answer to these questions came from Saint Augustine of Hippo, a great Catholic theologian living from 354 A.D. to 430 A.D. He essentially said that God created *time* when he created the world, so you may *not* ask what happened *before* time was created. To this one could also add that God created *space* then to put the world in.

Suppose we accept this philosophy. We can then ask whether a physical theory can be invented in which both space and time can be created like photons and gravitons, or electromagnetic fields and gravitational fields. In more technical terms, this is to ask whether a theory can be written down wherein space and time are treated as dynamical variables in a local quantum field theory.

If we think about it, we will see that the question may not even make sense. After all, a dynamical theory tells us how dynamical variables change with time. One writes down differential equations to describe its dynamical behavior. If we take away *time* as the parameter whose unfolding leads to the dynamical changes, and promote it to be a dynamical variable, then what do we even mean by 'dynamics'? Where is that independent parameter that we need to make sense out of that word?

Similarly, quantum field theories also regard the spatial coordinates (x, y, z) as parameters. A field is usually regarded as a function of (x, y, z), as well as t. Equations of motion in field theories are partial differential equations of x, y, z, and t. If we take away x, y, z as parameters, then no partial differential equations can be written, and it is not clear what we will still mean by the word 'field theory'.

There is one way out of this impasse. If all that we need are parameters to describe the evolutions of dynamical variables, we may simply introduce them into the theory. Call these evolution parameters *artificial time* (τ) and *artificial spatial coordinates* $(\sigma_1, \cdots, \sigma_n)$. Of course they have nothing whatsoever to do with *real time and real space*; they are simply there to enable us to write down (partial) differential equations and make sense out of the word 'dynamics'.

At first sight it seems that we are not going to get anywhere with this silly game. There are just too many equations of motion one can write down. Without any physical input and without any guiding principle, it is impossible to go any further than that.

The amazing thing is, we do have a *guiding principle*.

As the name indicates, these newly-introduced parameters are artificial. They have absolutely nothing to do with the real time and the real space that we know. Nobody but nobody has ever seen or detected the artificial space-time. Therefore, in writing down a dynamical theory involving them, we must be extraordinary careful. We must ensure that *nothing about the artificial space-time is observable*. This then is the guiding principle that we are looking for. We shall refer to that as the *unobservability principle*.

Can dynamical theories be written down which obey the unobservability principle? The answer is 'yes', as we shall explain. The principle actually turns out to be so powerful that the theory we can write down is essentially the string theory discussed in Sec. 7.5 of Chapter VI.

To understand how this comes about let us turn to general relativity, where a similar *equivalence principle* exists. This demands that physics depends on the geometry of the (physical) space-time, but not on what *coordinates* are used to parametrize it. However, the curvature of space-time is geometrical and is important in the theory, for it measures the strength of the gravitational field at that point. Well known techniques in differential geometry are available to construct dynamical theories obeying the equivalence principle. Such theories are also said to be *reparametrization invariant*.

Now the unobservability principle deals with the artificial space-time and not the physical space-time. Nevertheless, the mathematical technique is the same. There is no problem in writing down theories which are *reparametrization invariant*, and there are many of them. But the unobservability principle *goes beyond that*. It demands that *nothing* about the artificial space-time be observable. This includes geometry and curvature.

Now by curvature, we mean the *intrinsic curvature*. A one dimensional rope winding and curving in space actually has no intrinsic curvature, for a one-dimensional animal confined on it will never be able to tell whether it is curved or straight. This is not so for two dimensions. A two-dimensional animal living on a surface can tell whether the surface is intrinsically curved or flat. All that it has to do is to draw a triangle on the surface and measure the sum of its three internal angles. If it comes out to be 180°, then the surface is flat. Otherwise it is curved. For an animal living in higher dimensions, it can do so for each two-dimensional subspace, and hence at each given point it can talk about several (sectional) curvatures corresponding to two dimensional surfaces facing different directions.

So far we leave the dimension n of the artificial space rather arbitrary. If $n = 0$, then there is an artificial time but no artificial space. Theories obeying the unobservability principle can be written down just like theories obeying the equivalence principle. But with no artificial space, these theories are not field theories, and they turn out not to be interesting.

If $n = 1$, the artificial space-time is two dimensional, and at every point there is a curvature. Curvatures, like the inverse square of the radius of a circle, have a dimension (a non-trivial unit). To make curvatures irrelevant and unobservable in a physical theory, we must make all dimensions irrelevant. This can be accomplished if we can write down a theory which is invariant under any change of scales used in the units of measurements. In that case any quantity involving a non-trivial unit must not be measurable, for how else could its numerical value remain unchanged when we change the unit of measurement? Such theories exist and are called *conformal field theories*. The fact that the artificial space is one-dimensional ($n = 1$) says that all dynamical variables are functions of τ and a single σ. At every given (artificial) time, the dynamical variables are parametrized by one continuous variable σ, which can be thought of as parametrizing along the length of a string. This is why these theories are string theories.

If $n > 1$, then there are many (sectional) curvatures at every point, and we need to have a theory with many invariant properties at every point to remove the dependences of *all* these artificial curvatures in physics. This turns out to be impossible.

So we have come to the very interesting conclusion that in order to answer the age-old question about the beginning of the universe, we are more or less forced to the introduction artificial space-time. In order to prevent these artificial quantities from being physically relevant, we must insist on the unobservability principle. This then more or less leads to the string theory which is reparametrization and conformally invariant.

Summary

Modern theories of particle physics provide some answers and some problems to our universe. The cosmological constant computed from modern particle theories is invariably far too large, unless we resort to fine tuning. Objects of the topological types such as monoples, cosmic strings, and domain walls can be produced as a result of misalignment of the newly created condensates at the times of phase transitions. They may have important cosmological consequences, but they could have been diluted during inflation. The puzzle of having any nucleons left at all in the present universe can be solved if non-equilibrium interactions of the right kind existed in the early universe. Finally, the necessity of considering space-time as dynamical variables at the beginning of the universe is discussed, and string theory is shown to be the theory which can implement this goal.

---

8. INFLATIONARY UNIVERSE

The classical big bang theory is infected with a couple of diseases.

The first is known as the *horizon problem*. The cosmic background radiation which we measure today left its source back at the decoupling time. The source can be thought of as a spherical surface S. The radiation coming from different directions originated from different parts of S. This radiation is known to be isotropic at least to the accuracy of one part in ten thousand. Yet it can be calculated from the classical big bang theory that S can be divided into some 10^5 regions which had not been in causal contact with one another from the beginning of the universe to the time when the radiation left the surface. This is because the universe expanded so fast in the beginning, that light from one part of S had not had a chance to catch up to get to all other parts of S. As a result the surface S can be divided into 10^5 non-overlapping regions, so that light started at the center of a region at the beginning of the universe has not yet had time to reach into any other region. In spite of that the cosmic radiation coming from all these mutually non-communicative regions conspire to have the same magnitude. How can that possibly happen? This is the horizon problem.

The number of causally disconnected regions is inversely related to how far light can travel in a fixed amount of time t. Light propagates locally with a constant speed c. If the universe were static, then the distance light travels after a time t is ct. But the universe is expanding, and this expansion stretches every point away from every other point. Light can make use of this free ride to get further; how much further depends on the expansion rate of the universe. If $R(t) \propto t^n$, with $0 \leq n < 1$, then the distance covered is $ct/(1-n)$. For static universe, $n = 0$, this goes back to ct. A radiation dominated universe has $n = \frac{1}{2}$, so light goes twice as far but this is still not far enough to get over all S by the decoupling time. If $R(t) \propto \exp(Ht)$, however, then the distance covered becomes $(c/H)\exp(Ht)$, which can be very large indeed for a large t. Light would now be able to get all over S by the decoupling time, so there would no longer be any causally disconnected regions on it. This is how the *inflationary universe theory* solves the horizon problem: it demands the universe to grow exponentially over a sufficiently 'long' period of time.

The second problem of the classical big bang theory is the *flatness problem*. It is essentially the statement that the size of the universe is uncharacteristically big. The present radius of the universe is of the order of ten to twenty billion light years, which works out to be $R \approx 10^{28}$ cm. The corresponding quantity in inverse-energy units is $R/\hbar c = 5.1 \times 10^{41}$ GeV^{-1}. The statement that this is uncharacteristically large is the statement that this is very large compared to the inverse temperature T^{-1} of today. Indeed, $T^{-1} = (2.7\text{K})^{-1} = (2.3 \times 10^{-13}\text{GeV})^{-1}$, and this leads to $RT/\hbar c \approx 10^{29}$. This largeness of $RT/\hbar c$ implies that the density ρ became incredibly close to the critical density ρ_c in the early universe *: $|\rho - \rho_c|/\rho < 10^{-60}$ when

---

* To see how this comes about, we go to the energy equation given in Sec. 3.2: $(\dot{R}/R)^2 = \frac{8}{3}\pi G_N \rho/c^2 - kc^2/R^2$. The critical density ρ_c is by definition the solution of this equation with $k = 0$. Subtracting this equation

$T = M_P = 1.22 \times 10^{19}$ GeV. This requires a tuning of one part in 10^{60} to achieve, and is deemed to be extremely unnatural. This is the flatness problem.

A solution to these problems is given by the *theory of inflationary universe*, first proposed by Alan Guth in 1981, and subsequently modified by A.D. Linde and by A. Albrecht and P.J. Steinhardt in 1982.

The essential idea is the following. If for some reason the universe is stuck at a *constant density* ρ for some time, then during that time the universe will expand at an exponential rate. This gives rise to the inflation of the inflationary universe, and solves the horizon problem as mentioned before. It also solves the flatness problem for this inflation explains how the universe gets to be so uncharacteristically large.

We will now discuss why a constant ρ leads to an exponential inflation. Recall that in the radiation dominated universe, using the Stefan-Boltzmann law, $\rho \propto T^4 \propto R^{-4}$, and the expansion grows like $R(t) \propto t^{1/2}$. The density of the matter dominated universe goes down slower with R, like $\rho \propto R^{-3}$, and the radius $R(t) \propto t^{2/3}$ grows faster than $t^{1/2}$. Extrapolating these results, one might expect the expansion rate at a constant ρ to be faster still. This can actually be seen in the energy equation of Sec. 3.2 *.

To understand physically how the exponential inflation arises, let us go back to the energy equation and Fig. VII.1 and ask whether it is possible to *add* an extra mass m to the surface of the sphere. To do so we must conserve energy. The gravitational attraction of the universe gives this mass m a potential energy proportional to $-M/R \propto -\rho R^2$. For fixed ρ, its magnitude grows with R. The larger the universe becomes, the more can be gained from this potential energy. But the sum of the kinetic energy and the potential energy of this *additional mass* must be zero to conserve energy†. A growing

---

from the corresponding equation with $k = 0$, and using Stefan-Boltzmann law $\rho = aT^4$, one obtains the formula $(\rho - \rho_c)/\rho = \xi\epsilon/T^2$, where $\xi = 3kc^4(\hbar c)^2/8\pi G_N a$ and $\epsilon = (RT\hbar c)^2$. Since RT remains constant with time, this product is given by the present value of 10^{29}, thus giving rise to a very small ϵ, and consequently the flatness problem.

*   Treated as a differential equation in time, the energy equation has an exponential solution for $R(t)$ if ρ is constant.

†   The rest energy mc^2 is ignored because we need to use non-relativistic

potential energy therefore implies a growing kinetic energy, which increases the radius and drives up the inflation rate of the universe, this in turn increases the potential energy and hence the kinetic energy at an even faster rate, and so on and so forth. The result of all this positive feedback is an exponential inflation rate.

Other than solving the horizon and the flatness problems, there are also two other very interesting consequences of this inflationary scenario. First, the mass M of the universe grows! In fact, it is possible for M to grow to the present value, including the masses of you and me and all the galaxies, from a seed of mass as little as 10 kg. This really looks like magic; we seem to be getting something from nothing. Once again, this is not true, because that 'nothing' is actually the potential energy. It is the same energy that a meteor falling through earth's atmosphere derives from to produce the photons illuminating the shooting star.

Secondly, the universe not only grows, but grows with an ever increasing rate. At the end of the inflationary era, when the classical Friedmann-Robertson-Walker universe takes over, the universe is endowed with a large outward velocity, which can be interpreted as the explosion in the original big bang.

We will now consider how the universe can get stuck with a constant density ρ to trigger this exponential growth.

The universe can do so in at least two ways. First, suppose for some reason there is a large density fluctuation which takes a very long time to die away. Then before it dies, the density is stuck at this large value and an exponential growth results. This mechanism is called the chaotic universe mechanism.

Another mechanism is to make use of *supercooling* during a phase transition. To understand how that goes let us first examine the familiar case of condensing steam into water at a temperature of 100 C.

When steam condenses an energy called the *latent heat* is released. This is the energy that drives the wind and feeds the storms in our weather system, and is what your electric kettle has to provide to boil your water. Let us next examine how condensation takes place. It starts by having small water droplets form on some nuclei (seeds). Then these droplets grow and coalesce

---

kinematics for this heuristic derivation of the energy equation.

until finally a body of water is formed. If there are no nuclei present to start the water droplet, it may remain in the steam phase for some time even after the temperature drops below 100 C. The steam is then said to be *supercooled.*

Let us now return to the consideration of the universe. The universe in the low temperature phase (like water) is the standard Friedmann-Robertson-Walker universe, so we will call that the *FRW phase.* The high temperature phase (like steam) is the one that may be supercooled. Its energy has two components: the thermal energy of the radiation in the universe, and the latent heat energy it still retains when supercooled. The first gives rise to an energy density decreasing with the temperature, but the energy density of the second is constant. As the universe expands and cools, at some point the first becomes negligible compared to the second, and from there on the universe will have a constant ρ as long as it remains supercooled. During that period the universe expands exponentially and is called a *de Sitter* universe. We will therefore call the high temperature phase the *de Sitter phase.*

In the original scenario of Guth, droplets of the FRW phase are formed amidst the bath of the de Sitter phase background. The de Sitter phase background expands exponentially in time, whereas the FRW phase expands only like a power. As a result, the de Sitter background grows much faster than the FRW droplets, and it is impossible for these droplets to find one another to coalesce into a big FRW universe at the end.

The modified scenario of Linde-Albrecht-Steinhardt (called the 'new inflationary universe theory') resolves this difficulty by assuming that the droplets are still in a de Sitter phase, but in one that has a slightly lower energy density than the original de Sitter background. Now the droplets also grow exponentially with time just as the background. As the droplet grows, its energy density decreases. Suppose the decrease is very gradual in the beginning to give the droplets time to grow to a large size. When a sufficient size is attained, the energy density in the droplet is allowed to decrease quite suddenly by releasing its latent heat. One single droplet then develops into the FRW universe in which we live. The latent heat so released reheats the universe to a very high temperature.

For the new inflationary scenario to work, parameters have to be carefully tuned to make everything happen at the right time with just the right amount. This fine-tuning of parameters can be avoided if the dynamics of droplet growth is made more complicated. This is sometimes known as the 'extended inflationary universe theory'.

Whatever particles and objects that may have been produced before the inflation will now be greatly diluted. If magnetic monopoles were produced rather copiously before the inflation, after the inflation they will be so diluted as not to be observable at the present time. This may explain why they have never been found.

Summary

The classical big bang theory suffers from a couple of difficulties known as the horizon problem and the flatness problem; they can be resolved if the universe expanded exponentially at the beginning. This is called the inflationary phase of the universe. The theory proposed by Guth, modified by Albrecht, Steinhardt, and Linde, explains how this can come about in the particle physics context. The inflation also provides the mechanism for the initial 'explosion' in the classical big-bang theory.

APPENDICES

A. GENERAL CONCEPTS IN CLASSICAL PHYSICS

B. GENERAL CONCEPTS IN QUANTUM PHYSICS

C. THERMAL PHYSICS

   AND STATISTICAL MECHANICS

APPENDIX A

GENERAL CONCEPTS

IN CLASSICAL PHYSICS

1. THE PHYSICAL UNIVERSE

As its name indicates, *physics* in its broadest sense is the study of nature. It is the study of all inanimate forms of matter and energy, encompassing the smallest constituents of matter and the largest bodies in the Universe, even the whole Universe itself. It is the attempt to observe the formation and evolution of all objects, to probe their inner structure and their interactions with one another. It is also the search for the basic rules, rules with three basic functions: to correlate seemingly disparate facts in a logical structure of thought, to focus thinking along new directions, and, finally, to predict future observable events.

The scope of physics is illustrated by the three charts in Fig. A.1, which show the scales of size, mass, and age (or characteristic time) of some physical objects. We express these fundamental quantities in the *cgs units*: length in centimeters (cm), mass in grams (g), and time in seconds (s or sec). This is one of the two systems of units most often used by physicists, the other being the *mks system* in which the units of length, mass, and time are meter (m), kilogram (kg), and second (s). To deal with measurements that may be very small or very large, it is convenient to adopt the exponent notation of numbers — as in 10^n (written as 1 followed by n zeroes) and 10^{-n} (a decimal point followed by $n-1$ zeroes then 1) — some of which are assigned names with suggestive prefixes (such as kilo $= 10^3$, mega $= 10^6$, giga $= 10^9$, milli $= 10^{-3}$, micro $= 10^{-6}$, nano $= 10^{-9}$, pico $= 10^{-12}$, femto $= 10^{-15}$, etc.). It is evident from the figure that the physical universe englobes objects differing widely in nature and separated by huge factors in size, mass, and age. Less evident, but no less significant, is the fact that these dry numbers give more than just the numerical values of things; they epitomize the fruit of labor by several generations of scientists, and represent important signposts for efforts by new generations.

We characterize size, mass, and time as 'fundamental' because all other physical quantities can be considered as derived quantities, expressible in units that are various combinations of powers of the base units, which are cm, g, and s in the cgs system. Among the more familiar examples, the cgs unit of energy is erg $=$ g cm^2 s^{-1} and the cgs unit of electric charge is esu $=$ g$^{1/2}$ cm$^{3/2}$ s^{-1}. In atomic and subatomic physics, one often prefers a smaller, more appropriate unit of energy, the electron volt (eV), defined as $1\text{eV} = 1.60 \times 10^{-12}$ erg. The temperature is a measure of heat, and heat is a form of energy (Appendix C). Thus, the unit of temperature (degree Kelvin or K) can be considered as a derived unit; it is related to the unit of energy by a numerical factor which one may identify with a numerical constant called the Boltzmann constant k.

The Boltzmann constant is one of the many *fundamental constants* which recur again and again in physics. They are considered as universal, i.e., unchanging in any physical circumstances and, of course, constant in time. Their numerical values are the results of many delicate and laborious experiments (and therefore subject to observational errors) and are as yet unexplainable by theories. Some of the more important physical constants are given in Table 1.

The popular belief that scientific information doubles every seven years may well be a myth, but it is nevertheless true that a vast body of facts has been accumulated by mankind ever since the ancient Greeks started pestering each other with endless discourses and questions about earth and heaven, thus embarking on a great intellectual adventure that modern scientists are still pursuing. But facts alone do not necessarily give understanding, and knowledge by itself is not science. Explanations of specific phenomena in

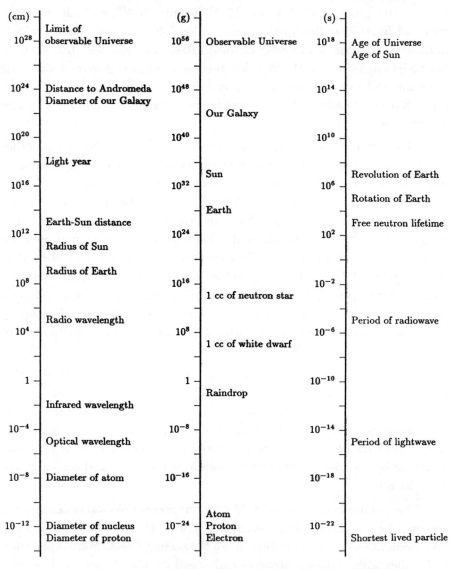

FIG. A.1 *Scales of length, mass, and time of the physical universe.*

TABLE 1
Physical constants (cgs units)

Speed of light in vacuum	$c = 3.0 \times 10^{10} \, \mathrm{cm \, s^{-1}}$
Gravitational constant	$G = 6.67 \times 10^{-8} \, \mathrm{cm^3 \, g^{-1} \, s^{-2}}$
Planck constant	$h = 6.63 \times 10^{-27} \, \mathrm{erg \, s} \; (\mathrm{erg} = \mathrm{g \, cm^2 \, s^{-2}})$
Boltzmann constant	$k = 1.38 \times^{-16} \, \mathrm{erg \, K^{-1}}$
Stefan-Bolztmann constant	$\sigma = 5.67 \times 10^{-5} \, \mathrm{erg \, cm^{-2} \, s^{-1} \, K^{-4}}$
Elementary charge	$e = 4.80 \times 10^{-10} \, \mathrm{esu} \; (\mathrm{esu} = \mathrm{g^{1/2} cm^{3/2} s^{-1}})$
Electron mass	$m_e = 9.11 \times 10^{-28} \, \mathrm{g}$
Proton mass	$m_p = 1.67 \times 10^{-24} \, \mathrm{g}$

terms of known facts must be generalized into rules that apply to larger classes of phenomena, and rules must further be distilled into a few universal brief statements, or *laws* that cover not only observed cases, here and now, but also unobserved cases, anywhere and at any time. Finally, laws are further organized into a coherent and logical structure, called *theory*.

In physics, existing theories can be divided in three broad categories: *mechanics*, the *theories of matter and interactions*, and the *physics of large systems*. Mechanics, the study of motion, plays a unique role central to all of physics. On the basis of the oldest of its three components, *classical mechanics*, modern physicists have constructed *quantum mechanics* and *special relativity*, two superb theories that underpin all contemporary physics and serve as guiding principles for further research. Loosely speaking, the fundamental dynamical theories deal with forces. But, in one of the greatest developments of this century, it was discovered that *forces* (or interactions) are in fact intimately related to *matter*. All matter is composed of a few species of fundamental, or elementary particles, and these particles interact via four fundamental forces: the gravitational force, the electromagnetic force, the strong force, and the weak force. Finally, when we deal with macroscopic bodies, the presence of a very large number of component parts requires specialized techniques. The study of the behavior of such systems is the domain of both *thermodynamics* (a theory based on a set of four general rules deduced from observations of controlled experiments on macroscopic bodies) and *statistical mechanics* (a statistical treatment of the mechanics of a very large number of particles).

In this Appendix, we give a very brief discussion of some general aspects of classical mechanics. Quantum mechanics will be the subject of Appendix B, while thermal physics and statistical mechanics will be considered in Appendix C. Other more specialized parts of physics will be dealt with at appropriate places in the main body of the book.

2. MATTER AND MOTION

2.1 Space and Time

Whenever physicists describe their observations or formulate their theories, they cannot fail to make use of certain quantities called *space* and *time*, two basic physical concepts that are intuitively evident to us all and yet cannot be formally defined in terms of any simpler entities. Since space and time cannot be defined, ways of measuring them must be devised — a distance in space is measured by comparing it with some unit length, and a time interval is measured in terms of the period of some cyclical phenomenon. The concepts of distance and time are ultimately defined by the operations carried out in making the measurement. This measurement then is equivalent to the missing formal definition.† Such a measurement, which is absolutely essential to the development of phyics, may appear trivial to us now, but had not always been possible in the past. It was perhaps not a coincidence that the first detailed study of the pendulum and the first systematic experiments on motions were carried out by the same person, Galileo Galilei.

It is a fact of experience that the space in which we live is three dimensional and, to a very good approximation, Euclidean. That our space is *three dimensional* — i.e., any point can be located relative to another by specifying three and only three coordinates — is intuitively clear. The assertion that the geometry of our space is *Euclidean* — i.e., the postulates of Euclid are valid for our world — is less evident but nevertheless true. A simple way to

---

† In his book *Logic of Modern Physics*, P. W. Bridgman, physicist and philosopher, writes: 'The true meaning of a term is to be found by observing what a man does with it, not what he says about it.'

check is to measure the sum of the interior angles of a plane triangle; it is always found to be 180° to within the measurement error. The Euclidean geometry is valid on or near the Earth but breaks down in the immediate vicinity of very massive bodies (e.g., the Sun).

Day to day experience tells us that time flows in one direction only, from past to present to future. This is indeed confirmed by experiment; time reversal never occurs in the macroscopic world. Time has another important property: it is *absolute*, that is, it flows at a rate that does not depend on position and velocity. Suppose we have two identical well-made clocks, initially at rest and perfectly synchronized. Suppose one of them is transported in motion along some path at varying moderate speeds, then brought back to its initial position. If we can verify that the two clocks are still synchronized, we say that time is absolute. Experiments show that time is indeed absolute to a very high precision in ordinary circumstances on Earth. But this notion of absolute time is not exact in any circumstance; in particular, it fails when very rapid motions or effects of gravity are involved: clocks moving at high velocities or exposed to strong gravity are found to lose or gain time relative to stationary clocks.

Classical (or Newtonian) physics is based on the assumption that space is Euclidean and time absolute. Because Newton's theory rests on postulates that do not exactly hold in all situations, we admit that it is not an exact theory. It is only an approximation to the real world, but nevertheless an excellent approximation, quite adequate in most circumstances we encounter on Earth or in space. Furthermore, its formulation is restricted to certain special frames of reference, called the *inertial reference frames*, i.e., those frames that are at rest or, at most, moving at a constant velocity relative to 'fixed' stars. Thus, a reference frame fixed on the ground is an inertial frame, but one fixed on an accelerating train is not. As any good theory, Newtonian physics has few, simple, and plausible assumptions, which are general enough to let it 'grow' to maturity and in various directions, and eventually blend itself with new, improved theories.

2.2 Three Laws of Motion

With the notions of space and time understood, one can define velocity and acceleration in the usual way. The only other concept we need is that of *force*. Again, this concept comes naturally to us, as part of our everyday experience. We can instinctively perceive a force wherever it exists and can gauge its strength. We can feel it as the airplane carrying us on board is taking off or landing; we know the amount of push or pull we must exert to set an object at rest in motion. Therefore, we can consider force as a basic entity and define it by measuring the effect it would have on some standard device (for example, the amount of stretching of a standard spring). With those elementary definitions on hand, we can state Newton's three laws of motion, which together form the basis of classical mechanics and dynamics.

First Law: *A body at rest remains at rest and a body in a state of uniform linear motion continues its uniform motion in a straight line unless acted on by an unbalanced force.*

This statement (often referred to as the law of inertia) simply means that a body persists in its state of rest or motion unless or until there exists an unbalanced force (i.e., when the sum of forces acting on it does not vanish). What happens then is described by the second law.

Second Law: *An unbalanced force (*F*) applied to a body gives it an acceleration (*a*) in the direction of the force such that the magnitude of the force divided by the magnitude of the acceleration is a constant (*m*) independent of the applied force. This constant, *m*, is identified with the inertial mass of the body.*

Stated algebraically, this law reads: $F = ma$, which is referred to as 'Newton's equation of motion' because it describes, for a given force, the motion of the body via its acceleration. Three points should be noted. First, it is a vectorial equation; it relates the three components of a vector (force) to the three components of another vector (acceleration). The second notable feature of the equation is its general character; it is applicable to any force, independently of its nature. On the other hand, to solve the equation for the body's acceleration, one must know the exact algebraic expression of the force, which is a separate problem. Finally, the Second Law plays a dual role: it gives a law of motion and also states a precise *definition of mass* by indicating an operational procedure to measure it. At this point, turning things around, one could choose mass as a basic quantity, define a mass unit, and use the Second Law to derive the force.

Third Law: *If a body exerts a force of any kind on another body, the latter exerts an exactly equal and opposite force on the former.*

That is, forces always occur in equal and opposite pairs in nature. We walk by pushing backward with our foot on the ground while the ground pushes the sole of our foot forward. The pull of the Sun on the Earth is equal to the Earth's pull on the Sun, and the two forces are along the same line but opposite to each other. Similarly, the positively charged nucleus of the hydrogen atom exerts an attractive force on the orbiting electron, just as the electron exerts an equal and opposite force on the nucleus.

As long as we are describing the kinematics of an isolated body, we just need to know how its velocity changes. But when we deal with the dynamics of the particle, we must take into account its mass as well. It is then more meaningful to deal with mass and velocity together in a single combination, called the *momentum*, $\mathbf{p}$. The momentum of a body is given by the product of the mass, m, and the velocity, $\mathbf{v}$, of the particle. It is a vector pointing in the same direction as the velocity. Since the acceleration of a particle is the rate at which its velocity changes, the Second Law simply tells us that the force exerted by the particle is the rate at which its momentum changes. It follows then: (1) in the absence of unbalanced forces, the momentum remains constant, and (2) the total momentum of an isolated two-particle system is a constant of motion. We recognize in these results a simple restatement of the First and Third Laws, which, however, has the virtue of making a generalization to systems of more than two particles almost trivial: if a system is isolated so that there are no external forces, the total momentum of the system is constant (*law of conservation of momentum*). If there are external forces, the rate of change of the momentum is vectorially equal to the total external force. The law of conservation of momentum has a more general validity and a deeper meaning than its derivation given here would indicate; it is a manifestation of the existence of a symmetry of space (Chapter III).

2.3 Angular Momentum

So far, we have discussed motion as if it involved only translational motion, i.e., a motion that keeps the relative orientation of different parts of the body unchanged. But pure translation rarely occurs in nature; there is always some degree of revolution or rotation in any motion. In revolution, a particle is acted on by a force that continually changes its direction of action, making it move along some curved path. In rotation, no straight line (except one) connecting any two points in the body remains parallel to itself; the line that does is called the body's axis of rotation. In pure rotation, a body changes its orientation without changing its position, and therefore must be acted on by something other than simple force.

Suppose we nailed a long rod loosely at one end to the ground. If we pushed or pulled the rod near its free end, the applied force would generate an equal and opposite reaction of the rod at the fixed end. Two equal and opposite forces, which are not on the same line and therefore remain uncompensated, act on the rod and make it rotate about the fixed end. Thus, a correct measure of the effects observed cannot be given by the applied force alone, but rather by that combination of forces, called a *torque* (meaning 'twist'). Clearly the amount of torque depends on where the force is applied. It also depends on the orientation of the force with respect to the rod (or generally on the position vector that defines the location of the application point relative to a fixed point); in particular, it vanishes if the force is parallel to the rod. A nonvanishing torque is a vector perpendicular to both the force and the position vector. Just as force produces change in (linear) momentum, torque produces change in *angular momentum*; if no torque exists, the angular momentum does not change; it is said to be conserved.

The angular momentum ($\mathbf{L}$) of a particle is a vector perpendicular to both the position vector ($\mathbf{r}$) and the momentum ($\mathbf{p}$) of the particle relative to some origin of coordinates, and whose magnitude is $rp\sin\theta$, where θ is the angle between the two vectors $\mathbf{r}$ and $\mathbf{p}$. As we have seen, in the absence of forces, a particle moves along a straight line at constant velocity . The direction of its angular momentum relative to an arbitrary reference point is perpendicular to the plane defined by the line of motion and the reference point. Its magnitude is simply the product of its momentum (p) and the shortest distance between the origin of coordinates and the line of motion ($r\sin\theta$). Thus, it is a constant vector for the chosen origin: in the absence of forces both vectors $\mathbf{p}$ and $\mathbf{L}$ are constants of motion. If the origin changes, $\mathbf{L}$ changes to a new constant vector. In particular, we can make $\mathbf{L}$ vanish by locating the origin on the line of motion ($r\sin\theta = 0$). Now, take a particle in uniform circular motion, i.e., a motion with constant velocity along a circle. Its momentum is constant in magnitude but continuously changing in orientation. There is a force (*centripetal force*) directed toward the center of the circle continuously acting on the particle to keep it on its circular orbit.

The angular momentum of the particle, defined with respect to the center of the circle taken as the origin of coordinates, is simply $L = rp$, the product of the particle momentum and the circle radius; it is perpendicular to the plane of the circle. So, again, there is no torque, and $\mathbf{L}$ is a conserved vector although the momentum is not. But, in contrast to the previous case, there is no way we can make $\mathbf{L}$ vanish by some choice of the reference point. Thus, angular momentum is a property characteristic of rotational motion.

Kepler's Second Law ('*the radial line segment from the Sun to a planet sweeps out equal areas in equal times*') can be seen as an example of application of the principle of conservation of angular momentum. Since the gravitational force of the Sun on a planet is along the Sun-planet line, there is no torque, and the angular momentum of the planet (which is essentially equal to the area swept out in a unit time) is conserved. Just as the conservation of momentum implies a symmetry of the laws of nature, so does the conservation of angular momentum. This symmetry — the rotational symmetry — implies that space is isotropic; the geometry of space is the same in all directions.

2.4 Work and Energy

Let us now turn to the concepts of *work* and *energy*. To simplify the discussion, we take the case of one-dimensional motion. If a constant force F acting along the x-axis causes a particle to move some given distance x, then the work done is defined by $W = Fx$. This is a reasonable definition, because it corresponds precisely to the effort we would provide and what we would be paid for to push the object that distance, regardless of the nature of the object. If the force applied is not constant over the distance being covered, then a careful summation (*integration*) over all small segments composing the distance must be made to get the correct result. When we do work on a body, the body changes by acquiring 'energy', a quantity defined such that the *acquired energy* equals the work done on the body by the force. If we apply a force to a particle of mass m at rest on a smooth (frictionless) surface in a vacuum, the particle will be moving at a definite speed v after the force stops acting on it. The force applied is $F = ma$, and the work done is $W = \frac{1}{2}mv^2$, which we equate with the acquired energy. Similarly, if the particle has initial speed v_1 and acquires a final speed v_2, the work done is $W = \frac{1}{2}mv_1^2 - \frac{1}{2}mv_2^2$. This is the acquired energy. Since the energy appears in the form of motion, it is called the *kinetic energy* of the particle, and defined as $KE = \frac{1}{2}mv^2$.

But energy may appear in other forms as well. When it represents the capacity of the particle to do work by virtue of its position in space, it is called the *potential energy*. A simple example is that of a particle moving under the influence of a constant gravitational force (e.g., the Earth's). To lift vertically a body of mass m a height z above the surface of the Earth, the work we must do on the particle is $-mgz$, where g is the gravitational acceleration and the minus sign comes from the downward direction of the force. The function $V(z) = mgz$ is called the gravitational potential energy. It is defined relative to the surface of the Earth, that is, the particle's energy at height z is larger by mgz than it was on the Earth's surface. If now the particle is released to fall back freely from this height, it will reach a speed v when it hits the ground. If gravity is the only force acting on the particle, the work done by gravity on the particle must be the same in both directions. Thus, the sum $E = KE + V(z)$, called the *mechanical energy*, has the same value at any height although the two energy components may be converted back and forth. This is the law of *conservation of mechanical energy*.

So far we have ignored dissipative forces, e.g., friction. If now we attempt to push a heavy object on a rough surface, we must do more work to obtain the same result because not all work done will be expended into changing the body's mechanical energy; some energy will be dissipated in heat, for which we do not usually get paid. In this case, the mechanical energy clearly is not conserved. However, the loss is only apparent because we are taking the macroscopic point of view, i.e., working on the level of the whole body, ignoring its microscopic components. As the body is rubbing against the surface, the energy lost by the body is in fact transmuted into disorderly kinetic and potential energy of the atoms in the body and the surface. This disorderly mechanical energy of atoms is what is meant by heat. No energy has been lost, only part of it has been transformed into another kind of energy. To be fair, we should be paid for work done on both the body and the atoms it contains.

Energy can manifest itself in many other ways as well. Some of the basic forms are the following. Chemical energy is the term often used for molecular energy or atomic energy, which are, respectively, kinetic and potential energy of atoms in molecules or electrons in atoms. Nuclear energy is kinetic and potential energy of particles (neutrons and protons) contained in the nuclei of atoms. Electromagnetic energy is the form of energy carried by all sorts electromagnetic radiation, such as moonlight and TV signals. Another important form of energy is mass, as expressed by the famous equation discovered by Einstein: $E = mc^2$. This equation means that an amount m of mass is equivalent to an amount E of energy given numerically by product of m and the square of the speed of light c. The equation can be read in either direction: energy can be transformed into mass, and mass can be converted into energy; energy and mass are equivalent.

It is a general property of our physical world that the total amount of energy of any isolated system never changes in time, although the form of energy may change. This is the *general law of conservation of energy*. Again, this conservation law arises from a symmetry of space, the irrelevance of the absolute measure of time. To summarize the above paragraphs, let us restate that the laws of classical mechanics depend neither on the origins of space and time coordinates nor on the orientations of the coordinate system. Also, they are insensitive to the state of motion of physical events, as long as this motion is uniform, rectilinear, and free of rotation. All these invariance properties are products of experience, not *a priori* truths. It is only because regularities such as these exist in nature that we can hope to discover its secrets.

As a simple application of the above discussion, consider a particle in uniform circular motion with constant speed v along a circular path of radius r under the influence of some force F. The situation just described can be seen as an idealization of the motion of an artificial satellite around the Earth (F=gravitational force), or of an electron around the atomic nucleus (F=electric force). One can show (with the help of calculus) that the magnitude of the centripetal acceleration is $a = v^2/r$. It follows from $F = ma$ that $mv^2 = rF$, which represents work done to move the particle to infinity (where it has zero potential energy by convention). Hence the kinetic energy of the particle, $KE = \frac{1}{2}mv^2$, satisfies the equation $KE = -\frac{1}{2}PE$, where PE stands for the potential energy of the particle. Although the derivation relies on a particular situation, the relation obtained gives the statement of an important general result of classical mechanics, the so-called *virial theorem*. The theorem is statistical in nature in the sense that it involves time average of various mechanical quantities. It is applicable to a large class of physical systems (those in quasistationary motion, in which coordinates and momenta always remain within finite limits).

As another application, let us consider how the *law of universal gravitation* could be derived. To begin with, we observe that planets must be under the action of some net (attractive) force because otherwise they would be moving in straight lines instead of curved paths, as required by Newton's First Law. Such a force acting on a planet must be directed at any instant toward the center of motion, otherwise Kepler's empirical Second Law could not be satisfied. For an elliptical orbit, the center of motion is one of the foci of the ellipse; for a circular orbit, the center of the circle. Now, Newton has proved mathematically that the centripetal force acting on a body revolving in an ellipse, circle, or parabola must be proportional to the inverse square of the distance of the body to the focus.

What is the origin of the force? Newton suggested that the force governing the motion of the planets around the Sun, or the revolution of the Moon around the Earth is of the very same nature as the gravitational attraction that makes an apple fall to the ground. Whatever their makeup, all heavy bodies experience the same basic force. Consider any two objects, for example the Earth (mass m_1) and a stone (mass m_2). We know by experience that the stone is pulled toward the Earth by its weight, which is proportional to its own mass m_2; but, similarly, there must be a weight of the Earth pulling it toward the stone and proportional to m_1. By the symmetry of action and reaction, we see that the gravitational force is proportional to both m_1 and m_2. Assuming this force to depend only on the masses of the two bodies in interaction and the distance between them, we come to the conclusion that the gravitational attraction between any two objects of masses m_1 and m_2 separated by a distance r is given in magnitude by $F = Gm_1m_2/r^2$. It is a vector directed from one body's center-of-mass to the other's. This is Newton's famous *law of gravitation*. The constant of proportionality, G, has come to be known as the universal gravitational constant.

3. WAVES AND FIELDS

The physical universe of the late nineteenth century was so dominated by Newton's ideas that it must have appeared to be endowed with a superb unity. Not only could the motion of all bodies, whether on earth or in space, be explained by the same laws of motion, but the invisible world of the atoms was also made part of the realm of mechanics when heat was treated as a mere mechanical phenomenon by the kinetic theory of gases. Certainly, there would not be any problems, most physicists of the time must have thought, that could not be ultimately solved within the existing framework.

But problems there were, and not necessarily anodyne. First, what is the nature of light? Is light composed of something similar to microscopic particles, as Newton believed, or of something immaterial akin to wavelike impulse, as Christiaan Huygens, Thomas Young, or Augustin Fresnel advocated? Then, how to incorporate into the existing mechanical framework the increasingly large number of new electric and magnetic phenomena observed by Hans Oersted and Michael Faraday, among others? Finally, what to do with the clearly unacceptable action-at-a-distance interpretation — i.e., that two particles interact even though they are not touching — as suggested by Newton's law of gravitation and the newly discovered Coulomb's law of electric force? It turns out that the answers to these questions are not unrelated, all relying on two new concepts, *waves* and *fields*, which are to play key roles in the physics of the twentieth century.

3.1 Waves and Wave Propagation

Before discussing the nature of light, let us pause briefly to describe what turns out to be an analogous phenomenon, the vibrations of a string that is fixed at one end. If we take hold of the free end of the string and move it rythmically up and down, imparting to it both *motion* and *energy*, successive segments of the string, from the free end down the line, also move up and down rythmically. A crest (maximum displacement) is generated, then a valley, and both travel down the string, soon to be followed by another maximum, then another minimum, and so on. Thus, we have a periodic wave traveling along the string, carrying with it energy (due to the motions and positions of the particles) from one end to the other. The speed with which a crest travels may be taken as the speed of propagation of the wave, v; the distance between two successive crests is called the *wavelength*, λ; and the time needed for a crest to cover that distance is called the *period* of vibration, T. The three quantities are related by $v = \lambda/T$, and the reciprocal of T gives the *frequency* of vibration. Here, the wave propagates horizontally while segments of the string move up and down; such a wave is called a *traveling transverse* wave.

Now, if we have a long rubber cord or a row of particles connected by massless springs and if we snap the free end of the string back and forth horizontally, a disturbance is forced on the string, and is transmitted from one particle to the next. The disturbance begins traveling along the string and will produce successive variations of particle density — condensations and rarefactions of particles, which oscillate in the same direction as the motion of the wave itself. Such a wave is called a *traveling longitudinal* wave. Let us imagine we have a number of such identical strings, of either kind, fixed to the same wall and forced to vibrate together by the same disturbance. We may then define a *wavefront* as an imaginary surface, generally perpendicular to the direction of propagation of the wave, which represents the disturbance as it travels down all the strings.

Waves have a characteristic property that distinguishes them from particles. When two or more waves propagate in the same medium, they fuse together to form a new wave. We call this phenomenon an *interference*; it can be either *constructive* (if the incident waves reinforce) or *destructive* (if the incoming waves tend to cancel out). Streams of (classical) particles do not seem to exhibit such behavior under normal conditions.

3.2 Nature of Light

To the question, is light particle-like or wavelike?, experiments have given a clearcut answer: transmission of light exhibits all the properties characteristic of propagation of waves. One of the first decisive experiments which helped to establish the wave theory of light is the famous Young's double-slit interference experiment, to be described in Chapter I. But we will find it instructive to repeat here the historical argument. It hinges on explaining why a light ray crossing the interface between a rare and a denser medium is *refracted*, that is, deflected toward the normal to the interface. If light is particle-like, one would naturally expect the light ray to be slowed down by the denser medium (more precisely, its velocity's component perpendicular to the interface to be reduced, and the component parallel to the interface unchanged) and hence to be deflected *away from the normal*, contrary to observations. However, if the 'light particles' are accelerated at the interface by some unknown force, then their path would be bent *toward the normal*,

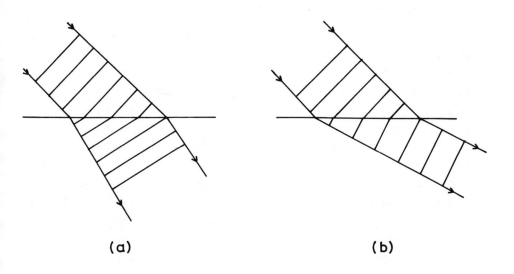

(a)   (b)

FIG. A.2 *Refraction of a light pulse going from air to water; wavefronts are at right angles to the light ray and delineate successive maxima. In (a) it is assumed that the speed of light is smaller in water than in air, and so wavefronts travel closer together in water; in (b) light is assumed to move faster in water.*

as observed.

On the other hand, if light is wavelike, one would expect that when a light pulse strikes an interface at an oblique angle, the wavefront is split into two, one traveling in the first medium with the old speed, the other in the second medium with a new speed. From Fig. A.2, one sees that light will be bent toward the normal if its speed is *smaller* in the new medium, and will be deflected away from the normal if its speed is *greater*. This means that if light is wavelike, it is slowed down in going from air to water, just the opposite of the behavior of a light particle. The predictions of two competing theories could not be more unambiguous. To decide between the two theories, it suffices to compare the speed of light in air with that in water or glass. It was not until mid-nineteeenth century that such a delicate experiment could be carried out, with the result by now well-known: light travels faster in air than in water.

3.3 Electric and Magnetic Fields

The *electric force* between two *stationary*, charged particles was experimentally discovered by Charles-Augustin de Coulomb in 1785. Its magnitude is given by $F = Kqq'/r^2$, where q and q' are the two interacting charges, r is their separation distance, and K a numerical constant; its direction is along the line joining the two particles. The electric force is repulsive if the charges are of like signs, and attractive if the charges are of unlike signs. When the particles are set *in motion*, there is an extra force acting on these particles; this extra force is called the *magnetic force*. The first concrete evidence that moving charges induce a magnetic force was given by Hans Oersted in 1820 when he showed that a strong electric current (flow of charges) sent through a wire aligned along the north-south direction caused a magnetic compass needle, originally set parallel to the wire, to rotate by 90° and settle in the east-west direction. When the direction of flow of the current was reversed, the needle immediately rotated by 180° and aligned itself perpendicular to the wire. The behavior of the needle indicated that it was acted on by a force quite unlike the electric force, emanating from the current and perpendicular to it.

Anybody who, like every physics student, has seen the regular alignment of small bits of thread between a pair of charged plates or the pattern of iron filings near a magnet, is convinced that space is modified by the presence of charges or magnets. Something new must have appeared in the intervening space.

To discover what it is, let us perform this experiment. Let us attach a small body to the end of a nonconducting massless string, and give it a small charge q, so small that it will not affect any system of charges we want to study (we assume the charges to be at rest for the moment). As we place this device at every point in space, we realize that our probe experiences an electric force which depends on the location of the body, the source charges that set up the action, and the probe charge q itself. To obtain a quantity intrinsic to the system, independent of q, we define at any given point P the *electric field strength* (call it $\mathbf{E}$) as the net force $\mathbf{F}$ on q at P divided by q. $\mathbf{E}$ is a vector, as is $\mathbf{F}$. We could fill the whole three-dimensional space with imaginary arrows to a given scale to represent all such vectors. We call the full set of the $\mathbf{E}$-vectors in a given region the *electric field* in that region. As

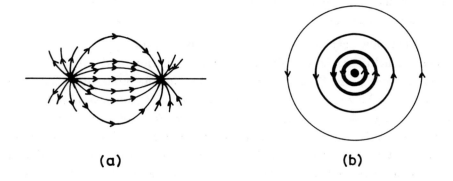

(a)   **(b)**

FIG. A.3 *(a) Representative electric field lines generated by two charges equal in magnitudes and opposite in signs. (b) Representative magnetic field lines around a straight wire carrying an electric current. The current emerges perpendicular to the plane of the page.*

an aid to visualization, we may represent the electric field by *electric field lines*. At every point along such a line, the electric field is tangential to the line. Furthermore, the density of the lines is proportional to the field strength. When lines crowd together, we have a relatively stronger field; when they spread out, we have a relatively weaker field (Fig. A.3).

The *magnetic field* (**B**) can be similarly represented by *magnetic field lines*. Repeating Oersted's experiment by placing a compass needle at various points around a current-carrying wire, we discover that the force exerted by the current on the magnet is circular. We can mentally picture the wire as surrounded by concentric lines of force. The tangent to a field line indicates the direction of the field **B**, and the density of field lines, the strength of the field.

What Oersted had shown was that a *steady* electric current produced a constant magnetic field around the current-carrying circuit. Inversely, one may ask, could a constant magnetic field or a steady current flowing in one circuit generate a current in another circuit nearby? The answer was given by Michael Faraday, who discovered the general principle of *electromagnetic induction*: (1) a constant magnetic field or a steady current in one wire cannot induce a current in another wire; (2) only *changing* lines of magnetic force can induce a current in a loop; the change of the force lines can be caused either by moving a magnet relative to the loop, or by suddenly varying the current in a wire nearby.

What now transpires from these results is that: (1) the electric and magnetic fields are sensitive to both space and time; and (2) they are interdependent; if one changes in any way, so does the other. It remained for James Clerk Maxwell to discover the exact physical laws governing the electric and magnetic fields, which he stated in the form of a set of differential equations for **E** and **B**. These equations, which form the basis of his electromagnetic theory, describe how the space and time variations of the electric and magnetic fields can be determined for a given electromagnetic source, or distribution of charge and current.

At points far removed from the source, Maxwell's equations are reduced to two separate equations, one for **E** and one for **B**. They are called the wave equations because they show that the two fields propagate together as periodic oscillations, both perpendicular to each other and perpendicular to the direction of propagation of the waves. Thus, the electric and magnetic fields behave as transverse waves. They cause any electric charges or magnetic poles found anywhere in their region of action to oscillate with a characteristic frequency, in the same manner as a wave transmitted along a string would force particles on its path to fluctuate.

To appreciate fully the importance of Maxwell's theory we must realize that his *electromagnetic wave propagates with the same speed as light*. Generally, a wave equation relates the spatial variations of a field amplitude, A, to its time variations. The spatial variations of A are essentially A divided by the square of a small distance, symbolically A/d^2, and the time variations are essentially A divided by the square of a small time interval, or A/t^2. These two quantities are dimensionally different, one has the dimension of the reciprocal of a squared length, the other the dimension of the reciprocal of a squared time. They cannot enter as two terms in the same equation unless the denominator of the second term is multiplied by the square of a speed. This speed is precisely the speed of propagation of the wave, which Maxwell showed to be numerically equal to the speed of light. It was a most remarkable result with far-reacing implications. His theory also predicted that a beam of electromagnetic waves would be reflected on metallic surfaces and that it would be refracted on entering a layer of glass. In other words, electromagnetic radiation is expected on theoretical grounds to behave in every way like light.

Maxwell has thus unified not only electricity and magnetism but also light in a single theory, the electromagnetic theory of radiation, which ultimately encompasses the whole electromagnetic spectrum, from the longest radio waves to the shortest gamma rays. The correctness of his views was eventually confirmed by Heinrich Hertz, who demonstrated by experiment that an oscillating current indeed sent out electromagnetic waves of the same frequency as the emitting oscillations, and that these waves carried momentum and energy that could induce a fluctuating current in a wire nearby. Maxwell gave the electromagnetic theory mathematical rigor, Hertz gave it physical reality.

To summarize, the concept of *force* can be conveniently replaced by the concept of *field* of force. Instead of saying that a particle exerts a force on another, we may say that a particle creates a field all around itself which acts on any other particle placed in its zone of action. While in nonrelativistic physics, the field is merely another mode of describing interactions of particles, in relativistic mechanics, where the speed of light is considered finite, the concept of field takes on a fundamental importance and, in fact, acquires physical reality of its own. The picture of particles interacting at a distance gives place to a picture of interaction by contact, in which a particle interacts with a field, and the field in turn acts on another particle, such that there is overall conservation of energy and momentum. Applied to electricity and magnetism, the concept of field is essential to a unified treatment of these two phenomena. The unified field thus introduced, — the *electromagnetic field* — generates *electromagnetic radiation*, which behaves in every way like *transverse waves* in regions far from the source that produces it . Light is but one form of such radiation. The field concept, introduced by physicists of the nineteenth century, will blossom to full significance in the physics of the twentieth century.

APPENDIX B

GENERAL CONCEPTS

IN QUANTUM PHYSICS

1. INTRODUCTION

Quantum mechanics is now known to be the correct framework theory for describing all physical phenomena in detail, that is at the microscopic level. Its predictions generally disagree with those of classical mechanics (See Appendix A). The numerical disagreement is, however, pronounced only in the domain of the very small, which is roughly the case for objects having very small inertia, such as electrons and many other atomic and sub-atomic particles. For the commonly experienced objects of sensible magnitudes, the deviations are practically negligible. It is this circumstance that tends to hide from us the otherwise profoundly different character of quantum mechanics. This Appendix briefly acquaints you with some of the remarkable features of Quantum Mechanics, which is undoubtedly the profoundest of all contemporary scientific thoughts.

2. HEISENBERG'S UNCERTAINTY PRINCIPLE

In Classical Mechanics, the state of a particle at any given instant of time is specified completely by giving simultaneously its position and momentum (or velocity) at that instant of time. (The statement can, of course, be generalized to a system of particles). This, together with the knowledge of all the forces acting on the particle will completely determine its trajectory which is then calculable through *Newton's laws of motion*. In general the accuracy of our prediction will be limited by the accuracy of the initial data which is always beset with the ubiquitous errors of measurement, by our incomplete knowledge of the forces acting on the particle, and not a little by our computational capability. But nothing in principle prevents our predicting the trajectory of the particle with absolute precision. Besides, unknown and unobserved as it may be to us (the observers), the particle is assumed to have a well defined trajectory, existing as an unobserved objective reality. Now, all this has to be renounced in Quantum Mechanics — the position and the momentum cannot be determined simultaneously with arbitrary high degrees of precision, even in principle. More specifically, if the instantaneous particle position, its x-coordinate say, is determined to within an interval of uncertainty Δx and if the corresponding x-component of its momentum p is determined simultaneously to within an interval of uncertainty Δp, then the *Heisenberg uncertainty principle* asserts that the product of these two uncertainties will be greater than or equal to an irreducible minimum: $\Delta x \Delta p \geq \hbar/2$, where $\hbar$ is Planck's constant divided by 2π ($\hbar \simeq 10^{-34}$ Joule-second). The very act of measurement involved in determining any one of them perturbs the other sufficiently so as to satisfy the above inequality, often by a wide margin, no matter how carefully and cleverly we design the measuring apparatus. This reciprocal latitude of fixation, or shall we say the 'frustration of errors' has observable consequences. For a particle of mass 'm' localized within a box measuring Δx on the side, the uncertainty of momentum would be of the order of $\hbar/\Delta x$. The associated 'zero-point' kinetic energy would be about $(\hbar/\Delta x)^2/2m$. Thus, for $\Delta x \sim 1$Å, the 'zero-point' energy for an electron is about 3 eV (1 electron-Volt = 1.6 x 10^{-19} Joule). This will exert a pressure on the walls of the cube of several million times the atmospheric pressure. The lighter the particle, the greater is the zero-point energy. It is precisely this zero-point energy that prevents the inert gases of *light* atoms from solidifying even at the absolute zero of temperature. Thus helium remains liquid down to the lowest temperatures known. True that hydrogen, the *lightest* of all elements, does form a solid at low enough temperatures, but the reason is that the attractive potential between the two hydrogen atoms is sufficiently strong to 'contain' this zero-point motion. For the inert atoms like those of helium, the attraction is relatively much too weak. Again, it is this zero-point energy that prevents the electron in a hydrogen atom from collapsing on to the proton and staying stuck there. The uncertainty principle holds for other pairs of dynamical variables too, for example, energy and time, or angle and angular momentum etc.

3. WAVE FUNCTION AND PROBABILITY: STATE OF THE SYSTEM

Having renounced the classical mechanical idea of sharply defined trajectories, the state of the particle is now given by a quantum mechanical *wave function* $\psi(r,t)$ which is a complex function of position and time, i.e., $\psi =\mid \psi \mid e^{i\theta}$. The wave function has a statistical interpretation. Thus, $\mid \psi(r,t) \mid^2$ gives the probability (density) of finding the particle at point r at time t, *if* such a measurement is made. The associated probability current (density) is given by $(\hbar/m) \cdot (\mid \psi \mid^2) \cdot$ (gradient of phase θ). The wave function is to be determined in any specific case by solving a differential equation, the Schrödinger equation that now replaces Newton's equations of motion. The point is that ψ contains all the observable information about the particle motion. Thus, the classical observables like energy, momentum, angular momentum, etc., retain their usual meaning, but are now obtained from the wave function through certain unusual set of rules prescribed by *quantum mechanics*.

For the simplest case of a freely moving particle, the wave function is a plane wave, $\psi \propto e^{i\vec{p}\cdot\vec{r}/\hbar}$. It has a well defined momentum $\vec{p}$ and therefore, as demanded by *quantum mechanics*, its position is completely uncertain, i.e., $\mid \psi \mid^2$ is the same for all points. This 'matter wave', or the *de Broglie wave*, carries an energy $p^2/2m$ as in classical particle mechanics. It has a wavelength $\hbar/p$. In point of fact, except for its probabilistic significance, this wave is like any other wave motion. Thus, for a free electron of energy 1 eV, the de Broglie wavelength is about 12 Å, much shorter than the wavelength of ordinary light. The shorter wavelength means higher spatial resolution. This is the idea underlying the high resolution electron-microscope where light is replaced by high energy electrons.

For the not-so-simple case of a hydrogen atom, where an electron is bound to the proton by the Coulomb attraction, much the same way as planets are gravitationally bound to the sun, the solution of the Schrödinger equation gives stationary states that roughly correspond to the elliptical orbits of the planets around the sun. The set of these stationary states is, however, discrete corresponding to only certain allowed values of energy, angular momentum and its component along a chosen direction. Accordingly, these states are conveniently labelled by certain integers, called the *quantum numbers*. Thus, we write ψ_{nlm}. Here n $(= 1, 2, \cdots)$ is the principal quantum number giving the energy $E_n = -1/n^2$ in Rydberg units (1 Rydberg = 13.6 eV); ℓ $(= 0, 1, 2, \cdots, n - 1)$ is the angular momentum quantum number and gives the angular momentum in units of $\hbar$; m $(-\ell, -\ell + 1, \cdots, \ell - 1, \ell)$ is the component of angular momentum along any arbitrarily chosen axis! In addition to these negative energy (i.e. bound) states, we also have the positive energy 'scattering' states having a continuous range of energy from zero to infinity — these would correspond to the case of some comets that are merely deflected by the sun into open hyperbolic orbits, and not bound by it.

The same principles, of course, apply to more complex atoms, molecules, bulk matter and the sub-nuclear matter. Quantum mechanics has unrestricted validity. Thus, the electromagnetic waves (light), classically described by Maxwell's equations, must also be 'quantized'! The resulting quantum is a 'packet' of energy called *photon*. If λ is the wavelength (with frequency ν) of light, then the photon carries energy hc/λ, or $h\nu$ where c is the speed of light. It also carries a momentum h/λ. Interaction of quantized radiation with matter involves absorption (emission) of a radiation quantum ($h\nu$) accompanied by an electronic transition from a lower (higher) energy state 1 (2) to a higher (lower) energy state 2 (1) such that $h\nu = E_2 - E_1$ ($E_1 - E_2$). This is the origin of energy 'spectrum' characteristic of atoms, molecules etc. Similarly, the sound-wave-like oscillatory motion of atoms in a solid also must be quantized — the resulting quantum is the *phonon*.

Certain processes, forbidden classically, can take place quantum mechanically. Thus, a particle having an energy which is less than a potential barrier will be classically reflected back by it — it cannot escape over the barrier. The all pervasive waviness (or fuzziness) of the quantum wave function enables the particle to take the barrier, so to say, in its stride — it can *'tunnel'* through it even at sub-barrier energies. This is what enables an electron to jump across thin insulating layers separating metallic or superconducting electrodes.

4. SUPERPOSITION OF STATES
AND WAVE INTERFERENCE

This is a characteristic feature of quantum mechanical waves. If ψ_1 and ψ_2 are the two possible states, then their linear combination (superposition) $a_1\psi_1 + a_2\psi_2$ is also a possible state. It is clear that $|\psi|^2$ will then contain a cross-term (interference term) whose sign will depend on the relative phases of the two complex components, leading to constructive or destructive interferences. Thus, these probability waves interfere and diffract just as any other wave. Hence the phenomenon of electron, or neutron diffraction which is of great practical use in studying crystal structures — the latter act as diffraction gratings. It is important to note that a particle-wave interferes with *itself* — in a Young's double-slit experiment the electron can propagate through the two slits as alternatives, and the interference is between these two alternatives.

5. INDISTINGUISHABILITY OF IDENTICAL PARTICLES

Classically, identical particles remain distinguishable even if only in virtue of their being spatially separated. For, in principle we could keep track of which is which by following their *trajectories continuously*. This is obviously *not* allowed in quantum mechanics. Thus, if $\psi(r_1, r_2)$ is the wave function of two identical particles then $|\psi(r_1, r_2)|^2$ only tells us the probability density for finding one of them at r_1 and the other at r_2 without specifying which one. It follows then that under interchange of the labels 1 and 2, the wave function either remains the same (symmetrical) or changes sign (antisymmetrical). The particles obeying the 'symmetrical statistics' are called *bosons*, and those obeyng the 'anti-symmetrical statistics' are called the *fermions*. For identical *fermions*, not more than one particle can occupy a given one-particle state. For *bosons*, any occupation number is permitted. It turns out that particles having intrinsic spin angular momentum half-integral $(1/2, 3/2, \cdots)$ in units of $\hbar$ are fermions, e.g., the electron, the isotope ^{3}He, etc. Particles with integral spin are bosons, e.g., photon, ^{4}He, etc. Here spin refers to an intrinsic angular momentum that a particle may have. This may be likened to the particle spinning about its axis just as the earth spins about its axis.

This indistinguishability (symmetry or antisymmetry) under interchange gives rise to purely quantum effects which are pronounced when the amplitude of 'zero-point' motion is comparable with the inter-particle spacing. This happens for electrons in metals, or for helium atoms in liquid helium. We call these *quantum fluids*. Superconductivity or superfluidity of these fluids is an important consequence of *quantum statistics*. The same indistinguishability also gives rise to 'exchange interaction' responsible for magnetic ordering.

6. QUANTUM MECHANICS AND NAIVE REALISM

Quantum mechanics not only forbids simultaneous measurement of position and momentum (and, therefore, trajectories), it even *denies* their objective existence, independent of our measurement, as physically meaningless. Thus, prior to measurement, an electron in a room remains in a state of 'potentiality' (the probabilistic wave function), which 'collapses' to a certain point when it is detected on a photographic plate, say. All this is philosophically very disturbing. Attempts have been made to introduce variables 'hidden' from our reckoning that make the observed particles behave apparently probabilistically while the entire system of particle-plus-hidden variables is *deterministic* in the spirit of classical statistical mechanics. But crucial experiments have ruled out all such 'hidden variable' theories so far. There are paradoxes. There is philosophical uneasiness. But the detailed agreement of quantum predictions with experiments has muted much of the criticism so far.

APPENDIX C

THERMAL PHYSICS

AND STATISTICAL MECHANICS

1. THERMODYNAMICS

Heat is a form of energy. It flows from a hot spot to a cold one. Heat energy in a macroscopic body is carried by the *random motion* of its microscopic constituents, which may be molecules, atoms, elementary particles, or whatever. The constituents at a higher temperature possess more energies than those at a lower temperature. These constituents collide with one another as they move about, as a result of which energy is constantly being exchanged, thus causing those possessing more energy to share the extra amount with those with less, and this transfer of energy is interpreted as heat flowing from hot to cold. In this way energy is eventually shared equally by every part of the system, the temperature of the whole body becomes identical, and the system is said to reach a state of *thermal equilibrium*. These collisions also change the direction of motion of the constituents, so that any constituent could be moving in any direction at any given time. For that reason these motions carrying heat are called 'random motions'.

Temperature can be measured in any of the familiar scales, but for scientific purposes it is usually measured in the Kelvin scale (K). The *Kelvin scale* differs from the familiar Celsius scale (C) by 273.15°: the freezing point of water at 0° C is 273.15 K; the room temperature of 26.85 C is 300 K. The temperature measured in the Kelvin scale is called the *absolute temperature*. It is 'absolute', because it can be shown that no temperature can ever be lower than 0 K. At 0 K, all random motions stop and there is not the tiniest amount of heat enregy left in the body, hence it is not possible to lower the temperature any further.

So a system tends to have less random motion at a lower temperature, but the ordered nature of it is not completely determined by its temperature. Look at H_2O at 0° C, where water and ice can both exist. The hydrogen and oxygen atoms inside a piece of ice are arranged in an orderly crystal structure, whereas those in liquid water, because water can flow and move freely, clearly are not. So although water and ice are at the same temperature, ice has a more orderly structure than water, and therefore there must be another thermodynamic quantity which signifies the orderliness (or the lack of it) of the constituents of matter. It is called *entropy*, and is usually denoted by S: a substance in a state with a larger amount of entropy is *less* orderly than when it is in a state with less entropy. Looking at the example of ice and water, we can also understand entropy in another (more quantitative) way. Fix the temperature T, in this case at 0° C. To freeze water into ice, you must put it in a freezer to extract out the *latent heat* in the water. As a result, the final product (ice) becomes more orderly with less entropy, so there must be a connection between the decrease of entropy S at this fixed temperature T (and a fixed volume) and the heat that is extracted from the liquid water to turn it into ice. In fact this amount of heat is precisely TS, and this can be used as the definition of entropy: at a fixed temperature T, the *decrease* of entropy (S) is given by the amount of heat extracted from the substance divided by the temperature T.

If we ignore the influence of heat, as we usually do in studying mechanics, an object becomes stable when it settles down at the bottom of the potential well, where its potential energy U is minimal. When heat is taken into account, then we must require the total energy in a given system, *viz.*, the sum of the potential energy U and the *inflow* of heat enregy $-TS$, to be mimimal for thermaodynamic stability. This sum $F = U - TS$ is called the *free energy*, it replaces the mechanical energy U in the absence of heat.

This example of water and ice illustrates another important aspect of thermodynamics, that a substance can have different *phases*. For H_2O at the normal atmospheric pressure, it is in the steam phase above 100° C, in the water phase between 0° C and 100° C, and in the ice phase below 0° C. The temperatures 100° C and 0° C at which a phase changes are called the *transition temperatures*. As a rule, the phase at a lower temperature is usually more orderly than the one at a higher temperature. Another way of saying the same thing is that the phase at a lower temperature usually has less *symmetry* (see Chapter III) than the one at a higher temperature. This last statement may seem contradictory but it is not. A drop of water is spherical and it has spherical symmetry — or symmetry upon rotation for any amount about any direction. A piece of ice is a crystal, so we must *confine* the rotation to some specific amounts about some specific directions in order to move one atom to the position of another atom, and hence its symmetry is smaller.

This phenomenon of *phase transition* occurs not only in water, but also in most other substances and systems. For example, the change from a normal metal to a superconductor at a low temperature is a phase transition (see Chapter II). As the universe cools down from its big-bang beginning towards its present temperature of 2.7 K (see Chapter VII), various phase transitions are believed to have taken phase, so that the universe we see today is not in the same phase as the universe that was in the very beginning; at the early epoch the universe was much more symmetrical. Phase transitions are important not only in superconductivity (Chapter II), and in cosmology (Chapter VII), but are also important in the modern theory of elementary particle physics (Chapter VI).

2. STATISTICAL MECHANICS

At thermal equilibrium, each constituent carries the same amount of energy *on the average*, but this does not mean that the energy of every constituent is identical at any given time—only the average over time is the same. To obtain the energy spread about the mean, or the *energy distribution*, we must employ the full apparatus of *statistical mechanics*, which is beyond the scope of this book. Nevertheless, it is not difficult to describe and to understand the outcome of these calculations.

Energy distribution depends on whether quantum effects (see Appendix B) are important or not. According to quantum mechanics, a particle in a given volume with energy lying within a specific range can occupy only one of a finite number of *quantum states*. How these quantum states may be occupied depends on what type of particles they are. Particles of this universe are of two types: they are either *bosons*, or *fermions*. Two bosons of the same kind (*e.g.*, two photons) have an additional tendency to occupy the same quantum state. It is this remarkable property that leads to the feasibility of a laser (see Chapter I) and the remarkable phenomena of superfluidity and superconductivity (see Chapter II). On the other hand, fermions of the same kind (*e.g.*, two electrons) are absolutely forbidden to occupy the same quantum state. It is this special property that leads to the distinction between a conductor and an insulator, and prevents the collapse of a white dwarf or a neutron star (see Chapter V).

If the total number of particles in the macroscopic system is small compared to the number of quantum states available, then it would be very unlikely for two particles to occupy the same quantum state anyway, in which case whether these particles are bosons or fermions is immaterial. The energy distribution then follows what is known as the *Boltzmann distribution*. At a given (absolute) temperature T, the *average energy* per *degree of freedom* for a non-relativistic particle is $\frac{1}{2}kT$, where $k = 1.38 \times 10^{-23}$ J/K (Joules per Kelvin) is known as the *Boltzmann constant*. What this says is that the average energy per degree of freedom at $T=1$ K is 0.69×10^{-23} J; at $T = 300$ K is 2.07×10^{-21} J. Sometimes units are chosen so that $k = 1$, as is done in Chapter VII. In that case we simply identify each degree Kelvin as an energy unit of 1.38×10^{-23} J, and the energy per degree of freedom becomes simply $\frac{1}{2}T$. Since one Joule is equal to 6.24×10^{18} eV (electron-Volt), each degree of temperature can also be thought of as carrying the energy of 8.617×10^{-5} eV, or 86μ eV (micro-electron-Volt). Thus at a room temperature of 27 C, or $T = 300$ K, the energy per degree of freedom will be $\frac{1}{2}T = 150$ K$=0.013$ eV.

So far we have not explained what a 'degree of freedom' is. A point particle can move in any one of the three spatial directions and is counted to have three degrees of freedom; a rigid body of finite size can do so as well as rotating about each of the three axes pointing at the three possible directions, so it has six degrees of freedom. The average energy per point particle is therefore $\frac{3}{2}kT$, and that for a rigid body is therefore $3kT$.

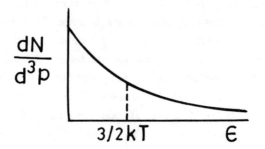

FIG. C.1 *Boltzmann distribution, depicting the number of particles per unit momentum volume (dN/d³p) with an energy ε at a given temperature T. The average energy per point particle is (3/2)kT.*

We have also not explained what a 'Boltzmann distribution' is. This distribution is shown in Fig. C.1, in which the number of particles per unit momentum volume is plotted as a function of the particle energy*. The average energy of each non-relativistic point particle, as mentioned before, is $\frac{3}{2}kT$. The Boltzmann distribution of other objects is qualitatively similar.

We have so far considered only the Boltzmann distribution, which is valid when the number of particles N in the system of volume V is so *small* that it is improbable for any two of them to occupy the same quantum state. If this is not so, then quantum effects become important and we must distinguish bosons from fermions.

In order to decide whether quantum effects are important or not, we must first ask: 'how small is *small*'? That depends on the temperature T of the system, because the number of quantum states N_0 present depends on it. To see that, recall that the energy of the particle depends on the temperature. Energy is related to momentum, and according to quantum mechanics, momentum is related to the wavelength λ of the particle[†], which

---

* The mathematical formula is $dN/d^3p = C\exp(-\epsilon/kT)$, where dN is the number of particles in the momentum volume d^3p whose energy is ϵ. C is a normalization constant determined by the total number of particles in the system

[†] The formula for λ is: $\lambda = \hbar(2\pi/mkT)^{\frac{1}{2}}$, and m is the mass of the

measures the effective size of that particle. Thus the effective volume λ^3 occupied by one such particle is temperature-dependent, and so is the number of quantum states $N_0 = V/\lambda^3$.

Thus whether quantum effects are important or not depends on whether N is large or small compared to N_0, and this in turn depends on the temperature T of the macroscopic body. At high temperatures, the wavelength λ is small, the number of quantum states N_0 is large compared to N, and quantum effects are relatively unimportant. At low temperatures, the opposite is true.

The temperature below which quantum effects become important is called the *degeneracy temperature*, and will be denoted by T_0. The energy distribution below the degenerate temperature will be said to be *degenerate*. From the discussion above, T_0 must increase with increasing N/N_0 *. As a result, it increases with the number density N/V of these particles and decreases with the particle mass m. For example, a *proton* (hydrogen nucleus) or a hydrogen atom has a mass of $m \simeq 2 \times 10^{-24}$ g. At a normal density of $N/V \sim 10^{22}$ cm^{-3}, the degeneracy temperature is $T_0 \simeq 1$ K. A *photon* has $m = 0$ so a photon gas is always degenerate, as $T_0 = \infty$ in this case. For *electrons* in metals, typically $N/V \simeq 10^{22}$ to 10^{23} cm^{-3}. With the electron mass being $m \simeq 10^{-27}$ g, the degeneracy temperature is $T_0 \simeq (16 - 20) \times 10^3$ K, hence the electronic distribution in a metal at room temperature is degenerate.

We will now proceed to discuss the distributions when the quantum effects are important. For bosons, the distribution is known as the *Bose-Einstein distribution*; for fermions, the distribution is known as the *Fermi-Dirac distribution*. These two distributions are different, but for $T \gg T_0$, both approach the Boltzmann distribution discussed before.

Electrons, protons, and neutrons are examples of fermions. They obey the Fermi-Dirac distribution, and two of these particles are not allowed to

---

particle. This formula may be understood as follows. The average energy of a particle at temperature T is of order kT, so its average momentum is of order $(2mkT)^{\frac{1}{2}}$, and its average wave length according to quantum mechanics is of order λ given above. The exact numerical coefficient appearing in the expression for λ can be obtained only through detailed calculations.

   * The exact formula is $T_0 = T(N/N_0)^{\frac{2}{3}} = 2\pi(N/V)^{\frac{2}{3}}\hbar^2/mk$.

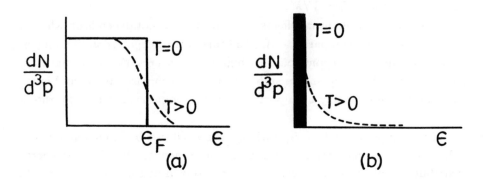

FIG. C.2 *The number of particles per unit momentum volume (dN/d^3p) with an energy ϵ at temperature T. (a) Fermi-Dirac distribution, with Fermi energy ϵ_F; (b) Bose-Einstein distribution, where the heavy vertical line shows Bose condensation.*

occupy the same quantum state. Therefore, at absolute zero temperature, the N particles in the system must simply fill out the N quantum states with the lowest energy. The distribution is therefore given by the solid line of Fig. C.2a; the energy of the last occupied state is called the *Fermi energy* and is denoted by ϵ_F**. When T rises above absolute zero, the distribution smears out a bit as given by the dashed line in Fig. C.2a, but the amount of smearing is still relatively small as long as $T \ll T_0$. Finally, as $T \gg T_0$, the amount of smearing becomes so large that eventually it approaches the Boltzmann distribution given by Fig. C.1.

Helium atoms (^{4}He) are bosons. They obey the Bose-Einstein distribution and these particles have the tendency to occupy the same quantum state. Thus at absolute zero temperature, these particles *all* occupy the lowest possible quantum state, as indicated by the heavy vertical bar in Fig. C.2b. This is the complete opposite of the fermions, no two of which is allowed to be in the same state. As T rises from 0, the available thermal energy pumps more and more of these bosons to quantum states of higher energy, as shown by the dashed line in Fig. C.2b. However, as long as T is lower than some

---

** $\epsilon_F = \frac{\hbar^2}{2m}(3\pi^2 N/V)^{2/3} \equiv kT_F = 1.31\ kT_0$.

condensation temperature T_c, a finite percentage of the particles will remain in the lowest energy state. This unusual phenomenon, where a macroscopic number of particles occupy a single quantum state and therefore exhibits quantum mechanical properties such as phase coherence, is known as the *Bose-Einstein condensation*. The condensation temperature T_c is of the order of the degenerate temperature T_0 [‡]. For liquid helium, this turns out to be 3.1 K, if all the inter-particle interactions are ignored as we have done so up to the present. The actual value for T_c is 2.19 K, because of the presence of these interactions. Below this condensation temperature, all sorts of interesting quantum mechanical behaviors (*superfluidity*) happens. Another interesting example of these Bose-Einstein condensation is the phenomenon of *superconductivity*. This happens because a *pair* of electrons also behaves like a boson. See Chapter II for detailed discussions of these topics.

Photons are also bosons, but they differ from helium atoms in two respects: photons have mass $m = 0$, and they can be created and absorbed by the walls of the container or any other material body in the box. As a result of their masslessness, the degenerate temperature T_0 is infinite. However, photons of zero energies cannot exist, so there are no Bose-Einstein condensation of the photons—these photons must disappear and be absorbed by the walls of the container or other material bodies.

The Bose-Einstein distribution of photons are known as *Planck's distribution*. This distribution law is also called the *black-body radiation law*. It was discovered in 1900 by Planck way before the general Bose-Einstein distribution was worked out; it is through this law that the Planck's constant was first introduced, and the seeds for quantum mechanics sowed.

Planck's distribution is sketched in Fig. C.3 for two temperatures. The peaks of the curves at various temperatures are connected by a dashed line.

Note that this differs from Fig. C.3 because we are now plotting the *energy density* of photons per unit *wavelength* (which we denote by u) against the *wavelength* λ of the photon, rather than the photon *number* per unit *momentum volume*[*]. Since the distribution is fixed for a given temperature, a measure of this distribution can determine the temperature of a body even

---

[‡] $T_c = 0.527 T_0$.

[*] Momentum p is related to wavelength λ by $p = 2\pi\hbar/\lambda$, and frequency ν is related to the wavelength by $\nu = c/\lambda$. The energy of a photon is $\epsilon = 2\pi\hbar\nu$.

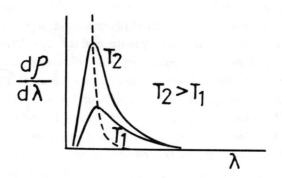

FIG. C.3 *Planck distribution, showing the photon energy density per unit wavelength $(d\rho/d\lambda)$ as a function of the wavelength λ at a given temperature T. The dotted line joind the peaks at different temperatures and is the curve for Wien's law.*

when this body is unreachable as long as its light reaches us. This is the case if we want to know the temperature inside a steel furnace, or the temperature of a distant star (see Chapter V). It is useful to learn more about two special aspects of this distribution. This is the *Wien's law*, and the *Stefan-Boltzmann law*.

Wien's law states that the wavelength λ_{max} at the peak of Fig. C.3 is inversely proportional to the temperature: $\lambda_{max}T = 0.2898$ cm K. This is so because quantum mechanically, the energy carried by a photon of wavelength λ is inversely proportional to λ (energy= $2\pi\hbar c/\lambda$). Thermodynamically, this energy is derived from heat and so is proportional to T. Consequently, $\lambda_{max} \propto T^{-1}$ as required by Wien's law. The proportionality constant ($\simeq 0.3$), however, cannot be obtained without a detailed calculation.

To illustrate this relation, let us just approximate the number 0.2898 by 0.3, and consider the radiation at a room temperature of $T = 300$ K. The peak wavelength works out to be $\lambda_{max} = 10^{-3}$ cm=10 micron (1 micron= one millionth of a meter). The wavelengths of the visible light being from 0.4 to 0.7 microns, this radiation at the room temperature is too long to be visible. Nevertheless, it is there in the form of the infrared. To emit visible light the body would have to have a higher temperature. The surface of the sun, for example, has a temperature of 6000 K. Its peak wavelength of 0.5

micron is right in the middle of the visible light region. That is why sunlight is visible, and why it contains all these beautiful colors of the rainbow visible at sunrise and sunset.

The *Stefan-Boltzmann* law states that the total amount of power density, or the radiation energy of all wavelengths emanating from a body per unit surface area and per unit time, is given by $f = \sigma T^4$, where $\sigma = 5.67 \times 10^{-5}$ erg/cm^2/s/K^4 is called the *Stefan-Boltzmann constant*. Wien's law tells us what 'color' the radiation has; this law tells us how bright the radiation is. The fact that it is proportional to the fourth power of the absolute temperature is the reason why hot objects are so bright.

A related and a more useful quantity for our purpose is the energy density of the radiation. If one imagines putting the heated body and the radiation it emits in a big box, then the radiation and the body will eventually reach an equilibrium state. The energy density ρ_γ of the radiation in the box is then given by $\rho_\gamma = aT^4$, where $a = 7.56 \times 10^{-15}$ erg/cm^3/ K^4 = 4.73×10^{-3} eV/cm^3/ K^4.

Another version of the Stefan-Boltzmann law concerns the number of photons per unit volume, n_γ. It is proportional to T^3: $n_\gamma = bT^3$, with $b = 3.2 \times 10^{13} T^3/\text{eV}^3$ cm^3 = $20.4T^3$/ K^3 cm^3.

3. DIMENSIONAL ANALYSIS

To understand the various versions of the Stefan-Boltzmann law, let us resort to a very useful tool known as *dimensional analysis*.

Every quantity in physics carries a *dimension* (*i.e.,* unit) made up of three fundamental units: second (s) for time, centimeter (cm) for length, and gram (g) for mass. Dimensional analysis is simply the statement that whatever quantity we want to consider must have the right units (dimensions). This by itself is of course trivial, and is not going to get us anywhere. However, if we also know on physical grounds what that quantity may depend on, then quite often this will yield a very valuable information as we shall illustrate below with the 'derivation' of the Stefan-Boltzmann law.

To begin the analysis for the photon number density n_γ, we must first decide what it can depend on. It will clearly depend on the temperature T. It must also depend on the dynamics and the kinematics governing the emission and the absorption of photons, the balance of which generates the eventual photon density in a volume. The details of these do not concern us, as long as we realize that they are controlled by two fundamental physical constants: the speed of light $c = 3 \times 10^{10}$ cm/s, and the Planck's constant $\hbar = 1.054 \times 10^{-27}$ erg s. It depends on c because anything that has to do with electromagnetic radiation involves c, and it depends on $\hbar$ because we are in a regime of degenerate representation where quantum effects are important. The numerical magnitudes of these two constants do not even matter for our purpose, but their units do.

So the photon number density n_γ is a function of T, c, and $\hbar$: $n_\gamma = f(T, c, \hbar)$, where f is some appropriate function which we seek to determine. The unit for n_γ is number per unit volume, say cm^{-3}. We must now choose the function f above so that $f(T, c, \hbar)$ comes out to have this unit cm^{-3}. The unit of T is energy (with units chosen so that $k = 1$), say erg=g (cm/s)^2; the unit of c is cm/s, and the unit of $\hbar$ is erg s. The only combination of these three variables, $T, c, \hbar$, to yield something of a unit cm^{-3} is clearly $c^{-3}\hbar^{-3}T^3$, which gives a unit (cm/s)$^{-3}$(erg s)$^{-3}$erg^3=cm^{-3}. Thus we must have $n_\gamma = b_0 T^3/(\hbar c)^3$, and the proportionality constant b_0 is dimensionless (*i.e.*, without any unit, or is a pure number). To calculate this number, we must know the technical details of quantum electrodynamics and statistical mechanics. Nevertheless, as we have just seen, the very fact that n_γ is proportional to T^3 can be deduced simply from dimensional analysis.

The fact that the energy density ρ_γ is given by $\rho_\gamma = a_0 T^4/(\hbar c)^3$ can be obtained in exactly the same way through dimensional analysis, once we note that the dimension of energy density is erg/cm^3.

The beauty of these arguments is not only that it is simple, but that it is also quite general. We used the arguments above to obtain n_γ and ρ_γ, but nothing in the argument above says that we are dealing with photons γ rather than some other relativistic particles. The appearance of the speed of light c, which is the speed of travel of a massless particle, however suggests that this argument is not valid for massive particles unless they are at such a high temperature to render the mass m negligible ($mc^2 \ll kT$) and their velocities approach that of light. Indeed, it is easy to see that if we had to assume f to be a function of m as well as T, c, and $\hbar$, then the dimensional argument would fail because there are infinitely many combinations of these four variables to yield the right dimension. No useful conclusion can then be drawn. However, as indicated before, the argument for photon is equally valid for any *relativistic particle* (meaning $m \ll kT/c^2$), in which case its number density is still given by $n = b_0 T^3 / (\hbar c)^3$, and its energy density is given by $\rho = a_0 T^4 / (\hbar c)^3$.

The proportionality constants a_0 and b_0 cannot be obtained without detailed dynamical inputs. With them, they are calculable and we shall quote the results. For bosons, the values are $a_{0B} = (\pi^2/30)g = 0.33g$, and $b_{0B} = (1.202/\pi^2)g = 0.12g$, where g is the *statistical degeneracy factor*, or the number of spin orientations the particle can have. This number is $g = 1$ for pions and $g = 2$ for photons. For fermions, the two proportionality constants are $a_{0F} = \frac{7}{8}a_{0B}$ and $b_{0F} = \frac{3}{4}b_{0B}$. That in both a_0 and b_0 the values for fermions are smaller than the values for bosons is understandable. It is due to the fact that fermions tend to avoid one another while bosons tend to stick with their own clans. As a result, there must be more bosons per unit volume than fermions, everything else being equal. Hence $b_{0B} > b_{0F}$, and similarly $a_{0B} > a_{0F}$.

If we write the energy density for the photon as $\rho_\gamma = aT^4 = a_{0B}T^4/(\hbar c)^3$, and substitute into it the known values of a_{0B}, $\hbar$, and c, we get $a = 4.73 \times 10^{-3}$ eV/cm^3/ K^4. Similarly we can calculate the constant b defined by $n_\gamma = bT^3$ from the value of b_{0B} and get $b = 20.4/\text{cm}^3/\text{K}^3$.

Note that the photon density is fixed at a given temperature. This could not possibly be the case for the *electron* density at low temperature because the electron density depends on how many electrons we put in the volume, so it is a number which could in principle be anything. This is not so for photons, because photons can be *created* and *absorbed*. If we force more photons in the volume at a given temperature, they will simply be absorbed by the walls. If we take out photons, then the walls will radiate enough photons so that the Stefan-Boltzmann law remains valid.

We also mention that Stefan-Boltzmann's law is valid for any relativistic particle, and that the proportionality constants a_0 and b_0 are completely fixed as well. That this is so is again because all these particles can also be created and absorbed *if* the temperature is *high enough*. For electrons, if the temperature is considerably larger than 1 MeV (million eV), then there is sufficient heat energy present to create electron-positron pairs. At low temperatures, such creations are energetically forbidden (electron and positron both have a mass of 0.5 MeV), and such automatic adjustments of their numbers cannot occur.

It should also be mentioned that the pressure P these particles exert on the walls of the box and on one another is given by $P = \rho/3$, both for bosons and for fermions. The fact that P is proportional to ρ can be obtained from dimensional analysis, and the factor 3 in the denominator can be obtained only through detailed calculations.

GLOSSARY

absolute zero the lowest possible temperature. It is by definition the zero of temperature on the Kelvin scale, or about −273 degrees on the Celsius scale. At this temperature, thermal random motions completely disappear, but not quantum fluctuations.

amplitude the maximum displacement of a wave motion above its average value. It determines the energy carried by the wave, the energy being greater for a greater amplitude.

angular momentum a vectorial quantity characteristic of rotational motion, analogous to momentum in linear motion.

association (star) a loose group of mostly young stars that are still close in an interstellar cloud from which they were formed.

asymptotic freedom a principle in quantum chromodynamics which says that the gluon-exchange forces acting between quarks become weaker as the quarks move in very close, as they often do in very high-energy collisions, so that the quarks become *free* of the forces at *asymptotically* high energies.

attractor a stable end-state toward which a system with appropriate dynamical properties will ultimately approach.

β-decay (beta-decay) transmutation of a neutron into a proton, accompanied by the production of an electron and an electron-antineutrino; also similar transmutation of a neutron-rich (radioactive) nucleus. It is the most familiar manifestation of the weak interaction.

baryons heavy particles of half-integral spins made up of three quarks, such as the proton and other, higher-mass particles. They can interact through any of the four known fundamental forces.

beauty one of the quark flavors; same as *bottom*.

bifurcation splitting of an equilibrium state in *two*, progressively, as a control parameter is varied. Hence bifurcation diagram.

big bang theory describes the origin of the universe as started from a big explosion some ten to twenty billion years ago.

black hole a region of space-time which cannot be seen or otherwise observed by distant observers because everything, light or matter, is trapped by the strong gravitational field existing there. Black holes are believed to be the end state of very massive stars after gravitational collapse; there exists some observational evidence for their existence. Quantum, or mini black holes are much less massive and as yet hypothetical objects, which might have been formed at the big bang.

boson name for particles obeying the *Bose-Einstein statistics*, which favors the occupation of the same state by many particles. In practice, a boson always has an integral-valued intrinsic spin (measured in units of $\hbar$, the Planck constant divided by 2π). Photons, gravitons, gluons, W and Z — all particles associated with the transmission of forces — are bosons. Composite objects made up of even number of fermions (particles possessing half-integral spin) — e.g., mesons, ^{4}He, ^{16}O — also behave as bosons. They also exhibit a phenomenon called *Bose condensation* observed at a temperature near the degeneracy temperature, where an appreciable fraction of particles in a system of bosons begin to congregate in a single one-particle state.

Cepheid variable a type of star with variable luminosity due to pulsations (regular oscillating changes in size). The luminosity can be deduced from the period of pulsation, making these stars celestial standard candles, useful for distance determination.

chaos aperiodic and apparently random behavior of a dynamical system despite governing laws being deterministic. Hence, deterministic chaos.

charm one of the quark flavors.

color a quantum number for quarks and gluons.

conduction band closely-spaced energy levels partially occupied by current carrying electrons in a metal, and the lowest normally empty band in an insulator or a semiconductor; electrons excited to the conduction band in an insulator or a semiconductor can carry current.

conservation fact or principle according to which the total amount of some physical quantity in an isolated system always remains the same, whatever internal changes may occur in the system.

Cooper pair stable complex of two fermions having equal and opposite momenta and opposite spins in a fermion degenerate system; such a pair behaves as a condensed boson if an effective attractive force exists between the two fermions in the pair.

cosmic rays high energy particles (mostly protons, electrons, and helium nuclei, with a few percent of heavy elements) present throughout our Galaxy. Their origin is not well known. They can be detected above the Earth's atmosphere. Cosmic rays are the object of study of *cosmic-ray astronomy*.

cosmic string a string-like object with a high energy density that could have been formed at the beginning of the universe. It can serve as a seed to accrete matter into galaxies.

cosmological constant a constant that can be incorporated into Einstein's gravitational theory (general theory of relativity). It describes the energy density of a matterless universe. This number is observationally known to be either zero or very small, though present theories of elementary particle physics almost invariably predict a large value.

cosmological principle states that the universe looks the same to every observer anywhere in the universe if local matter fluctuation is averaged out. It is the modern version of what Copernicus advocated some 500 years ago.

critical phenomena singular thermodynamic behavior of a system close to a second-order phase transition. Divergence of the magnetic susceptibility of a magnet at the *critical point* (Curie temperature), for example.

degeneracy temperature the characteristic temperature of a system of identical particles below which the quantum effects of indistinguishability (Bose or Fermi statistics) become observable.

degenerate particles   a system of identical particles is said to be degenerate when they occupy the lowest possible one-particle levels consistent with their respective statistics. For bosons, this means the same one-particle state and for fermions, all one-particle states up to a certain energy (the Fermi level), while all states with higher energy are empty.

dielectric material   an electrical insulator or a material in which an electric field can be sustained with a minimum of energy dissipation. If the material is a solid, the valence band is full and is separated from the conduction band by at least 3 eV.

Doppler shift   the change in wavelength of waves as seen by an observer when the source and observer are in relative motion. When they are moving apart the wavelength increases and, conversely, when they approach the wavelength decreases . With sound, the Doppler effect causes a shift in pitch, and with light, a shift in color of the source.

electron gas   a collection of electrons which interact with each other weakly enough to be considered as practically free, subject only to the exclusion principle.

energy band   a continuous range of energies in a solid in which there are allowed quantum states for electrons. Energy bands may be occupied or empty, and are separated from one another by *energy gaps*, which are energy regions where no quantum states for electrons exist.

entropy   a thermodynamic property which gives a measure of the degree of disorder of a macroscopic system.

event horizon   an imaginary surface surrounding a black hole where outwardly traveling photons would barely be able to escape to infinity. Light or matter emitted inside it is trapped by the gravitational field.

exclusion principle   *see* Pauli's exclusion principle.

fermion   name for particles obeying *Fermi statistics*, or the Pauli exclusion principle, which forbids two identical fermions to occupy the same one-particle state in a given system. Fermions always have half-integral intrinsic spins (measured in units of $\hbar$). Quarks and leptons — all particles associated with matter and antimatter — are fermions. Composite systems of odd numbers of fermions, such as baryons, ^{3}He, ^{13}C, and so on, also behave as fermions.

Feynman diagram   a pictorial representation of the quantum field theoretical predictions of particle and of many-body dynamics.

field   a physical quantity that extends over space, to be contrasted with a particle, which is localized at a point in space. Examples are electric field, magnetic field, and temperature distribution in a spatial region.

flavor   the quality that distinguishes the different varieties of quarks — up, down, strange, charmed, top (or truth), and bottom (or beauty). This property is also used to distinguish the various types of leptons (electron, muon, tau-lepton).

four-wave mixing   combination of three waves in an optical medium to generate a fourth.

fractal geometry   geometry of highly irregularly shaped objects having fractional dimensions, e.g., jagged coast-lines, sponges, foams, etc. Studied first by Benoît Mandelbrot.

fundamental interactions   forces acting between the basic constituents of matter, to be contrasted with the *effective* interactions, which are simplified forms of interactions between composite particles. The four known fundamental interactions are: the gravitational force, the electromagnetic force, the weak force, and the strong force.

gamma rays   a very energetic electromagnetic radiation, with wavelengths shorter than 10^{-11} m, or energies greater than 1 million electron-volts. The branch of astronomy devoted to this waveband is called the *gamma-ray astronomy*.

gauge symmetry   invariance of the form of the dynamical equations with respect to certain (space-time dependent) changes (re-definitions) of various fields.

gauge theories   theories of fundamental interactions based on gauge symmetry. In such a theory a spin 1 *gauge field(s)* is always present, whose longitudinal polarization is decoupled by the gauge symmetry. Couplings are universal; particles having the same gauge quantum numbers always couple the same way to the gauge field(s).

general relativity   theory of gravitation developed by Einstein. It gives the necessary equations to determine the geometry of space-time in terms of a distribution of matter and energy. It is indispensable for a correct description of situations dominated by a strong gravitational field.

globular cluster   a gravitationally bound system of (typically 10^6) stars with a well-defined spheroidal or ellipsoidal shape. These stars are among the oldest known objects in our Galaxy and are characterized by very low abundances of heavy elements. They are found in the halo of the Galaxy.

gluon   massless boson of spin one and agent for the strong interaction of quarks.

graviton   massless boson with a spin value of two, associated with the gravitational force.

hadron   any strongly interacting particle, made up of quarks and antiquarks. Mesons and baryons are all hadrons.

Hawking radiation   radiation emitted by a black hole when a virtual pair of particle-antiparticle, produced near the event horizon, splits into two, one falling through the event horizon and the other escaping to infinity.

Heisenberg's uncertainty principle   a general proposition in quantum mechanics that not all quantities in a quantum system may have simultaneously definite values. The uncertainty relations are mathematical formulas (due to Werner Heisenberg) that describe that irreducible level of uncertainty of certain pairs of dynamical variables when they are observed together.

Hertzsprung-Russell diagram   the plot of luminosity versus temperature for stars. Ordinary stars lie in band called the main sequence band; other types of stars are found in specific places on the diagram.

Higgs field   a spin zero field, some of whose components have non-zero values in the vacuum at a low enough temperature. This *vacuum condensate* is instrumental in giving non-zero masses to the gauge bosons (fields) and fermions.

Higgs boson particle associated with a Higgs field which is not needed to provide the longitudinal polarization degree of freedom for a massive gauge boson.

holography photography by reconstruction of light waves. The two parts of a laser beam, split into two by a half-silvered mirror, are directed at a photographic plate, one directly, the other after being scattered off the subject. The photographic plate records the interference pattern caused by the recombination of the two beams. When the *hologram*, as this photographic record is called, is illuminated by light from a laser identical in characteristics to the laser used for the recording, a three-dimensional image of the original object is reconstructed.

Hubble's law the observation, first made by E. Hubble, that distant galaxies are receding from us with a velocity proportional to their distance.

inflationary universe a theory of the very early universe constructed to overcome certain difficulties of the classical big bang theory. According to this theory, the universe underwent a huge inflation in its very early life, a period during which much of the matter of the universe and the energy for the explosion in the classical big bang was produced.

interference a characteristic property of waves whereby a new wave is formed when two waves overlap. When the two interfering waves are in step, an amplified wave motion is produced (constructive interference); otherwise, an attenuated wave motion is obtained (destructive interference).

insulator a poor conductor of electricity.

interstellar medium the medium between the stars. It is composed of three constituents: gas in all its phases (atomic, molecular, and ionized), dust (grains with sizes about 1 μm), and high energy particles (mainly electrons and protons). There is also a weak large-scale magnetic field present in the interstellar medium.

laser acronymn for Light Amplification by Stimulated Emission of Radiation. It is a device composed of three elements: an optically active medium, a pumping mechanism to inject energy into the medium, and a resonant cavity to amplify the radiation. It generates intense directional beams of coherent light through stimulated emission of radiation.

lepton a generic name for a class of fermions (electron, muon, tau particle, and neutrino) that respond to weak and electromagnetic forces, but not to the strong force.

luminosity the amount of electromagnetic radiation emitted by a celestial object.

magnetic monopole a particle with a non-zero *magnetic* charge. All the known macroscopic and microscopic magnetic objects are *magnetic dipoles*, with equal and opposite magnetic charges which total to zero. A magnetic monopole has never been found, but is postulated to exist in certain theories.

maser acronymn for Microwave Amplification by Stimulated Emission of Radiation, the immediate predecessor of the laser and operating along the same principles in the microwave band.

Meissner effect exclusion of magnetic flux from the bulk of a superconductor on cooling through the transition temperature. Hence a superconductor is a perfect diamagnet.

meson bosonic particle composed of a quark and an antiquark. Examples are the π-meson and the K-meson.

microwave background radiation thermal radiation with a temperature of about 3 K uniformly distributed in the Universe. The radiation is believed to be the cooled remnant of the hot radiation that was in thermal equilibrium with matter during the very early phases of the existence of the Universe.

muon one of the three known kinds of charged leptons, a heavier analogue of the electron.

neutrino electrically neutral lepton. Its mass is either small or vanishing. There are three known varieties, one associated with each of the charged leptons (electron, muon, and tau lepton); their presence is always indicative of the action of the weak force.

neutron star a dense compact star supported against gravity by neutron degeneracy pressure. Neutron stars have radii of about 10 km and central densities comparable to the nuclear densities. The first neutron star was discovered in 1967 as a *pulsar*.

nonlinear optics study of optical properties of matter under intense radiation field. It is found that in such circumstances, the optical response of matter is nonlinear, i.e., strongly sensitive to the applied field, and manifests itself in some unusual forms, such as a nonlinear induced polarization, frequency mixing, harmonics generation, and so on.

nuclear fission division of a heavy-mass atomic nucleus into two nuclear fragments of nearly equal masses and possibly lighter particles. Nuclear fission can be induced by thermal neutrons and, as a result, can produce more neutrons, which in turn go on to provoke more fission events, leading to chain reactions. This process is the operational basis of nuclear fission reactors and atomic bombs.

nuclear fusion combination of two very light nuclei into a nucleus lighter than iron. This process is the main source of energy of stars, and also forms the operational basis of controlled fusion reactors and thermonuclear weapons.

nucleosynthesis synthesis of chemical elements in the Universe. Very light elements are synthesized during the hot early phases after the big bang. Other elements, not heavier than iron, are produced by nuclear fusion in the central regions of stars. Heavier elements are produced mainly during supernova explosions.

OB association a loose group, or association, of young, hot and massive stars and, therefore, of spectral types O and B.

open cluster a stellar cluster of irregular, not well-defined shape, younger than globular clusters. They are found in the disk of the Galaxy, typically contain 1000 stars, and have heavy-element abundances.

optical chaos an uncontrollable and unpredictable response exhibited by many nonlinear optical media under an intense applied field.

order parameter   a physical quantity that is a measure of the order of a thermodynamic state (or phase). Thus, magnetization is the order parameter for a ferromagnet. It can be scalar, vector, or tensor.

parity   left-right symmetry with respect to mirror reflection.

Pauli's exclusion principle   a general rule of quantum theory (found by Wolfang Pauli) that forbids two fermions of the same kind to occupy the same quantum one-particle state in a given system. At any space-time point, at most one such particle can carry a given complete set of quantum numbers.

phase   the phase of a wave is a measure of the state of the wave's motion. When a wavelength of a wave has passed at a given point in space, we may say that the wave motion has completed a cycle, or that it has changed its phase by 360°. Two waves are said to be in phase if their peaks coincide; otherwise, their phase difference indicates how far apart their peaks are. In a completely different usage, the thermodynamic state of a macroscopic system is also called a *phase*. Thus, ice, water, and steam represent different phases of H_2O.

phase transition   change of thermodynamic state as some parameter, e.g., temperature is varied. *First-order* transitions, like freezing/melting, are discontinuous and involve latent heat; *second-order* transitions, like magnetic transitions, are continuous and have no latent heat.

photon   massless spin-one boson and quantum of the electromagnetic force.

Planck constant   the fundamental physical constant (symbolized by h or $\hbar = h/2\pi$) which quantifies the scale at which the quantum effects become important.

planetary nebula   a small, gaseous nebula around a hot star at a late evolutionary stage.

polarization (of a dielectric)   electric response of molecules in a dielectric material; it is also a measure of this effect, i.e., the electric dipole moment per unit volume. Dielectric polarization may arise from the distortion of the electron distribution about the nuclei in an electric field, or from a change in dipole moment resulting from the stretching of electrical bonds between unlike atoms in molecules.

polarization (of waves) phenomenon observed when a transverse wave is *polarized*, that is, when the displacement of vibrations is completely predictable. A transverse wave is said to be *unpolarized* when the vibrations in the plane perpendicular to the direction of propagation appear to be oriented in all directions with equal probability; no preferred pattern of orientation can be observed over a long time period.

proton-proton chain the sequence of nuclear fusion reactions that transform four protons into an α-particle in the core of ordinary stars.

pulsar an astronomical compact object that emits regular pulses of electromagnetic radiation, now interpreted as a rapidly rotating neutron star.

quantum chromodynamics (QCD) the theory of strong interactions between quarks, in which the forces are mediated by gluons and the states characterized by color charges.

quantum electrodynamics (QED) the quantum theory of electromagnetic interactions between charged particles.

quark fundamental particle of all hadrons. There are six known flavors of quarks, which combine in twos (a quark and an antiquark) to form mesons and in threes to form baryons.

quasar a star-like, extragalactic object, now believed to be an extreme form of active nuclei of distant galaxies.

radiowaves electromagnetic radiation of wavelengths $10^{-3} - 10^4$ m. *Radio astronomy* is the branch of astronomy devoted to the radio observations of celestial objects. The radioband open to detection by ground based telescopes extends roughly between 1 cm and 30 m; the cosmic background radiation is a famous member of this band.

renormalization a procedure to rewrite a fundamental physical law in an effective form appropriate to a certain energy scale μ in terms of a number of measurable parameters.

renormalization group a mathematical relationship describing the fact that the arbitrary choice of μ should not affect the physical outcome of a measurable quantity. This relationship can be used to derive results not easily accessible by other techniques.

RR Lyrae stars a class of variable stars that can be used for distance determination.

scale invariance symmetry demanding invariance with respect to a change of scale (of length, of energy, etc.). Thus, a system at its *critical point* is scale invariant, whence all microscopic length scales such as lattice spacing become irrelevant.

space-time the four-dimensional space in which the three spatial coordinates and the time coordinate are treated on an equal footing, as required by the special and general theories of relativity.

special relativity the currently accepted description of space, time, and motion formulated by Einstein in 1905. A key result of the theory is no material body or physical signal can exceed the speed of light.

spin intrinsic angular momentum of a particle or system of particles. It is characteristic of the particle and independent of its motion. As a quantum entity, it can take on integral or half-integral values in units of $\hbar$. According to the Theorem of Spin and Statistics, particles with integral spins are bosons and particles with half-integral spins are fermions.

spontaneous emission emission of a photon by an excited atom (molecule, etc.) that is independent of external radiation.

spontaneous symmetry breaking the symmetry of the state becoming lower than that of the governing law. This takes place at a phase transition.

state function a mathematical object that contains the most complete physical information about a physical state of a quantum system.

stimulated emission emission of a photon by an excited atom or molecule under the influence of an external field. Stimulated emission proceeds at a rate that depends on the intensity of the incident light and is spatially and temporally coherent with the incident light.

strange attractor a region of phase space (*attractor*) to which the phase trajectories are attracted but on which they wander aperiodically. It has fractal geometry and describes deterministic chaos.

strangeness a quantum number measuring the number of strange quarks inside a hadron.

string theory a theory which holds that the elementary particles are different manifestations of the excited modes of a tiny string, of the order of 10^{-32}cm in size, and that all the four fundamental forces are produced by the joing and breaking-up of the strings.

superconductivity a low temperature phenomenon of zero electrical resistivity and perfect diamagnetism shown by many materials below their characteristic critical temperatures.

superfluidity total loss of viscosity exhibited by helium close to absolute zero of temperature, that enables it to flow through the finest capillaries.

symmetry invariance of an object with respect to certain operations like mirror reflection. The object may be an equation expressing physical law.

supernova the explosion of a massive star toward the end of its evolution. It leaves behind a supernova remnant composed of a gaseous expanding nebula and, in most cases, a central collapsed object which evolves eventually into a compact object.

tau lepton negatively charged lepton, similar but heavier than both the electron and the muon.

thermal convection heat transfer by moving masses of matter in gases and liquids.

truth one of the quark flavors; same as *top*.

tunnel effect a quantum property which allows particles to pass through regions which are energetically forbidden by classical physics.

turbulent flow fluid flow in which the velocity at any point of the fluid changes constantly in magnitude and direction, to be contrasted with a *laminar flow* which is characterized by a regular space and time dependence.

uncertainty relations *see* Heisenberg's uncertainty principle.

unified theory a theory unifying several kinds of fundamental interactions.

universality   the assertion that the critical behavior of a system (i.e., behavior close to second-order phase transition) is independent of the behavior of the microscopic details and depends only on symmetry and dimensionality. Confirmed experimentally.

valence band   the highest completely filled band of energy levels in a solid.

virtual particle   a quantum particle that exists only for very short durations and, because of Heisenberg's uncertainty relations, need not satisfy the usual relations between mass, energy and momentum, in contrast to the more familiar, long lived, *real* particles. Virtual particles may appear singly when exchanged between other particles, or in pairs of particle-antiparticle when spontaneously created in vacuum.

wavelength   the distance between two adjacent maxima in a wavetrain.

weak interaction   one of the four fundamental interactions. The weak force is responsible for β-decay and any interaction involving neutrinos or antineutrinos.

white dwarf   a dense compact star with mass less than 1.4 solar mass and radius of about 1000 km; supported against gravity by the pressure due to degenerate electrons; believed to be the final evolutionary stage of the less massive stars.

X-rays   electromagnetic radiation with wavelengths around 10^{-9} m, or energies around 1 keV. *X-ray astronomy* is the study of celestial X-ray emitters, which include stars, supernovas, and active galaxies.

zero-point motion   quantum vibrational motion that is always present in a system, even at absolute zero temperature. The energy due to these quantum fluctuations is a minimum, irreducible amount of energy called the zero-point energy. Its existence may be regarded as a consequence of Heisenberg's uncertainty principle.

Zeeman effect   splitting of atomic energy levels into two or several components when the atoms are subject to a magnetic field.

INDEX